YOUR PLANET IS SPEAKING

SOUND OF SIRENS

21ST CENTURY GLOBAL TEMPERATURE CHANGE

BOOK 1: TABULA RASA

G.E. JOHNSON

SOUND OF SIRENS: 21ST CENTURY GLOBAL TEMPERATURE CHANGE
BOOK 1: TABULA RASA

FIRST PRINTING 2011

for
you

ZUGANG ZUM INFOHIGHWAY
ENTRANCE TO THE INFORMATION HIGHWAY

- Things should be made as simple as possible...

...but not any simpler. -

-Albert Einstein

A NOTE FROM THE AUTHOR...

As we head full of force into the 21st century, humankind has entered the Global Information and Communication Age, where access, retrieval, and archiving of our dynamic global civilization and the planet we inhabit is available to the average citizen like never before.

Literally at the speed of light...or the fastest Internet connection.

The Internet has changed what we know and how we see the world around us, changed the way we communicate with those we share it with, changed our ways of commerce and consumption, and changed the way we inform, educate, and learn. The Internet has opened up our access to 'news' of all types and immensely expanded our ability to understand the global human condition and the state of the planet.

Along with other things in contemporary society, the Internet has also helped create a growing culture of 2-minute attention spans, immediate self-gratification, and acute self-absorption.

Generation X and the Y Generation have become Generation i.

But, this cultural trend is also strengthening the importance of the self, the value we put on our individual experience, and the value we put on the experience of others, allowing us a greater awareness and respect for our similarities and differences, opening our eyes to unacceptable injustices, and offering us a deeper understanding of the problems and solutions for our growing human settlements in an age of globalization.

The year is 2005 C.E. and we all live in settlements that make up a great global civilization. There are differences, there are similarities, there are disparities, and there is interdependence between members of our global family. Our settlements are sensitive to changes in the foundation they have been built upon and this foundation is predicted to change dramatically during the 21st century and beyond.

Three words: Global Temperature Change.

This book attempts to present a clear picture from the jumble of pieces that science has given us up to this point about our common future. The puzzle's border has been laid during the past 30 years and as more discoveries are made, as more research is completed, and as more advanced technologies are developed, a greater understanding continues to emerge.

The puzzle used to be full of large gaps which made it very difficult to envision the whole picture.

The gaps are what we call uncertainty, but the uncertainty of today is different than the uncertainty of yesterday. There is no longer uncertainty that global temperature change is happening, that greenhouse gases and global temperature change are intertwined, and even that human activity during the past 350 years has had a significant impact on the global systems of the Earth.

Today, the uncertainty lies in how the dynamic systems of the Earth will react, how quickly abrupt changes will occur, how human settlements will be affected, and how human settlements will adapt.

Climate change science and projecting future scenarios can never be 100% certain, there are too many variables. But past events give us a strong understanding of what might happen in the future and present trends can be trusted to fill in the gaps, recognize the signs of change, and guide us in our preparations to live under entirely different circumstances than we experience today.

You will notice that much of the material contained in these pages belongs to the scientists, journalists, organizations, and agencies researching and reporting the hard facts. I make no claim to this intellectual property as my own and in no way wish to infringe upon the copyrights of any individual or organization. I am simply a mouthpiece, a DJ mixing words, ideas, and facts. I have sampled their findings to show how our planet has behaved in the past, how it is presently behaving, and how it is predicted to behave in the future.

This is done in a purely fair and transformative way that is meant to frame our present situation.

I graciously thank all of those who I have sampled from and those who have and continue to support me.

Welcome to the world of 21st century global temperature change.

I hope you enjoy this embodiment of the ZEITGEIST.

We are all in this planet together, whether we like it or not.

Play nice and nobody gets hurt.

gej

CONTENTS..

ya'll ready for this?

SOMEDAY..

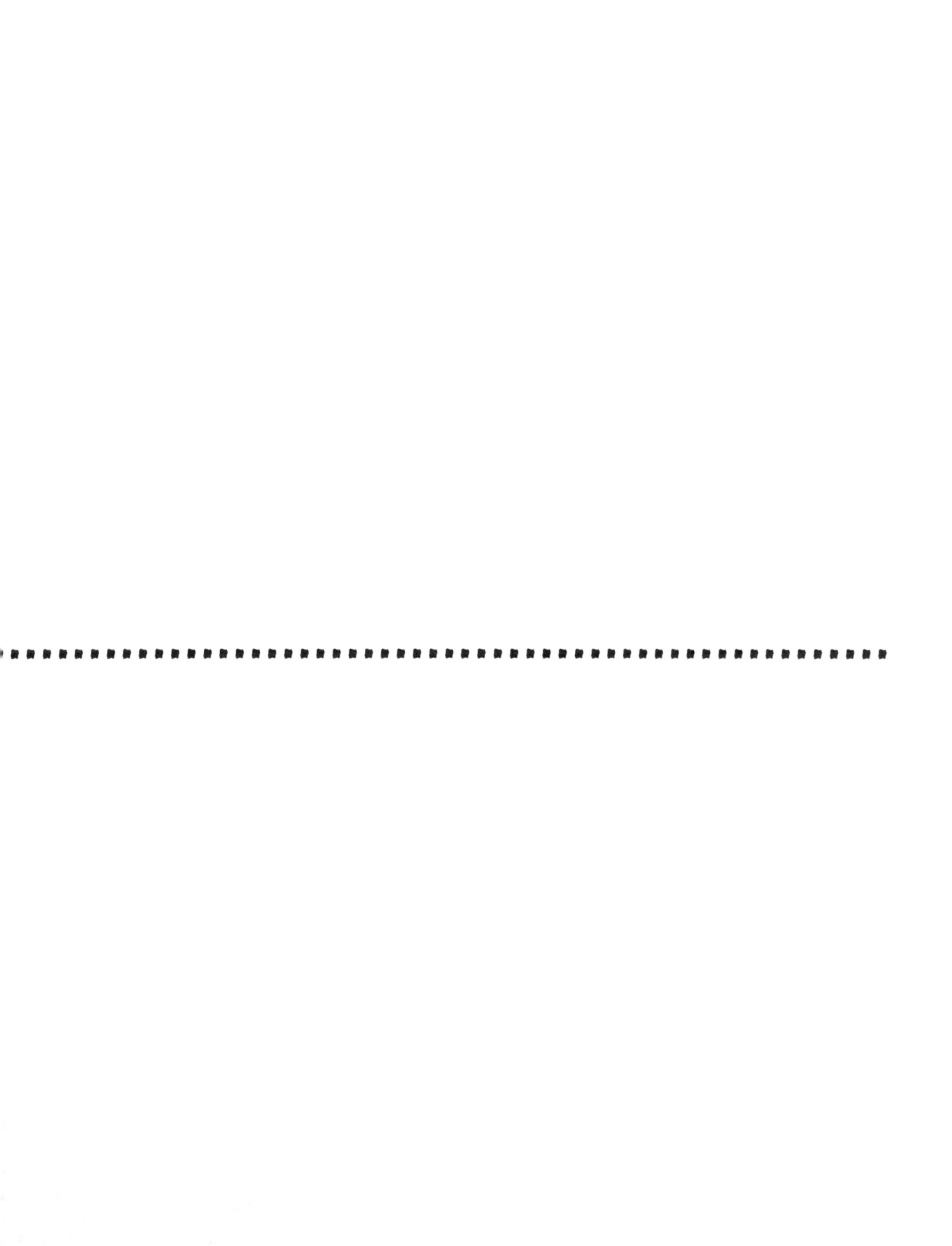

...IT WILL BE DISCOVERED THAT
EVERYTHING THAT IS ANYTHING BEGAN
AS NOTHING MORE THAN A COLLECTION
OF GASES, SET IN MOTION BY SOME
INITIAL FORCE AND LEFT TO GATHER
TOGETHER THROUGH RANDOM CHANCE,
ORGANIZED PRINCIPLES, LEARNING,
INSTINCT, FREE WILL, AND DESTINY...

...WE WILL LEARN THAT THE EARTH IS ABOUT 4,600,000,000 YEARS OLD AND GOES THROUGH PERIODS OF SPONTANEOUS GROWTH, SLOW EVOLUTION, DIVERSIFICATION OF LIFE, AND EXTINCTIONS OF MASS PROPORTION...

estimated and projected global species loss

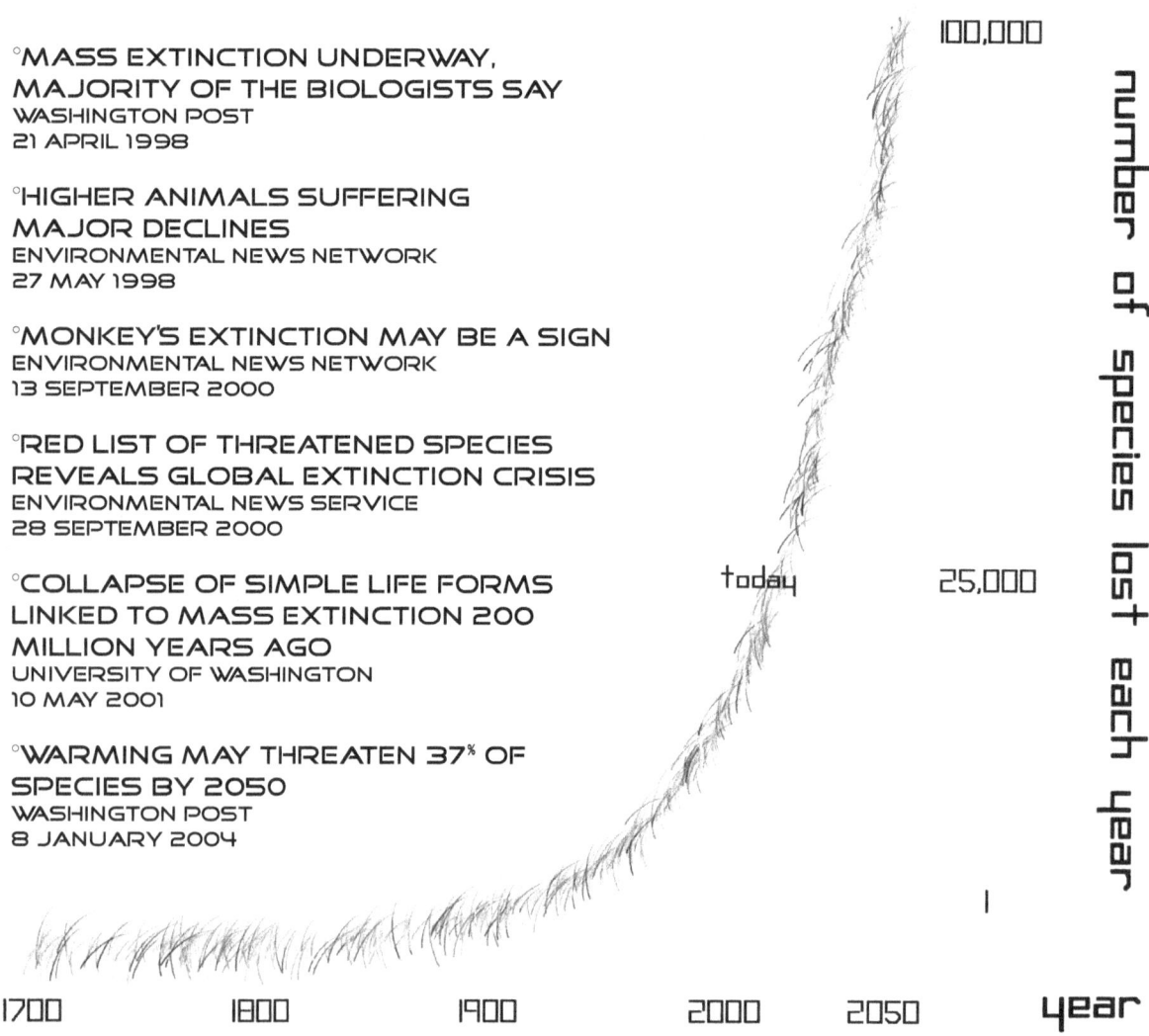

°MASS EXTINCTION UNDERWAY,
MAJORITY OF THE BIOLOGISTS SAY
WASHINGTON POST
21 APRIL 1998

°HIGHER ANIMALS SUFFERING
MAJOR DECLINES
ENVIRONMENTAL NEWS NETWORK
27 MAY 1998

°MONKEY'S EXTINCTION MAY BE A SIGN
ENVIRONMENTAL NEWS NETWORK
13 SEPTEMBER 2000

°RED LIST OF THREATENED SPECIES
REVEALS GLOBAL EXTINCTION CRISIS
ENVIRONMENTAL NEWS SERVICE
28 SEPTEMBER 2000

°COLLAPSE OF SIMPLE LIFE FORMS
LINKED TO MASS EXTINCTION 200
MILLION YEARS AGO
UNIVERSITY OF WASHINGTON
10 MAY 2001

°WARMING MAY THREATEN 37% OF
SPECIES BY 2050
WASHINGTON POST
8 JANUARY 2004

100,000

today   25,000

1

number of species lost each year

1700   1800   1900   2000   2050   year

...WE WILL DISCOVER
HUMANS WERE BORN 4-8
MILLION YEARS AGO FROM
A BRANCH ON A TREE OF
PRIMATES IN THE
AFRICAN STEPPE...

...THEY LEARNED TO
WALK AND TALK, SPREAD
AROUND THE EARTH,
REACHED THE MOON AND
TRAVELLED ON...

...THE GLOBAL HUMAN POPULATION WILL SURPASS SIX BILLION AND CONTINUE TO INCREASE...

...THE PRESSURES ON THE EARTH'S NATURAL RESOURCES AND STRAINS ON ENERGY, SHELTER, FOOD, AND WATER SOURCES WILL INCREASE AS WELL...

I drink **water**, because it is slowly becoming **scarce**

Ich trinke **Wasser** weil es langsam **knapp** wird

NORD-SÜD KONFERENZ
NORTH-SOUTH CONFERENCE

...SOME OF THE EARTH'S CHILDREN WILL CRY BECAUSE THEY LACK THE NEWEST TOY...

...400 MILLION WILL CRY BECAUSE THEY LACK CLEAN WATER...

...2.6 BILLION WILL CRY BECAUSE THEY LACK BASIC SANITATION...

THAT'S A LOT OF TEARS

...INSTEAD OF HELPING OUR UNKNOWN FAMILY MEMBERS ACQUIRE THE STUFF EVERYONE NEEDS TO SURVIVE, A FEW IN OUR GLOBAL FAMILY WILL SPEND TRILLIONS OF DOLLARS EACH YEAR TO KEEP MOST OF THE STUFF FOR THEMSELVES AND PREVENT IT FROM BEING TAKEN...

Peace Endangers Employment

Der Frieden gefährdet Arbeitsplätze

NORD-SÜD-GEFÄLLE

NORTH-SOUTH DIFFERENTIAL

...WHILE OVER 2 BILLION PEOPLE, 1/3 OF OUR GLOBAL FAMILY, WILL LIVE ON LESS THAN 1 DOLLAR PER DAY AND STRUGGLE TO SURVIVE WITH VIRTUALLY NO FOOD, WATER, SHELTER, SANITATION, MEDICINE...

...OR ANY OTHER STUFF AT ALL...

5

this is nuclear fission

neutron

plutonium-239

becomes
unstable

barium

krypton

neutrons

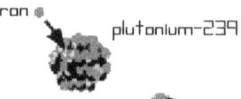
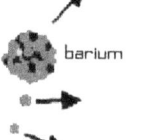

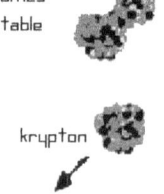

An atomic bomb, like the ones used by the United States to erase Hiroshima and Nagasaki during World War II, were fission bombs, containing a ball of plutonium-239 about the size of a grapefruit.

Plutonium-239 is usually derived from the recovered fuel of a nuclear reactor. A chain reaction starts in this ball when a stray neutron enters a plutonium nucleus. When this happens, the ball is suddenly compressed, squeezing the nuclei closer together, forcing more chain reactions and more compression.

This causes a burst of heat energy, equivalent to the reactions on the surface of the Sun. The amount of heat energy produced by modern atomic bombs equals 1 million tons of TNT and reaches temperatures over 50 million degrees Fahrenheit at the center of the fireball.

Many of the radioactive atoms produced by this explosion are carried up into the stratosphere, circling the earth and drifting back down to its surface as fallout, for you and your children to breathe.

Since 1945, there have been over 1500 explosive tests around the world.

There are over 20,000 active atomic warheads in the world today.

Hold your breath.

...AND DESTROY THE HOMES OF HUMANS...

...WE WILL DISCOVER THAT THE AVERAGE GLOBAL SURFACE TEMPERATURE IS INTIMATELY INTERTWINED WITH GLOBAL GREENHOUSE GAS LEVELS AND CYCLES...

...AND THAT GLOBAL GREENHOUSE GAS LEVELS AND CYCLES RISE AND FALL OVER TIME AS A RESULT OF ACTIVITIES UPON AND BEYOND THE EARTH...

change in atmospheric carbon dioxide concentration
change in atmospheric methane concentration
change in average global surface temperature

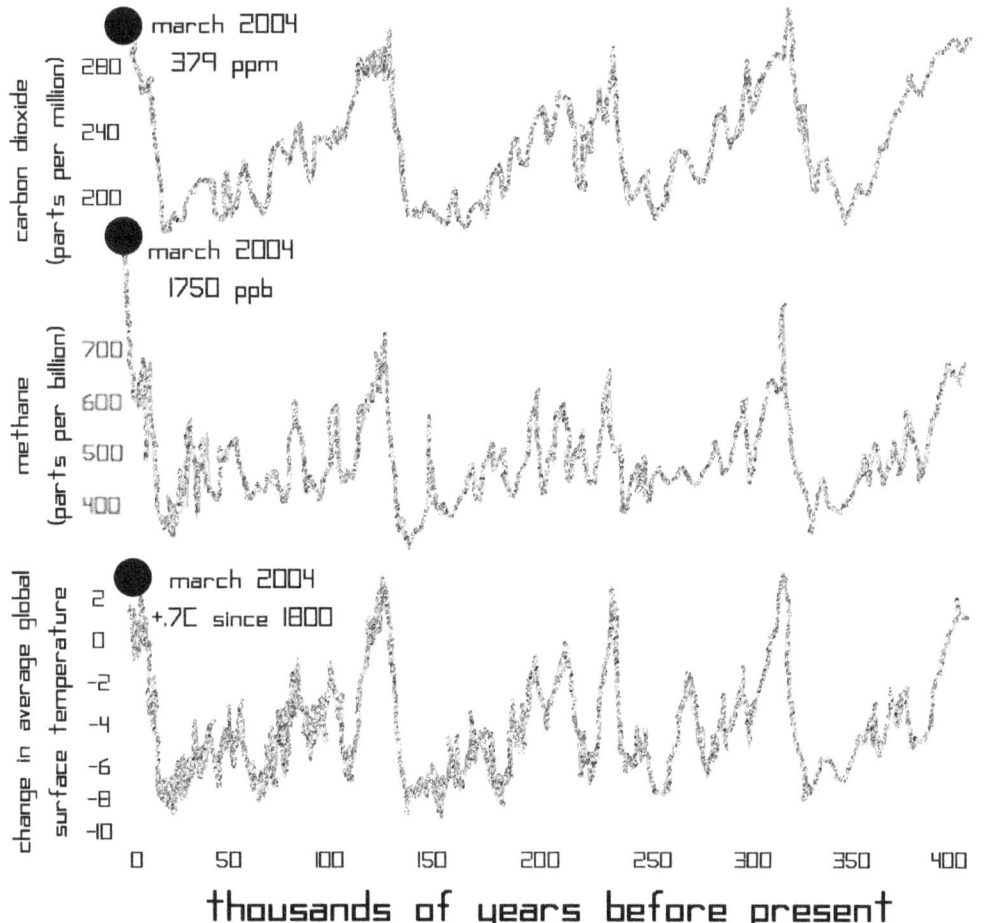

...WE WILL DISCOVER THE AVERAGE GLOBAL SURFACE TEMPERATURE HAS INCREASED DURING THE 20TH CENTURY AND CONTINUES TO RISE...

...WE WILL DISCOVER THE STABLE CLIMATE SYSTEMS AND WEATHER PATTERNS WE TAKE FOR GRANTED ARE BASED ON THE AVERAGE GLOBAL SURFACE TEMPERATURE...

...AS THE GLOBAL TEMPERATURE AND CLIMATE SYSTEMS CHANGE, NATURAL DISASTERS WILL BECOME MORE FREQUENT AND MORE INTENSE, AFFECTING MORE PEOPLE IN MORE PLACES IN MORE WAYS...

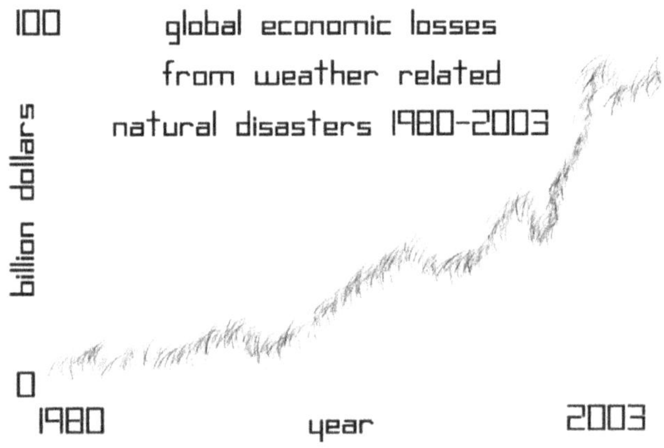

°20TH CENTURY GLOBAL WARMING IS UNPRECEDENTED
ENVIRONMENTAL NEWS NETWORK
22 DECEMBER 1998

°1999 SECOND-WARMEST YEAR THIS CENTURY, FOLLOWING LAST YEAR
REUTERS
28 DECEMBER 1999

°EARTH HITS '2,000-YEAR WARMING PEAK'
BBC NEWS ONLINE
1 SEPTEMBER 2003

°US EXPERTS SAY GLOBAL WARMING FASTER THAN THOUGHT
AGENCE FRANCE-PRESSE
24 JUNE 2004

°EARTH GETS WARM FEELING ALL OVER
NASA-GISS RESEARCH NEWS
8 FEBRUARY 2005

°1998 RECORD YEAR FOR WEATHER RELATED DISASTERS
ENVIRONMENTAL NEWS NETWORK
1 DECEMBER 1998

°FLOODS HIT 20 MILLION
BBC NEWS ONLINE
30 SEPTEMBER 2000

°WORLD DISASTERS SEEN AS GLOBAL WARMING OUTCOME
REUTERS
19 FEBRUARY 2001

°GLOBAL WARMING TO COST $300 BILLION A YEAR
REUTERS
30 OCTOBER 2002

100

billion dollars

global economic losses from weather related natural disasters 1980-2003

0

1980   year   2003

°EXTREME WEATHER ON THE RISE
CNN NEWS ONLINE
3 JULY 2003

...WE WILL DISCOVER AS GLACIERS AND POLES MELTED IN THE PAST, FRESH WATER FLOODED OCEAN SYSTEMS, HELPING TO CAUSE ABRUPT CLIMATE SHIFTS...

...AND WE WILL DISCOVER THAT IT COULD HAPPEN AGAIN...

-It would take no more than one-quarter of one percent more fresh water flowing in to the North Atlantic from melting glaciers in Greenland and Northern Canada to bring the northwards flow of the Gulf Stream to a shuddering halt.-

-Vittorio Canuto
Senior Scientist
NASA Goddard Institute for
Space Studies-

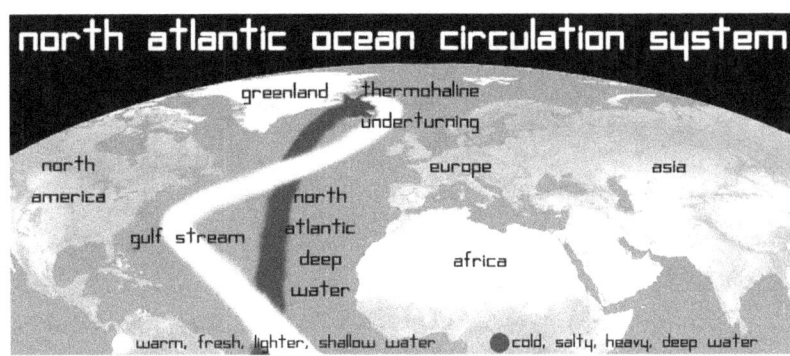

north atlantic ocean circulation system

greenland — thermohaline underturning

north america — europe — asia

gulf stream — north atlantic deep water — africa

warm, fresh, lighter, shallow water ● cold, salty, heavy, deep water

°EVIDENCE FOUND OF ANCIENT CLIMATE SWINGS
ENVIRONMENTAL NEWS NETWORK
24 APRIL 1998

°EARTH ENTERS BIG THAW
BBC NEWS ONLINE
7 MARCH 2000

°OCEAN CIRCULATION SHUT DOWN BY MELTING GLACIERS AFTER LAST ICE AGE
NASA-GSFC PRESS RELEASE
19 NOVEMBER 2001

°SCIENTISTS WARN OF ABRUPT CLIMATE CHANGE
REUTERS
12 DECEMBER 2001

°GLOBAL WARMING WILL PLUNGE BRITAIN INTO NEW ICE AGE 'WITHIN DECADES'
THE INDEPENDENT
25 JANUARY 2004

°SATELLITES RECORD WEAKENING NORTH ATLANTIC CURRENT
NASA-GSFC PRESS RELEASE
15 APRIL 2004

...THE AVERAGE GLOBAL SURFACE TEMPERATURE MAY CHANGE 6° CELSIUS WITHIN A FEW DECADES, AS IT DID 11,500 YEARS AGO...

...THE AVERAGE GLOBAL SURFACE TEMPERATURE MAY CHANGE 5° CELSIUS WITHIN A FEW YEARS, AS IT DID 14,600 YEARS AGO...

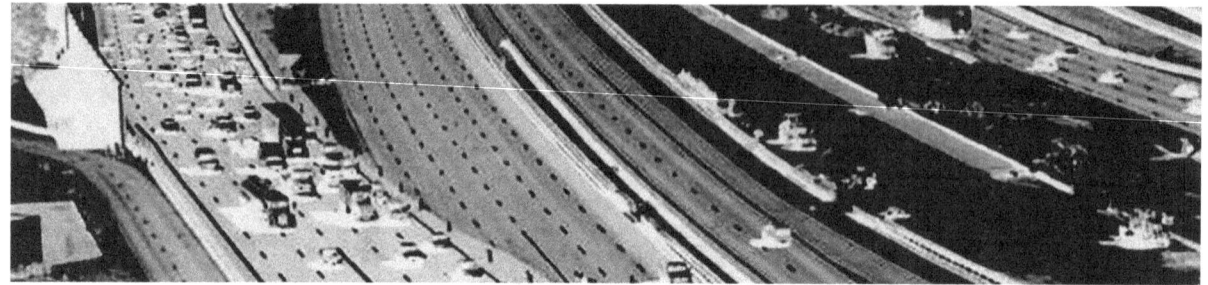

....HUMANS WILL BUILD THE PLACES THEY INHABIT TO ACCOMMODATE THEIR MACHINES, RATHER THAN THEMSELVES...

...OUR GLOBAL SPECIES WILL BUILD ADVANCED, COMPLEX AND GROWING SETTLEMENTS BASED AND DEPENDENT UPON AN AVERAGE GLOBAL SURFACE TEMPERATURE OF 14°-16° CELSIUS...

...AND THEN THAT TEMPERATURE WILL CHANGE...

°CATASTROPHIC CLIMATE CHANGE '90% CERTAIN'
THE INDEPENDENT
20 JULY 2001

°GLOBAL WARMING WILL CONTINUE FOR THE NEXT CENTURY
UNISCI.COM
20 FEBRUARY 2002

°CLIMATE CHANGE 'ENTERING THE UNKNOWN'
THE BALTIMORE SUN
3 DECEMBER 2003

°GLOBAL WARMING SPIRALS UPWARDS
THE INDEPENDENT
28 MARCH 2004

°WEST WARNED ON CLIMATE REFUGEES
BBC NEWS ONLINE
24 JANUARY 2000

°MORE HUNGER AND POVERTY MAY BE ENDURING IMPACT OF CLIMATE CHANGE
WORLD BANK PRESS RELEASE
14 JUNE 2001

°WORLD FORUM SCIENTISTS: 'CLIMATE CHANGE WILL BECOME A SECURITY ISSUE'
REUTERS
2 FEBRUARY 2002

°A QUARTER OF CHINA'S POPULATION AT RISK AS GLACIERS START MELTING
AGENCE FRANCE-PRESSE
13 MAY 2004

SOMEDAY..

...WE WILL PREDICT THE FOUNDATION OF OUR MODERN SETTLEMENTS WILL DRAMATICALLY CHANGE DURING THE 21ST CENTURY...

...WE WILL HAVE TO LIVE WITHIN THE EARTH'S LIMITS AND CANNOT CONTROL ITS THRESHOLDS...

...WE WILL LEARN THE FABRIC OF THE EARTH IS WOVEN FROM MANY INTERRELATED ACTIVITIES AND THAT WE ARE PART OF ITS FABRIC...

...WE WILL LEARN THAT OUR SURVIVAL IS DEPENDENT UPON COOPERATION BETWEEN MEMBERS OF OUR SPECIES IN THE FACE OF DRAMATIC GLOBAL CHANGES...

...WE WILL STOP KILLING OURSELVES...

...WE WILL STOP KILLING EACH OTHER...

...A RICHER, MORE SECURE, MORE STABLE, MORE PEACEFUL AND MORE BENEVOLENT ERA WILL EMERGE FROM THE ASHES OF AN INDUSTRIALLY CONSUMPTIVE AND WARRING GLOBAL CIVILIZATION...

SOMEDAY..

..IS TODAY.

WE ARE ALL ON THIS PLANET TOGETHER

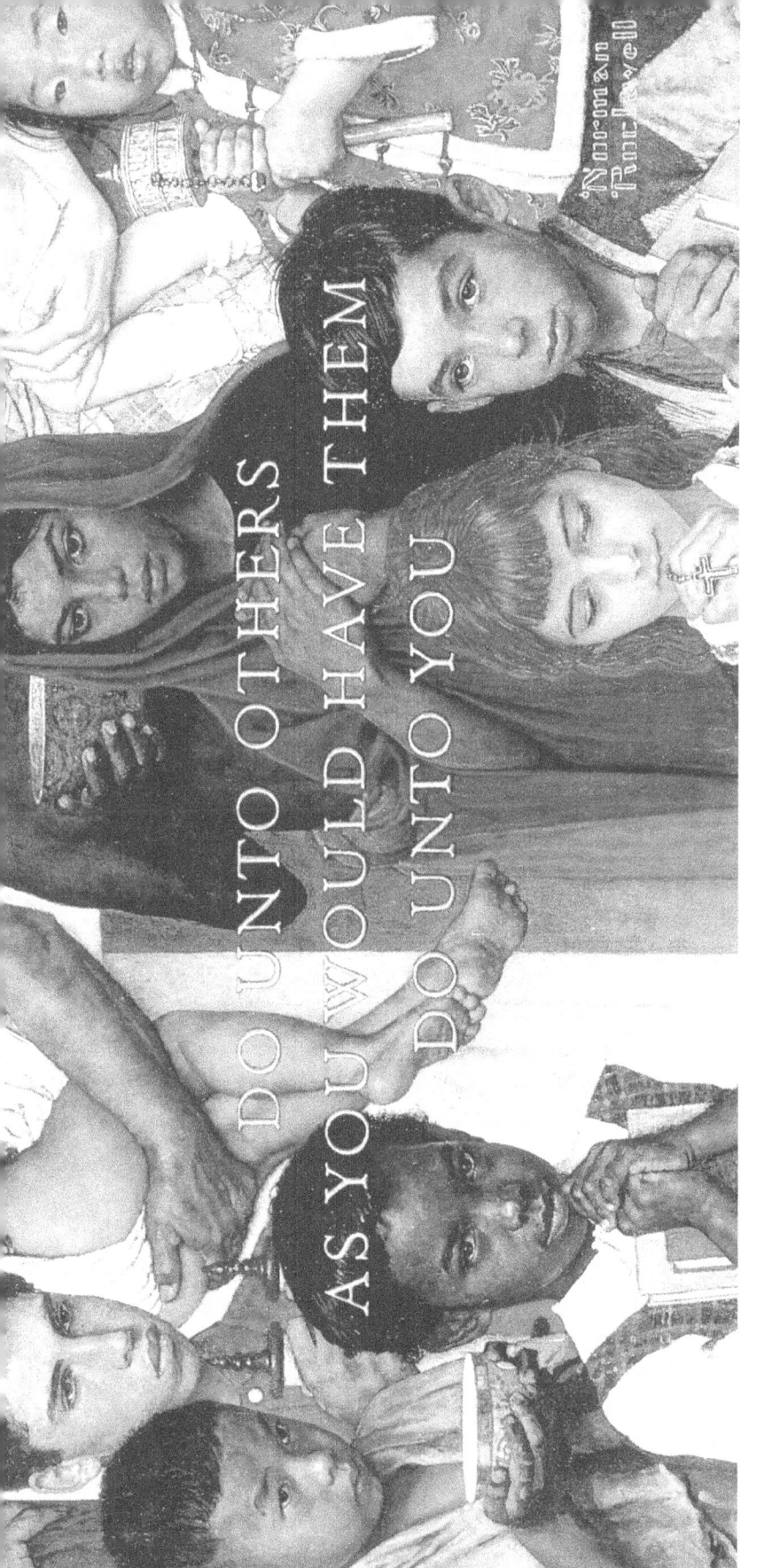

DO UNTO OTHERS AS YOU WOULD HAVE THEM DO UNTO YOU

GET WITH IT

OR GET OFF IT

PROLOGUE

AND NOW A WORD FROM ONE OF OUR
FELLOW HUMANS REGARDING RECENT
TEMPERATURE IRREGULARITIES IN
NORTH AMERICA...

-It's global warming, dude.
I don't care if the whole
planet burns up in a
hundred years. If I can get
me a fish today, it's cool
by me.-

-Pito Robles, age 28
New York Times
23 December 2001

WELCOME TO YOUR FUTURE
GREAT APES

ENJOY THE STRUGGLE

A
INTERVIEW WITH THE AUTHOR PART I

TELL ME WHERE WE ARE

"So G, what's this all about? Do you think people are actually interested in global temperature change? Isn't that kind of a dry subject?"

Well my friend, they say the consequences of 21st century global temperature change are the greatest threat to modern civilization in the world today. I find that fairly exhilarating.

A predicament on the magnitude of something like global temperature change and its consequences is unique and can only be assessed and properly remedied if it is properly understood. It can only be properly understood, if it is properly explained.

This book brings the heavily scientific subject matter to the people in a way they can access and understand what is happening with our planet and why the world's leaders are concerned.

I figure most people want to understand what is happening with the Earth and how their life might be affected in the near future.

"So what's with the title? What is Tabula Rasa and Sound of Sirens?"

Tabula Rasa is a term that was used by the English philosopher John Locke in the 1600's to describe the state of the brain upon birth; a 'blank slate' on which experiences are imprinted. I used this for my title to describe the blank slate of the reader when addressing the subject of global temperature change, because although most people have heard of global warming or climate change, they are not really sure what dynamics are involved with such global changes.

The title Sound of Sirens emerged pretty early on and has multiple meanings. When I observe our world today, I hear sirens from people and from places, sirens from the land and from the seas. Of course, not everything in the world today is doom and gloom, there is a bounty of beauty I see as well, but the state of the planet and our global civilization isn't as rosy as we may believe.

For this reason, it seemed appropriate to sound an alarm, to call S.O.S. (Sound of Sirens) to the issues I am discussing in the book.

The third meaning refers to the mythical Sirens from tales of ancient Greece, whose seductive songs lured sailors to their death along the rocky coasts. Similar songs of bewitching sweetness have been emanating during the recent period of industrialization and continue to be heard today in a world of economic globalization, resource intensive competition and material intense consumption. The Sirens sing to us of unrestrained and limitless production and consumption, ever-expanding and infinite economic growth, and the ability to have whatever we want, whenever we want, from wherever we want, as if there are no limits, boundaries or consequences of our actions.

All of our desires are at our fingertips in a consumption-based economy and this is the model all societies are pursuing.

We are following the Siren's song without seeing that they are leading us into a rocky coast of serious consequences. The Siren's song distract humankind from some very serious realities at hand:

°The reality of fossil fuel dependence and overuse.
°The reality of poisoning the air we breathe, the food we eat, and the water we drink.
°The reality of an increasing global population without effective means to care or provide for more.
°The reality of destroying self-reliant and self-sufficient lifestyles
°The reality of gaping 'have' and 'have-not' inequalities within our global family.
°The reality of committing severe crimes against humanity.
°The reality of degrading our common living space for present and future generations.
°The reality of causing disruptions to the Earth's biosphere and fostering a sixth mass extinction.
°The reality of destabilizing the global systems our settlements depend on to provide us with what we need to survive.

But the rocky coast that the international community is the most concerned with, is the very real possibility that global temperature changes, be they gradual or rapid, and resulting climatic and weather related disruptions during the 21st century will dismantle the systems and structures our modern settlements are based upon.

"Whoa, whoa, whoa…That sounds like a doomsayer's prophecy to me. Haven't we been hearing this sort of thing about the miserable state of the planet since the 1970's?"

We have, sort of. There have been piles of books written about global changes, many include catastrophic predictions based on the knowledge of that time. There have been alarming reports about problems like acid rain, biodiversity loss, human population explosions and crashes, dwindling resources, extreme pollution and ozone depletion.

We are still presented with reports of looming disaster, but nothing globally catastrophic has really happened as a result of these localized problems. We have found that the Earth is quite resilient in terms of how much we can take and use for our purposes, how much we can degrade it, and how much we can disrupt and transform natural cycles without collapsing natural or human systems. People are therefore skeptical and wary of those who exaggerate pessimism.

However, at the beginning of the 21st century, we have a multitude of extremely accurate, diverse, and advanced collaborative research that tells us, "Hey, something big is going down and we're sure of it." It is not just a small group of academics in a privately funded think tank who are saying this. It is thousands and thousands of the world's premiere scientists, working independently around the world, as well as international insurance agencies, political leaders, military and security specialists, businesses, and industries. It's not just a few NGO's raising concern, it's everyone else too.

"Is this book one of these doomsday forecasts?"

I have tried to be careful about being too catastrophe oriented and fatalist, but as you will see, there is an overwhelming amount of research that says global warming and climate changes are already happening, and the international community is making some pretty dramatic predictions for the near future.

Basically, the average global surface temperature and greenhouse gas levels are intertwined, rising and falling in tandem. Greenhouse gas cycles have been significantly disrupted during the past few hundred years, resulting from both human and natural activities; and this has risen atmospheric greenhouse gas levels. This rise is predicted to alter the global temperature, climate systems and weather patterns; which is all predicted to create a landslide of changes in the natural world and rattle our settlements to the core.

"This has been discovered by looking at events in the past and present trends?

The Earth is a big living body made up of a multitude of dynamic systems, like our own body, and by looking at as much research from as many areas as possible gives us a great holistic picture of the Earth and its activities in the past, present and probable future.

The idea is to listen to what the Earth is telling us about how it reacts to large disruptions within its systems.

One cannot rate the state of the planet by simply looking at mere numbers and statistics. Numbers and statistics are important, but in understanding the activities of the Earth, one has to take a much more sophisticated approach, integrating many fields in an interdisciplinary manner, and allowing connections and patterns to emerge.

"So what do you make of the present trends and future prognosis?"

Well, I think the facts speak for themselves: global temperature change is happening and is forecasted to cause greater disruptions to our settlements around the world if our settlements continue to function in their present fashion.

"Do you think humankind's activities have influenced the present trend of global warming?"

To be honest, I find the question of human involvement fairly irrelevant, because it does not matter what or who is causing the present trend. It is happening and our settlements will have to adapt.

No matter what the cause.

I think we like to discuss the possibility of whether or not human activity is responsible, because it brings the subject into our realm and allows us to exert some imaginary control over the predicament.

We are led to believe that we have such control, because God declared it so, science tries to prove it so, and industrialization tries to make it so. But the magnitude and complexity of Earthly changes we are dealing with are far beyond the human sphere of control.

Controlling greenhouse gas cycles is not like controlling ozone depleting chemicals, which are only made by humans and can only be reduced by humans. Naturally occurring greenhouse gases like water vapor, carbon dioxide, and methane are intimately woven into the fabric of everything that exists on the Earth, living and nonliving.

These are cycles of planetary proportion, which wee little humans cannot control; not today, not tomorrow, not ever. We cannot even control our own emissions of greenhouse gases, what makes us think we can control the Earth's?

°MAN HAS BEEN CHANGING CLIMATE FOR 8,000 YEARS
NATURE NEWS SERVICE
10 DECEMBER 2003

...Humans began altering the climate 8,000 years ago, long before the industrial revolution, claims a leading climate scientist.

Massive clearance and irrigation for agriculture released huge amounts of greenhouse gases into the atmosphere, says William Ruddiman of the University of Virginia in Charlottesville...

"But you do think human activities have had an impact on greenhouse gas levels during the Earth's recent past, don't you?"

Well, of course. It has been scientifically proven in many ways and I think it is just plain obvious that the collective activities of humankind during the past 8,000 years and more dramatically during the past 350 years have created a global scale disruption within the fabric of the Earth.

This disruption goes beyond natural variation and beyond the influences from the Sun, the Earth's orbit, or our descent out

of the last ice age. Humankind may only be one expression in the multitude of life on planet Earth, but we are a big, loud expression. Our activities during the recent past have had a global impact on the composition and integrity of the Earth's hydrologic, geologic, atmospheric, and biological systems.

Through the extensive extraction and use of fossil fuels like oil, coal and natural gas, through the removal of forest cover, through converting native lands into industrialized agriculture tracts, through pollution of the Earth's water, land, and air, through hunting animals to extinction, and through a general large scale disturbance of the natural world, we have significantly altered the face of the Earth.

Our countless independent activities have acted and are still acting as a powerful, cumulative disruptive force within the Earth's fabric of interrelated systems. We need to look no further than the present atmospheric concentration of the two most common and powerful greenhouse gases: the level of methane has doubled since pre-industrial times and the level of carbon dioxide is at its highest level in 420,000 years and probably in the past 20 million years.

This is a very significant event in the history of global greenhouse gas levels and the history of the Earth.

This event is similar to a sustained asteroid impact event that has been spread out across the Earth over a long period of time. There hasn't been one massive impact in one day, but rather countless impacts over hundreds of years every single day. Which is still just a blink in time for the Earth.

Look at it this way, the Earth functions on a geologic time scale, the time it takes for it to react to a disruption is stretched out, because it is a big celestial body. Let's say 10,000 years is equivalent to 1 minute, and 10 years is equivalent to .0001 minute. For illustrations sake, let's say the present trend of accelerated global warming is a result of countless activities during the past 300 years: 100 years + 100 years + 100 years = .003 minute. That's a dramatically quick event in geologic terms, but 300 years is quite a long time in our perception of time. The global temperature change that occurred at the end of the last ice age, roughly 11,500 years ago, took only 30 years (.0003 minutes) to rise 7° C, which is wickedly fast even in our perception of time.

LOOK AT IT THIS WAY

THE LAST TIME THE CARBON DIOXIDE LEVEL IN THE ATMOSPHERE WAS AS HIGH AS IT IS TODAY WAS 20 MILLION YEARS AGO:

10,000 years + 100,000 years + 100,000 years +

100,000 years + 100,000 years + 100,000 years + 100,000 years +
100,000 years + 100,000 years + 100,000 years + 100,000 years +
100,000 years + 100,000 years + 100,000 years + 100,000 years +
100,000 years + 100,000 years + 100,000 years + 100,000 years +
100,000 years + 100,000 years + 100,000 years + 100,000 years +
100,000 years + 100,000 years + 100,000 years + 100,000 years +
100,000 years + 100,000 years

CARBON AT 20 MILLION YEAR HIGH
BBC NEWS ONLINE
17 AUGUST 2000

Two British scientists say the level of carbon dioxide (CO_2) in the Earth's atmosphere is higher than for 20 million years. But their study of levels over the last 60 million years suggests that the gas was once even more abundant than it is today.

Carbon dioxide is the main gas caused by human activity that has been linked to global warming.

Concentrations now are about 360 ppm (parts per million), but will continue to rise as emissions increase...

+ 100,000 years + 100,000 years

As climatologists gather here this week to discuss new research on global warming, a disquieting idea has been gaining currency--the possibility that small shifts in global temperature could lead to sudden and abrupt climate changes.

...Newer, more sophisticated models suggest that the Earth's climate system is "nonlinear"--in other words, small changes can have large effects on everything from ocean and land temperatures to drought and monsoon patterns, icecaps and tropical rain forests...

Global climate change, often seen as a process stretching over thousands of years, could in fact occur abruptly and unexpectedly -- quickly pushing up temperatures by as much as 18 degrees Fahrenheit and wreaking havoc on human society, scientists warned on Wednesday.

"Climate change is not always smooth. Sometimes it is abrupt," said Richard Alley, a climate expert at Pennsylvania State University and lead author of a new National Academy of Sciences report on the threat of rapid climactic shifts...

THAT WAS A REALLY LONG TIME AGO

DRAMATIC GLOBAL TEMPERATURE CHANGES HAPPENED 11,500 YEARS AGO:

100 years + 100 years

AND THEY HAPPENED IN 30 YEARS:

10 years + 10 years + 10 years

THAT IS REALLY QUICK

"What do you think of the media coverage on global warming, climate change and possible abrupt shifts?"

I think the international news media has done a fantastic job at reporting the research findings, but one has to search out the news themselves; these issues are not usually on the front page of your daily paper. These topics have also made it into the mainstream, with million-dollar Hollywood movies about catastrophic and swift climate changes and popular books that weave the science of global warming into their story. But, these fragments of pop-culture are science fiction and often distort the facts of the matter, leaving the general public no more informed than they were before, probably more confused, and most likely very skeptical of facts they may read later.

A lot of eyes glaze over when these subjects are addressed, because the issues concerning global temperature change seem abstract, daunting, overwhelming, complicated, time consuming, and far off in a distant future. There are no easy answers and that is what we are used to.

Most people have heard about climate change and think, "Oh yeah, the planet is warming, the climate is changing...what does that mean? I don't know, so I won't worry about it. Besides, I won't be alive when it happens, the predictions might not come true anyway, and I probably won't be affected. So why should I worry, or even care, about the distant future, when I've got enough to worry about today?"

This is a totally normal response to something that appears incomprehensible and irrelevant upon first glance. But these issues are neither incomprehensible, nor irrelevant. One just has to learn how to connect the dots and see the picture that emerges; and people are beginning to see the emerging picture could dramatically affect our lives today and the lives of those for many generations to come.

Those living comfortably might have heard of global warming and climate change, but they are a non-issue for daily life. For those struggling out a daily existence, these global issues are not even present their minds, because they are concerned with finding water, food, and cooking fuel to keep them and their family alive one more day.

We focus on issues that are tangible and in our immediate reality. The weather in the year 2040 and disruptions related to it don't appear that important to our life today, because they don't affect what happens today.

It is also difficult to imagine how things might be done differently and what living under different circumstances in the near future might be like. We are comfortable with routine and familiarity, which is perfectly natural. Changes in the familiar make us uncomfortable and uneasy and nobody likes to feel this way.

But, the predicted consequences of global temperature change in store for our settlements are not some vague notion for sometime in the distant future; these things are real and are happening today. More people are being affected by natural disasters. In mountainous

°CLIMATE FILM 'FLAWED BUT USEFUL'
BBC NEWS ONLINE
12 MAY 2004

The blockbuster climate disaster movie *The Day After Tomorrow* contains badly flawed science and ignores the laws of physics, leading UK scientists believe.

But many of them have welcomed the film as a dramatic and popular way to raise people's awareness of climate change. Sir David King, the government's chief scientific adviser, said he hoped many ordinary Americans would see the film. And the former US Vice-President Al Gore said the risks the film portrayed were a threat to our common future.

Speaking in London, Sir David described The Day After Tomorrow as "a spectacular action film" which portrayed the switching off of the Gulf Stream and the Northern Hemisphere's subsequent plunge into a new Ice Age. The scientific consensus was that climate change might lead to a weakening of the thermohaline circulation (THC), the phenomenon that drives the Gulf Stream; but it was not expected to cause its complete halting, as in the film.

Sir David said the present global atmospheric carbon dioxide concentration of 379 parts per million, the highest for at least 420,000 years, was "very significantly higher" than during previous warm periods.

But that did not mean the THC, which keeps north-western Europe about 5C warmer than it would be otherwise, would switch off at all, and certainly not as quickly as *The Day After Tomorrow* suggested.

...Dr David Viner, of the Climatic Research Unit at the University of East Anglia, told BBC News Online: "The film got a lot of the detail wrong, and the direction of change as well - cooling of this sort is very unlikely with global warming. "But the fact that *The Day After Tomorrow* raises awareness about climate change must be a good thing."

°CARBON 'REACHING DANGER LEVELS'
BBC NEWS ONLINE
13 OCTOBER 2004

The UK government's leading scientist says levels of carbon dioxide in the atmosphere already represent a danger.

Professor Sir David King told a London audience the world had to adapt to prepare for significant changes ahead, and also to reduce greenhouse gases.

He said climate change was "the most serious issue facing us this century and beyond", needing global solutions.

On present trends, Sir David said, the world was just 60 years from triggering an irreversible climate disaster.

..."Is there a point where the melting becomes irreversible?" he asked.

"Yes, there is. When the temperature around Greenland is 2.7C above the pre-industrial level - that is the tipping point."...

regions, more precipitation is falling as rain, which affects the water source for drinking, industrial, agricultural, and energy uses for settlements downstream. Droughts are lasting longer and monsoon seasons are becoming more severe. Glaciers are melting and sea levels are rising. Coastal cities can only build their dykes so high and as for the island nations, well, they'll just have to find somewhere else to build; somewhere a few thousand miles away. These changes are already affecting state economies, food production, international insurance companies, health services, and security.

Regardless of wealth, level of development, economic and military might, religious beliefs, or political establishment, the disruptions predicted as a result of global temperature change will fundamentally affect everything and everyone, everywhere.

The Earth does not choose who is punished with what problems. The Earth knows no color, no language, no level of development, no sex, no religion, no ideas, no government, no geopolitical boundaries; it is simply reacting to a stimulus, and if it affects us, that's our problem.

"You obviously see the predicted consequences as a grave threat."

Well sure man. I think it's the greatest known threat to modern settlements that exists. And so do a lot of other people.

21st century global temperature change is greater than any other predicament facing humankind today and because it is something outside of the human sphere of control, it is a great equalizer, it is neutral, it has no agenda. For this reason, it is an issue that all of humankind can focus on and come together to do something about and at the same time, secure their own way of life without making the situation worse.

For this reason, I also think it provides a great opportunity.

I think it is a great opportunity for cooperative endeavors between North and South, East and West.

It is a great opportunity to remedy conflicts within our species by assisting our unknown brothers and sisters adapt to changes.

It is a great opportunity to change the way we acquire basic necessities, making our energy, food and water sources more healthy, stable, and secure in light of predicted consequences.

It is a great opportunity to restructure our settlements to function in a more sustainable and benign way within the fabric of the Earth.

The Earth is changing and will continue to change. We can either alter our thinking, our behaviors, and the foundations of society, or we can let global temperature and climate changes alter these things for us, which won't be quite as pleasant as us doing it ourselves.

I'm not talking about dismantling what we have, just making adjustments in the face of change. I'm also not talking about giving up our way of life, just figuring out different ways to supply the things we need to survive and prosper.

And in order to figure out what to do, we must first understand the predicament, the PROBLEMATIQUE related to global temperature change. So let's get into it.

"I guess we should let people read on and come to their own conclusions. Anything else?"

This book cannot answer all of the questions related to global temperature change. What it can do is give the reader a solid understanding of where humankind and the Earth have come, where we are, and where we are going at the beginning of the 21st century.

Global temperature change and the consequences predicted for our settlements is something that we all must comprehend and confront. This is not a matter of saving the Earth, it is a matter of saving ourselves.

This book is a guide through the labyrinth of global temperature change; a beacon of light, so that all may clearly see the rocky coast approaching. It would be a shame if we crashed because our eyes were blurred or we denied the rocky coast existed.

Welcome to spaceship Earth.

We're all in this together.

Better buckle up.

There are some bumps ahead.

to be and to become...that is the answer.

ACHTUNG BABIES

THIS IS YOUR CAPTAIN
SPEAKING

THANK YOU FOR
CHOOSING PLANET
EARTH FOR YOUR
TRAVELS THROUGH
THE COSMOS

WE HOPE YOU ENJOY
THE RIDE

TAKEOFF WILL
COMMENCE IN...

...5...4...3...2...1...

LIFTOFF

EARTH
RACING TO NOW+HERE

1

FEEL YOUR THOUGHTS

TOUCH THE GRASS, TOUCH THE SKY
TOUCH YOUR MIND, FEEL YOUR MIND'S EYE

There are probably a number of you reading this book who already have a strong understanding of the dynamics involved with global temperature change and climate change. But, for those of you who are delving into this subject for the first time, there are a few concepts that will aid in your understanding of how the dynamic systems of the Earth are believed to work. To understand these dynamics, one must summon the ability for complex thinking, which is basically looking at many related elements and pulling them together so a full picture emerges. Complex thinking involves investigation, reasoning, deduction, inference, even intuition and instinct; it takes all of the capacities which make our modern mind so creative and turns it into a brilliant problem solver.

For the past few hundred years, our educational learning has been highly **COMPARTMENTALIZED**, examining individual subjects within their individual fields. This has caused a **FRAGMENTATION** of knowledge by separating the pieces of scientific discovery without bringing them back together again, losing sight of the big picture and the interaction between pieces. Although it is important to examine each individual piece of evidence on its own within its own field, when dealing with global dynamics and change, it is very important to see how the pieces all fit together.

We are now moving toward an **INTERDISCIPLINARY** approach, in which many fields come together to share ideas and look at issues and problems from many different angles. By examining problems in an interdisciplinary way, one is less likely to leave out important factors; it is the only true way one can develop a comprehensive understanding of the issues involved.

In learning about global temperature change, one cannot adequately assess the issues in an isolated way using only a few sets of data from only a few areas of research and scientific fields. One has to look at as much information from as many sources and as many areas as possible. It is not only the climatologists who are doing important research, it is also the hydrologists, the atmospheric scientists, the paleoclimatologists, the biologists, the geologists, the chemists, the biogeochemists, the physicists, the meteorologists, the oceanographers, the anthropologists, and the archeologists. They are all working on the Great Puzzle, connecting the dots to produce a clear picture of the past, present, and future of global temperature change.

When observing Earthly dynamics and projecting their future behavior, we are looking at how the systems of the Earth act and react to one another on micro and macro levels. Looking at how the systems are related to one another is referred to as **SYSTEMS THINKING** and examines how large systems interact with one another, as well as examining the individual components of each system.

open your mind and say
aaaaaaaaaaaah

Systems thinking was an approach originally designed to address social problems and is commonly used by city planners when developing a city, ecologists when looking at habitat preservation, and doctors when addressing a patient's ailment. Understanding how the Earth works in this way allows us to adequately address and assess the consequences for the Earth and humankind as a result of global changes, which allows us to develop appropriate strategies for adapting the complex systems of our settlements to such changes.

This **HOLISTIC** way of thinking, a thinking which recognizes connections between life forms, between Earthly systems, between places, and between people is becoming increasingly prevalent in all aspects of science and policy. Holistic thinking pulls together relationships which were previously ignored and allows for a much better treatment of the situation at hand.

There are a number of names for holistic approaches that use systems thinking: **INDIGENOUS SCIENCE, INTEGRAL SCIENCE, ECOPSYCHOLOGY, EARTH SYSTEMS THEORY, AND THE GAIA HYPOTHESIS**. All of these perspectives see humankind and the Earth as integrated actors in a great global body. The activities of humankind affect the systems of the Earth in significant ways, just as the systems of the Earth significantly affect the activities of humankind as well. By approaching the situation in this way, consequences of actions are approached with greater foresight of how one activity affects a situation far down the line.

There are several other concepts which help us understand how the Earth works with regard to global temperature change. Most understand the concept of cause/effect and chain reactions. Fewer understand concepts like critical thresholds, synergetic momentum, positive feedbacks, steady-states, dynamic equilibrium, and linear and non-linear dynamics. These are concepts which help explain how certain phenomena work, how systems of the Earth can react so quickly to certain stimulus, and why scientists around the world are deeply concerned with the present state of the planet.

CAUSE/EFFECT is fairly straightforward. The cause is the action and the effect is what happens as a result. A cause would be pushing a ball, the effect would be that ball rolling. A cause would be global temperature rise, the effect would be global glacier retreat.

A **CHAIN REACTION** happens when one cause leads to one effect, which is the cause of another effect, and so on down the line. A chain reaction occurs when one fells dominos, or explodes a nuclear bomb, or starts an automobile. A chain reaction also occurs when, for example, the lowest species on a food chain dies off, affecting the next level species, and pretty soon the entire food chain has reacted itself into oblivion.

Sometimes it takes the Earth a long time to react to a stimulus that was caused very quickly. Sometimes it takes a very short amount of time for the Earth to react to a stimulus that has taken a long time to develop.

It depends when a **CRITICAL THRESHOLD** is reached.

A critical threshold can be imagined when one snaps a wooden stick. The stick is flexible when bent, able to withstand a certain amount of

°"FEEDBACK COULD WARM CLIMATE FAST: HOLISTIC MODEL HINTS NEXT CENTURY COULD GET EVEN HOTTER THAN WE THOUGHT
NATURE NEWS SERVICE
23 MAY 2003

The twenty-first century could see more warming, more quickly, than was previously estimated, hints a new approach to modelling the Earth's climate.

...Earlier climate models looked at a limited set of factors and often measured changes in the ocean and on land separately. The new approach, developed at the Hadley Centre for Climate Prediction and Research in Bracknell, UK, accounts for as many influences as possible, including volcanoes belching out millions of tonnes of carbon dioxide, fluctuations in the Sun's activity as well as changing levels of greenhouse gas and ozone. It also allows oceans to affect the land, and vice versa.

The Hadley Centre team first raised the alarm in 2000. They showed that, as increasing levels of carbon dioxide in the atmosphere warm the planet, more carbon dioxide would be released from vast reserves in oceans and forests. Even a slight drying of the Amazon rainforest, for example, would release billions of tonnes of carbon dioxide into the skies.

...Now the Hadley team balances the books with a new holistic climate model - dubbed the Earth systems approach.

..."It looks like [Jones' team is] about right," says [climatologist Jorge] Sarmiento. But he warns that modelling the importance of the feedback from stored carbon from oceans and forests of the future is a tricky business. "There is a lot of uncertainty associated with this stuff".

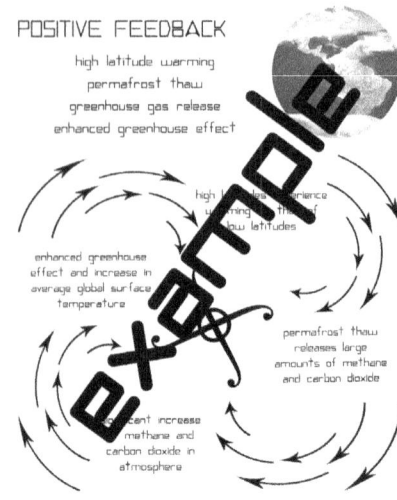

POSITIVE FEEDBACK

high latitude warming
permafrost thaw
greenhouse gas release
enhanced greenhouse effect

high latitudes experience
warming the of
low latitudes

enhanced greenhouse
effect and increase in
average global surface
temperature

permafrost thaw
releases large
amounts of methane
and carbon dioxide

recent increase
methane and
carbon dioxide in
atmosphere

°NON-LINEAR RESPONSES AND SURPRISES TO GLOBAL CHANGE
PROFESSOR IAN NOBLE
AUSTRALIAN NATIONAL UNIVERSITY
JULY 2001

We live in a world where rapid and unpredicted system responses can result from even very small changes in forcing conditions or from gradual and continuous environmental change. Such nonlinear dynamics is the product of complex interactions and feedbacks and/or from simple biochemical and structural threshold-like responses that cascade throughout a system.

This type of nonlinear behavior is characteristic of the way ecological and climatic systems function, yet, tremendous effort has been geared towards describing complex nonlinear responses as more readily tractable linear ones. Linear thinking is also very much entrenched in the way policy-makers perceive environmental change and, consequently, ways to manage it...

pressure and return to its original shape. However, there comes a point of pressure when the stick snaps in two. Just a fraction more of pressure exerted determines the difference between a broken stick and a non-broken stick. Once the threshold is breached, changes can be swift, dramatic, and long lasting.

SYNERGETIC MOMENTUM. I think it is a great name for an independent rock band. It is also a term used to describe how seemingly inconsequential activities combine to have a rapid cascading effect of immense magnitude. Synergetic momentum can be seen in a phenomenon like an avalanche. The effect of gravity on every flake of snow is relatively inconsequential until they are combined and after one more flake is added, thousands of tons of snow come crashing down.

A POSITIVE FEEDBACK is fairly self-explanatory: a cause produces an effect which reinforces the cause which again reinforces the effect; a chain reaction in a closed system which keeps reinforcing itself. The positive feedbacks that are of the most interest to those researching global temperature change dynamics are greenhouse gas feedbacks. In the present Earth system, one possible positive feedback is this: methane and carbon dioxide release from thawing Arctic permafrost. The tundra of the North Pole is frozen and contains a huge amount of frozen methane and carbon dioxide gas. As the global temperature rises, the permafrost thaws and releases these gases into the atmosphere. As more methane and carbon dioxide accumulate in the atmosphere, the greenhouse effect is enhanced, which increases the average global surface temperature, which warms the Arctic, which thaws the permafrost, which releases more methane and carbon dioxide, which enhances the greenhouse effect, which increases the global temperature...

It is believed that the Earth as a whole exists in a STEADY-STATE of DYNAMIC EQUILIBRIUM. This means that the Earth's dynamic systems are in relative balance and stay this way for long periods of time. Influences and actors come in and out of the system, but the system stays in relatively the same state, until something pushes it over a critical threshold and into a new steady-state.

The Earth is presently experiencing a more temperate, shorter lived and active INTERGLACIAL steady-state, compared to a longer lasting and less active ICE AGE. Steady-states also occur in individual Earth systems, for example, a tract of rainforest produces its own microclimate until something, like deforestation or forest fires cause the rainforest to dry out.

Planetary steady-states can change during the course of millennia or sometimes as quickly as decades or centuries, when the Earth experiences great system-wide disruptions to the balance of its present equilibrium. The Earth then settles into a steady-state of dynamic equilibrium which may be similar or different than the previous steady-state, a warmer or cooler planet than before.

Cause and effect, chain reactions, critical thresholds, synergetic momentum, positive feedbacks, steady-states, and dynamic equilibrium are all part of understanding how the Earth's systems function in LINEAR and NON-LINEAR ways. We are accustomed to thinking in a linear fashion: cause and effect in one straight line for now and forever. But, the complex systems of the Earth have numerous factors that are interdependent, acting and reacting upon one another in a sometimes random way which cannot be predicted with 100% accuracy. Non-linear thinking allows one to imagine the many possibilities available to a given existing state.

Here is one last example which ties all of these concepts together in a scenario that has happened during the Earth's past and may happen in the near future. This example takes us to the North Pole and the North Atlantic Ocean, a place considered to be the Achilles' Heel of the ocean's circulation system.

Areas in the North Atlantic Ocean act as a sort of pump, pushing deep, cold, saltier water under warmer, less salty surface waters. This pump eventually brings the warm Gulf Stream northward, where it unloads its heat, cools, sinks and flows southward again. As the global temperature increases, the ice on land and in the sea at the North Pole melts, the darker colored land is exposed, which increases open water, reduces the Earth's reflectivity, increases the absorption of solar heat energy, accelerates ice melt…another positive feedback.

As fresh water flow increases into the North Atlantic Ocean, it dilutes the colder, saltier water, weakening the thrust of these pumps. Should the amount of freshwater entering the North Atlantic reach a critical threshold and dilute the ocean enough to weaken the pumps and halt the movement of the Gulf Stream, there would be significant changes to the present steady-state of ocean circulation, the planet's climate systems, and the weather patterns our settlements are based and dependent on for life as we know it.

And that…is just the tip of the iceberg.

Trickle..trickle…trickle…

°**NATURE PLANTS DOOMSDAY DEVICES**
THE GUARDIAN
25 NOVEMBER 1998

…The world's ocean circulation seems to have two 'modes'. One maintains the Gulf Stream; the other does not. And geological evidence suggests that it switches between the two modes without warning, and virtually overnight. "The Gulf Stream has been flowing for 10,000 years. But we are close to the cusp of a change," says Stefan Rahmstorf of the Potsdam Institute for Climate Impacts Research in Germany.

Last week, Rahmstorf placed on the Internet new research showing that global warming could push the system over that cusp and shut down the Gulf Stream. The crunch could come within the next 50 years, though his best guess is that we have another century before the big freeze.

"If the system shut down, winter temperatures in the North Atlantic region would fall by 10 degrees Celsius within 10 years," says Wallace Broecker, a geochemist at Columbia University in New York. "London would experience the winter cold that now grips Irkutsk in Siberia."…

2
A GRAIN OF SAND IN THE BEACH OF TIME

THE BEACH OF TIME

Energy flows like a stream, through time, through matter, through each one of us, and through everything else in existence. The Universe is a big place, a really big place. If you consider that our existence on planet Earth is equivalent to living on a grain of sand in the beach of time, you feel small, really small.

In the grand scheme of things, that is about our place in it all.

The time we enjoy or despair on planet Earth is nothing more than a quiet peep and the Earth is nothing more than a tiny blip on the radar screen. This is why we have religions, to put things in perspective, to forget that our existence is more or less meaningless and to hope for a better life the next time around if we are good this time around.

This is a great paradox of life. Although our time here may be more or less meaningless in the grand scheme of things, our life and our immediate experience is the most important thing happening. This life is the only life we know and it is right here, right now. Those involved in our lives and the events that shape our lives are what make our world go around.

you are

Our existence on planet Earth may be more or less meaningless in the grand scheme, but it is ultimately meaningful to us as individuals during our time here. For some, anything that does not affect the immediate experience does not really matter. For others, anything that has happened in the past still affects them in the present. Still for others, anything that might happen in the future affects the actions they make today. Our experience influences how we go about our life and how we go about our life influences our present and future experience.

The human mind and body are very sensitive receptors to the stimulus of the world, whether we are consciously aware of it or not. The more aware we become of ourselves, our fellow humans, the rest of the living planet we inhabit, and the situation which bounds our immediate existence, the more likely it will be that we understand how to live in our immediate experience and in our not-so-distant future.

Supposedly the Universe began about 15 billion years ago with a tremendous burst of energy, originating out of a void of nothing to create everything. Whatever initiated it, the **BIG BANG** gave birth to all the energy, all the matter, and all the time that the Universe would ever contain; all happening in .0000000000000000 00000000000000000001 second.

Now that is a fast birth.

Most believe that a prime mover was involved, a divine force so awesome, that creating the known Universe was nothing more than blowing a kiss into space.

Some give the prime mover form, others do not.

Stories are stories and they all help us to go on.

The story of modern science says that after the Big Bang and over a long period of time, parts of atoms gathered into elements, which gathered into gases, which gathered into stars, planets, and black holes, which gathered into galaxies, galaxy clusters, everything else that we can see, and all of the rest we cannot see, but believe is there.

Do you believe we are alone in the Universe?

Do you believe there are no other life-bearing planets?

I find it hard to believe that the Earth might be the only life-bearing planet in the cosmos. It is possible, but highly unlikely, since the same elements that brought life to this planet are abundant throughout the entire Universe. There are a few hundred billion stars out there and probably just as many planets; just in our own galaxy!

The celestial bodies in the cosmos are not immutable objects, they go through changes, they are constantly in a state of becoming. These objects may appear to be unchanging and stable, because celestial beings of this size change slowly relative to our perception of time. However, as we look at the history of our solar system and the history of our planet, we see there have often been tremendous events which brought great changes in a relatively short amount of time.

There is a magic in the Universe.

There are many mysteries that are not fully explainable within the framework of understanding we presently have. Some of the principles within this framework are well established, some principles are not, and some are still waiting to be discovered and understood when the time is right. When the unknown principles are discovered and understood, the framework changes and the way we perceive the world around us changes as well.

The meaning of life, the creation of the cosmos, the evolution of a species; these things are truly marvelous subjects to ponder, allowing us to slip away from the grind of our daily struggle and daydream about the unimaginable.

Let us take a look at some of the brief moments of the past which have brought us to where we are today and may shed some light on what we can expect in the possibilities of tomorrow.

Let us begin the PARADIGM SHIFT.

here

GALACTIC MARBLES

°EARTH'S OFFSPRING:
THE COLLISION THEORY
SPACE.COM
1 SEPTEMBER 2000

In the decades since Apollo astronauts collected moon rocks, one theory has emerged over competing explanations of lunar birth – the moon formed when a big object whacked Earth.

Scientists believe the solar system formed 4.56 billion years ago from a cloud or nebula of gas and dust surrounding the newborn Sun. The nebula condensed into pebbles, rocks and larger "planetesimals," which collided and fused to create the planets.

A theory that the moon and Earth both condensed or accreted from the same material was rendered doubtful by a major difference in the moon's composition – the lack of a significant iron core like Earth's.

...The moon's composition was enough like Earth's mantle and crust to make some believe the moon formed from material hurled off of a rapidly spinning Earth. But hard evidence could not be found.

...And none of the old theories explained how the moon's oldest rocks solidified from molten rock about 4.44 billion years ago, roughly 100 million years after the solar system formed.

The giant impact theory proposed that at least 50 million years after the solar system formed, a large protoplanet whacked a perhaps still-molten Earth, heating and ejecting debris from both objects. Part of the debris then clumped together to form the moon, which was covered by molten rock from the heat of the collision...

Let us go back, way back, jump back, get deep, get back to where we came from, to where we once belonged. Not as far back as the beginning of the Universe, but back to the beginning of the Earth and Moon, around 4.5 billion years ago.

Although much of the evidence has been transformed or disappeared completely as the Earth has been folded, twisted, and molded over eons like a big ball of dough kneaded by the hands of God, there are remnants from that time still around today and they tell us a lot about how things were back then.

These fragments of the past tell us that 4.5 billion years ago was a time of planetary creation, when the Solar System contained hundreds of tinier planet-like objects, rather than the nine planets with their moons that are in our known Solar System today. These **PLANETESIMALS** circled the Sun for years and years in irregular orbits, crashing into one another, reforming, changing their orbits, and crashing into one another again, until all of the tiny planets and other objects gradually settled into their present forms and orbits we see today.

The fresh Earth during this hot, fiery Hadean period did not look like the blue-green Earth of today. At that time, the Earth was full of active volcanoes, seas of molten lava, and boiling sandy oceans. The Earth was just a big churning pot of gooey magmatic stew. Sure, it had an atmosphere, but it consisted primarily of methane and carbon dioxide, there was not much oxygen to be found.

Galactic Marbles was the only game in town and the Earth was one of the marbles still on the playing field, constantly being bombarded with an array of objects travelling around the foundling Solar System.

And then one day...PLOOOOPSSHH!

A littler planet-sized object slammed into the Earth, causing it to eject massive amounts of Earth goo far into the heavens. Just like tossing a stone into a pond, but on a larger scale.

Much of the material evaporated on impact and some material shot up in a giant plume and fell back to the Earth.

But one big molten glob splashed into the upper atmosphere, cooled, solidified, and got stuck; suspended in the Earth's gravitational pull.

That glob is our Moon.

Rocks from the Moon and the Earth dating back to this time have a very similar structure and composition. The Moon also has a very light weight and no atmosphere. It is more or less just a big ball of pumice created by an impact with the early lava filled Earth.

At least this is the way I imagine it happened.

so the moon was born

l
i
k
e

a

s
t
o
n
e

t
o
s
s
e
d

in the stream of time

complex crater formation

impact

melt

uplift

Since the formation of our present-day Solar System, the only objects that remain a threat of impact with Earth are asteroids, meteors and comets. These celestial bodies have and will continue to come crashing into the Earth. Sometimes these impacts cause great disruptions to the steady-state equilibrium of the Earth, sometimes they cause no noticeable changes at all.

There have been many collisions with such objects throughout the history of the Earth and when these objects make contact, they often leave craters as their signature. Humankind has discovered impact craters all over the world on land and because the Earth is 2/3 water, many more objects are believed to have fallen into the seas.

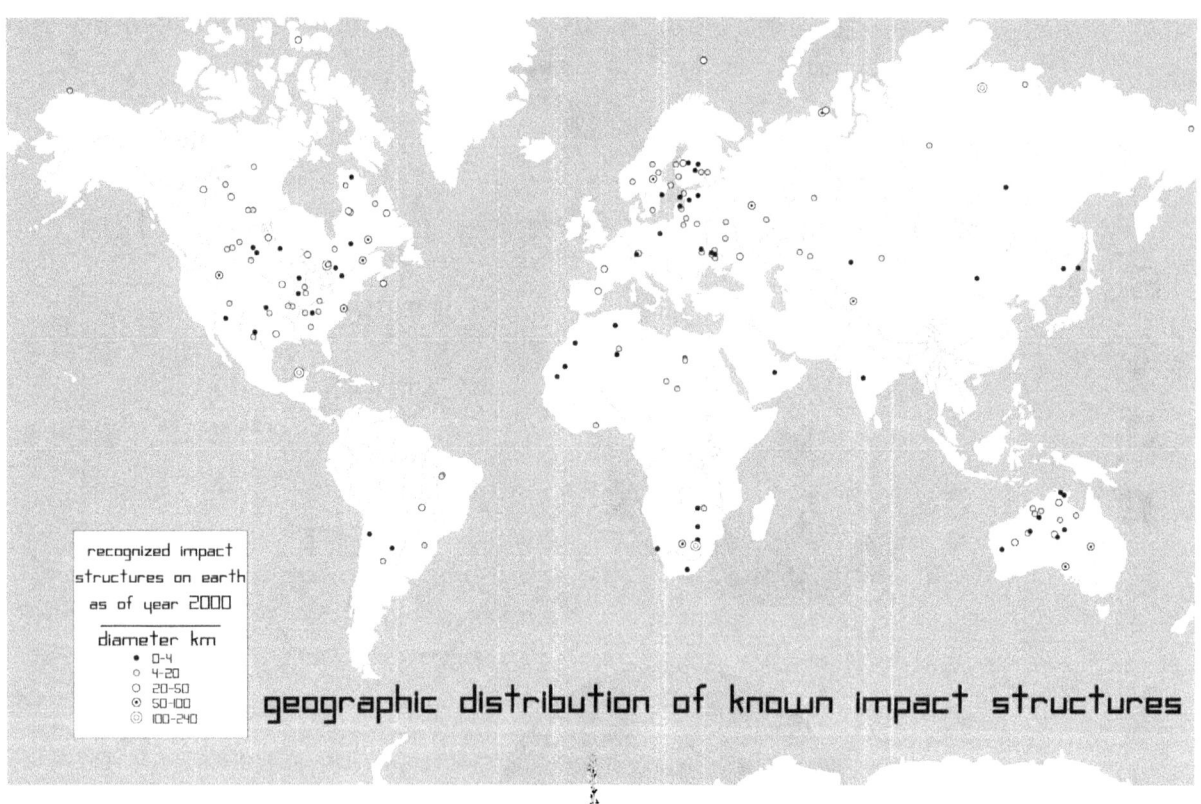

recognized impact
structures on earth
as of year 2000

diameter km
• 0-4
○ 4-20
○ 20-50
⊙ 50-100
⊙ 100-240

geographic distribution of known impact structures

Around 250 million years ago, the largest mass extinction occurred on planet Earth. This is an episode known as **THE GREAT DYING**, in which 75% of all plants and animals on land and 90% of all marine life died over the course of about 50,000 years.

Reasons for this mass extinction are varied. Most scientists believe the mass extinction was caused by an impact event with an asteroid the size of Mount Everest, around 15 kilometers in size. This event would probably have blotted out the Sun for months or years and caused a dramatic shift in local and global climate systems and weather patterns. The discovery of a massive undersea crater found off the coast of Australia adds weight to this theory, as does the existence of unique molecules of helium, argon, and iridium in rocks found in Antarctica and the European Alps that date back 250 million years. These chemicals are abundant in space rock, but rare on Earth.

Some believe The Great Dying was caused by a massive outpouring of greenhouse gases during a period of supervolcanism, at which time the Earth would have experienced accelerated global warming and reduced oxygen levels. This would also cause dramatic shifts in climates and weather patterns, and the lack of oxygen would have caused a global suffocation. This theory is partially based on the remains of a great ancient lava flow found in Siberia and corresponding paleoclimatic records that show a volcanic outpouring in this area from this time.

Others believe the violent volcanism was already occurring and an asteroid impact just made the situation worse. Still others believe the massive impact, whose shockwaves would have set off a chain reaction of earthquakes, stimulated supervolcanism around the world.

Regardless of which theory is right and which is not quite as right, disruptions of this magnitude, be their source extraterrestrial or terrestrial, will cause a multitude of changes for the systems of the Earth and that which lives within them. Events like these can initiate rapid and irreversible local and global changes, often taking the Earth a long time to settle into a steady-state again. The disruptions that take place during such a transition can cause species to die out and cause settlements to collapse.

These consequences have happened time and time again as a result of such geologic and biologic disturbances in the past and will happen as a result of such disturbances in the future.

That is just part of living within a dynamic body called the Earth.

Perhaps the most famous asteroid impact event happened around 65 million years ago. This event wiped out about half of all life on Earth, including the dinosaurs of the day, and may be the event that allowed mammals to succeed in becoming the dominant creatures on the planet.

65 million years ago, Tyrannosaurus Rex was the hot shot around, until a couple of big rocks flew into the Earth and disrupted everything the T-Rex was dependent on for survival. One asteroid, believed to have been about 10-15 km in size, left its mark on the border of the Yucatan Peninsula in the Gulf of Mexico. This is the site of the Chicxulub crater. Another crater has been found under the waters of the North Sea and is believed to have been created by an asteroid of relatively the same size landing the same time.

SPACE OBJECT THAT KILLED DINOSAURS BROKE THROUGH EARTH'S CRUST
SAN FRANCISCO CHRONICLE 18 DECEMBER 2000

The monstrous object from space whose impact on Earth doomed the world's dinosaurs and most other life forms to extinction 65 million years ago is yielding new insights into one of the world's great catastrophes.

Scientists studying drill cores from the impact crater's edges and analyzing the evidence from far-scattered rocks have determined that the object must have blasted all the way through the crust of the Earth, allowing molten material from the mantle beneath to well upward and releasing immeasurable quantities of deadly sulfurous gases and carbon dioxide.

...Evidence of the source of the blast later pointed to a crater known as Chicxulub (pronounced shix-shi-loob), centered on the seabed just off Mexico's Yucatan Peninsula...

°GIANT CRATER FOUND UNDERSEA

DISCOVERY.COM
1 AUGUST 2002

A quest for oil in the North Sea has turned up an ancient impact crater so well preserved that it could give scientists fresh insight into the effects of large meteorite impacts on Earth.

The 12-mile wide crater is buried under 120 feet of water and more than 900 feet of sediment, which has helped preserve features that on Earth's surface would have been eroded away.

The planetary scar is believed to have been caused by a meteor or comet impact 60 million to 65 million years ago, and is unlike anything that has been found on Earth or the moon. In fact, it is so pristine that it resembles craters previously found only on Jupiter's icy moons...

It is believed these asteroids originated from the Asteroid Belt between Mars and Jupiter, and were dislodged by a burst of energy emitted from the Sun. A burst so strong, that it knocked the asteroids out of orbit and affected the gravitational stability of Mercury and the Earth.

That is a serious starburst.

As these town-sized rocks hit the Earth, they hit with the force of a million nuclear bombs, all going off in one place at one time, sending shockwaves around the world, triggering tsunamis and earthquakes, and stimulating volcanism which blanketed parts of the planet with debris so thick it blocked the Sun for many seasons.

That is one big explosion with big consequences for all life on Earth and the systems that provided the Earth with its pre-impact equilibrium. After an impact like this, everything the dinosaurs were used to changed very dramatically, very quickly, and the changes lasted a long time. It is believed that the average surface temperature dropped significantly for many decades and that the climate systems never did return to their pre-impact state, taking the Earth over 1,000 years to settle into a different steady-state.

Those that could adapt did so.

You can see them in the form of birds, alligators, ferns and orchids.

And those that could not?

You can visit them in your nearest natural history museum.

SO WHEN WILL THE NEXT ONE PASS BY?

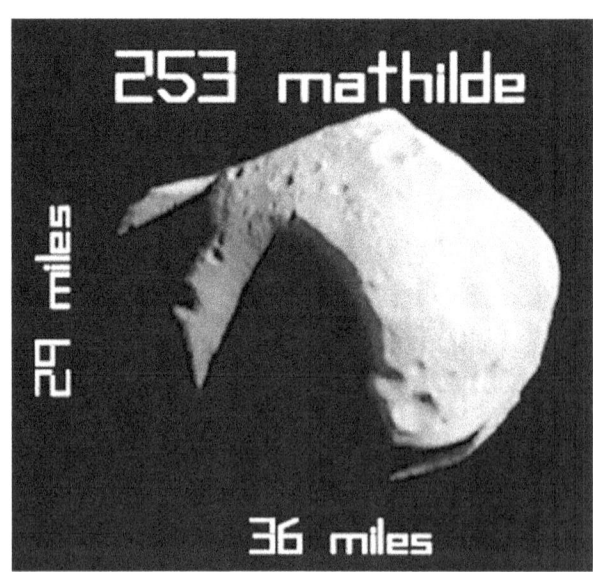

253 mathilde

29 miles

36 miles

WHEN WILL THE NEXT ONE HIT?

WILL WE EVEN KNOW?

°HUNT FOR POTENTIALLY DEADLY ASTEROIDS UNDERFUNDED, PANEL SAYS
SPACE.COM
10 JULY 2002

The U.S. government should invest more money in tracking near-Earth objects that might threaten Earth, said members of a space roundtable on Capitol Hill Wednesday. While the Air Force is not tasked with tracking near-Earth objects, U.S. Air Force Brig. Gen. S. Pete Worden said such a mission would be appropriate for the service and an assignment could occur "in the next few years," he said.

...Currently, NASA spends about $4 million per year on programs that track space objects larger than a kilometer in diameter (0.62 miles), said Colleen Hartman, director of NASA's Solar System Exploration Division.

NASA, however, does not track objects the size of the recently discovered 2002 MN2 an object between 50-100 meters in diameter (roughly 50-100 yards) that passed within 75,000 miles of Earth in June, the panelists said. The rock was found three days after it flew by...

SHOULD WE?

°'DON'T TELL PUBLIC OF DOOMSDAY ASTEROID'
THE TIMES ONLINE
15 FEBRUARY 2003

The public should not be told if scientists detect a huge asteroid on a collision course with Earth that cannot be deflected, a disaster expert said.

Geoff Sommer, of the Rand Corporation in Santa Monica, California, said that governments would be wrong to warn of an impending impact that could destroy all life if there was no realistic prospect of stopping it. The panic, misery and disruption that such a warning would cause would not be worthwhile, he told the association.

It would be better for the fewest people to know that mankind was about to become extinct in a fashion similar to the dinosaurs.

Mr. Sommer said: "It makes sense to warn if there's something you can do but if you can't intercept it, if you can't move people out of its way, it makes sense not to occasion further social costs."...

°UNSEEN COMETS MAY RAISE IMPACT RISK FOR EARTH
NATURE NEWS SERVICE
18 OCTOBER 2004

The Solar System could be teeming with almost invisible comets, according to some astronomers' calculations. If they are right, such extra comets would significantly increase the risk of a catastrophic impact with Earth.

These objects have never been observed, but the astronomers argue that 'dark comets' provide a likely explanation for an astronomical puzzle: we can only see a tiny fraction of the comets that theory predicts.

...The common explanation for this discrepancy is that the comets quickly disintegrate into smaller lumps after just one or two orbits, says Bill Napier, a recently retired astronomer who worked at the Armagh Observatory, Northern Ireland.

...The dark comets would present a major challenge to astronomers searching the skies for objects that might collide with the Earth. "They're so black you can't see the damn things," says Napier. "These things will just come out of the dark and hit you with no warning. It looks as if we're dealing with a substantial impact hazard that people haven't clicked into yet."...

°Human beings have transformed between 1/3 and 1/2 of the entire surface of the Earth.

°About 2/3 of the oceans fisheries have been exploited, over-exploited or depleted.

°1/4 of all bird species have been driven to extinction.

°1/2 of all surface freshwater is used by humankind.

°30% of tropical rain forests and woodlands have been removed since 1970.

1.5 acres of rainforest are removed each second.

°100,000 acres of wetland ecosystems are destroyed each year.

°Over 28,000 rivers have been dammed (or damned) since 1900.

°Levels of carbon dioxide and methane in the atmosphere have doubled since the industrial revolution.

Let us take a look at another time when the Earth experienced a significant impact event. A time of tremendous dislocation and transformation of matter. A time of tremendous disruption to local and global systems. A time of large scale extinction of life on Earth. A time that happened in a geologic split second.

It is a time much more recent than 65 million years ago. It is a time of bombardment that begun only a few thousand years ago and has intensified sharply during the past few centuries.

It is a time of bombardment by the collective activities of the human species.

That is what all of this talk about global warming and climate change really is; just the Earth reacting to a stimulus similar to that of many impact events during its recent history.

One can see the signature craters around the world that document these impacts from within.

There are hundreds of craters from limestone quarries, coal mines and empty oil wells.

There are a million square mile scars from deforested landscapes.

The Earth has experienced significant and rapid alterations in the composition of its atmosphere, water, land, and life.

These are tremendous tears in the fabric of the Earth's present equilibrium and have happened overnight in geologic terms.

The bombardment continues every day with every acre of rainforest removed, every barrel of oil pumped out, and every ton of coal dug up.

Billions and billions of unrelated impacts carried out by billions of actors, like a shower of meteors spraying the Earth for millennia.

We are all actors in this play spanning the length of modern civilization called the ANTHROPOCENE ERA, the epoch of global scale disruption as a result of human activity.

It is a dramatic play indeed.

The Earth does not care where the disruptions come from, be they asteroids, volcanoes, or humans. The Earth simply reacts to the disruptions it experiences. The Earth will react in a similar way tomorrow as it did when it experienced similar disruptions in the past.

Is that any surprise?

°Humans use an average of 300 million tons of fertilizer derived from a petroleum base each year.

°There are over 700 million motor vehicles travelling the planet.

°1/4 of the world's energy is used by 5% of the human population who reside in the united states.

°One out of every four vehicles in the united states is a 12-16 mile per gallon SUV.

°Humankind has depleted 40% of known oil reserves and peak production is being reached.

YOU CAN SEE THE OIL WELLS PUMP

...thump...thump...

YOU CAN FEEL THE COAL TRAINS RUMBLE

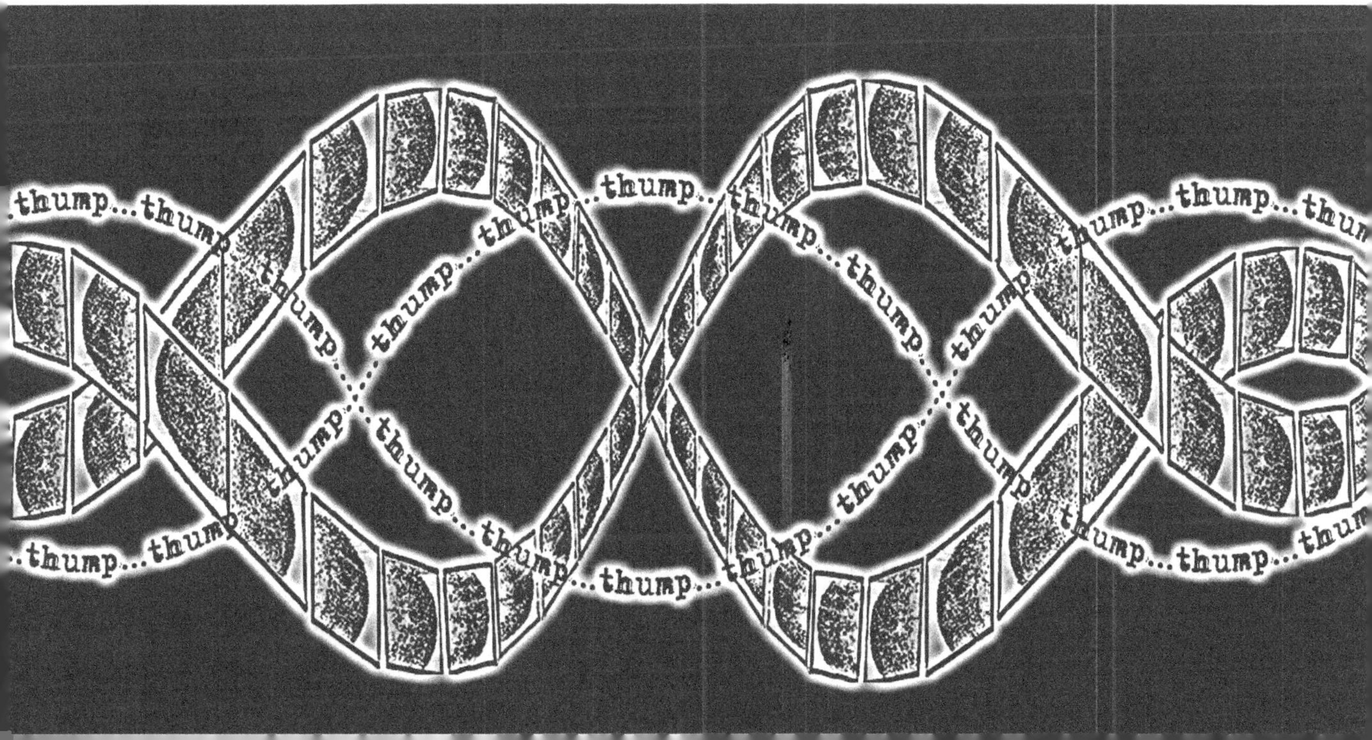

thump...

YOU CAN LISTEN TO THE CHAINSAWS WHINE
AND HEAR THE FORESTS CRY

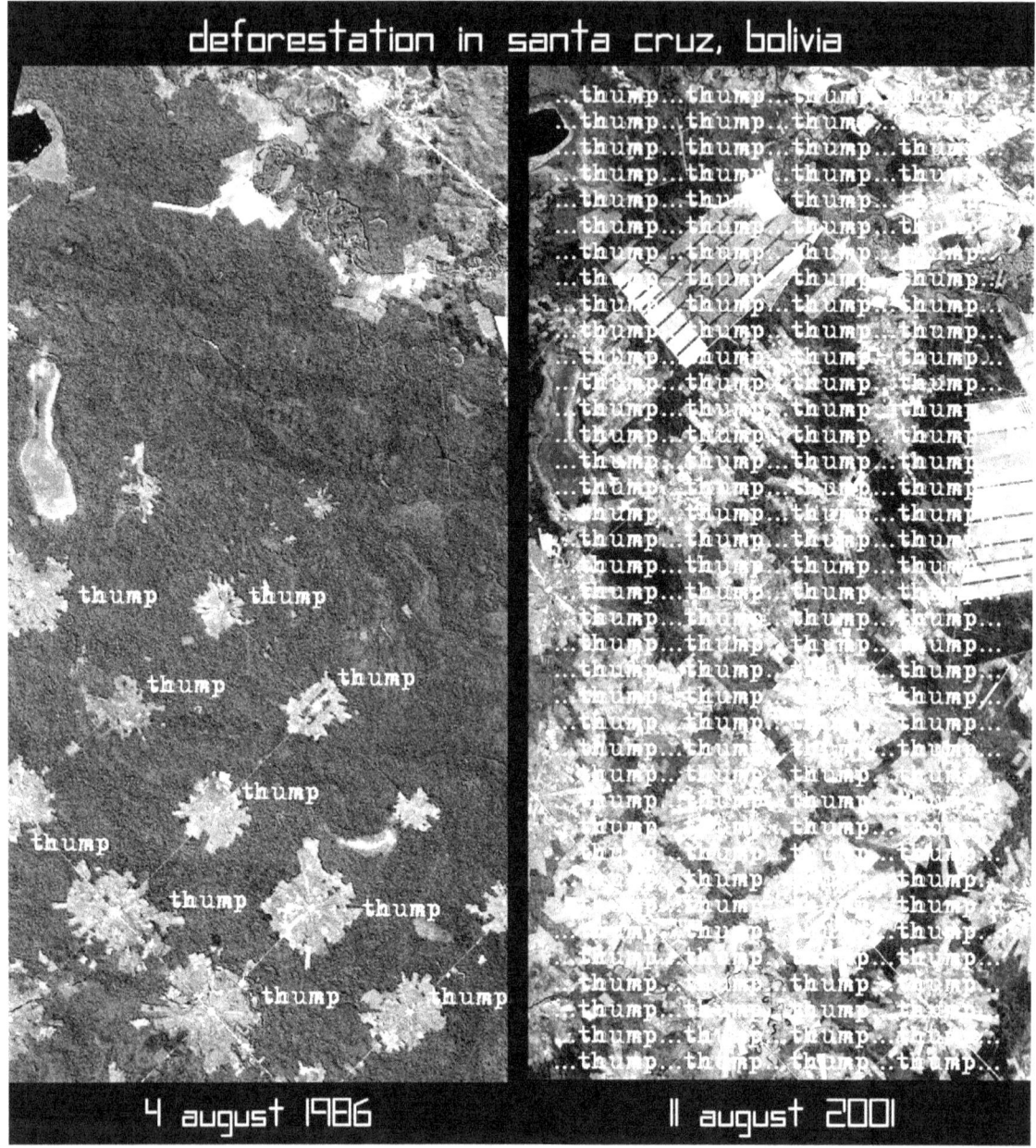

deforestation in santa cruz, bolivia

4 august 1986   11 august 2001

IF YOU LOOK...YOU WILL SEE
IF YOU TOUCH...YOU WILL FEEL
IF YOU LISTEN...YOU WILL HEAR
IF YOU THINK...YOU WILL KNOW

24 HOURS A DAY

365 DAYS A YEAR

IT'S ALWAYS RUSH HOUR SOME WHERE

KNOW WHAT I MEAN

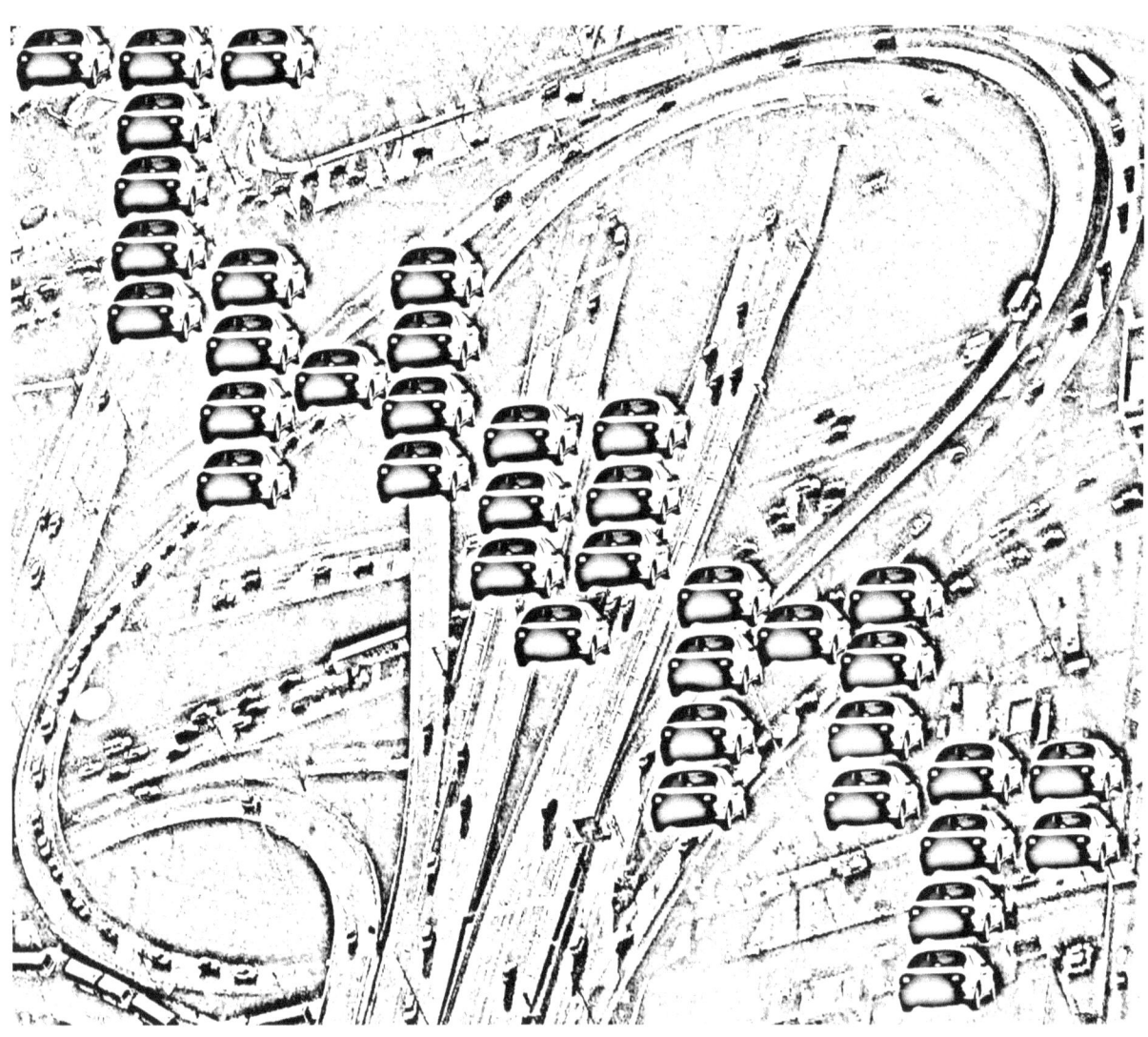

MOBILE MARBLE

The Earth moves. It spins completely around like a top every 23 hours 56 minutes and 4 seconds. It revolves around the Sun every 364.25 days. It changes its shape from spherical to oblong, depending on the amount of water in the oceans. The Earth changes its speed and rotation, depending on atmospheric pressure differences and the distribution of land mass and ice. The Earth also dances with the fabric of fourth dimension space-time, tugging it like a piece of chewing gum as it spins.

The Earth changes its tilt, spin, and orbital ellipse over the course of thousands of years. The theory that these variations could affect the Earth's global temperature is attributed to a Serbian astrophysicist named MILUTIN MILANKOVITCH, who in the 1930's documented that changes in the Earth's tilt, spin and orbit corresponded with global temperature differences between ice ages and interglacial periods. These three variables and their varying time scale cycles are called ORBITAL FORCINGS and can influence the global temperature.

°SPINNING EARTH
TWISTS SPACE
NATURE NEWS SERVICE
20 OCTOBER 2004

One of the last untested predictions of general relativity has been confirmed by the first reasonably accurate measurement of how the rotating Earth warps the fabric of space.

...The space warp is a consequence of Einstein's general theory of relativity, which describes gravity as a curvature in space produced by objects sitting in it. It also implies that a rotating mass will drag space around it like a spinning top placed in treacle - an effect known as the Lense-Thirring effect, or more commonly as 'frame-dragging'.

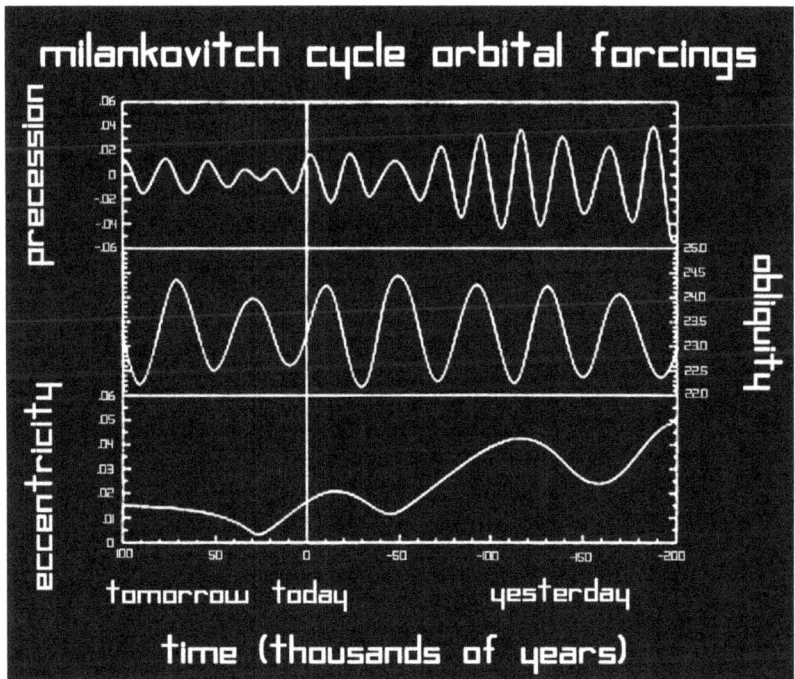

milankovitch cycle orbital forcings

ORBITAL ECCENTRICITY occurs on a cycle of about 100,000 years, as the Earth's orbit around the Sun changes from circular to oval and back again. When the Earth's orbit is more oval, the distance to the Sun is shortened during its approach and the Earth receives an exaggerated increase in the amount of solar radiation for half of the year. As the Earth moves away from the Sun, the distance lengthens considerably and

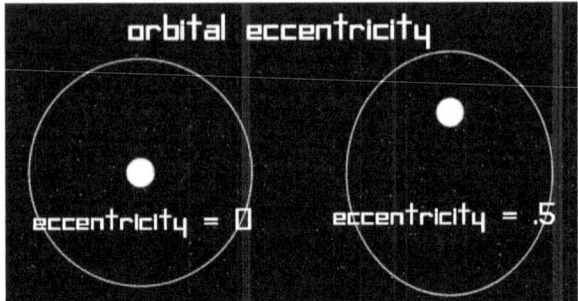

orbital eccentricity

eccentricity = 0

eccentricity = .5

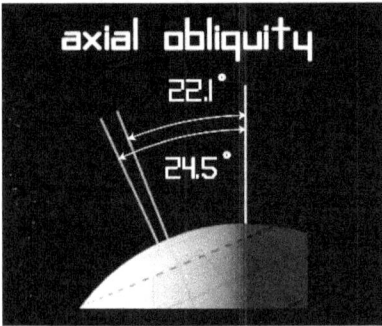

axial obliquity

22.1°

24.5°

precession

the Earth experiences an exaggerated decrease in solar radiation for half of the year. The present eccentricity of the Earth gives it an almost circular orbit at a fairly constant distance, so it receives a fairly even level of solar radiation throughout the entire year.

The change in **AXIAL OBLIQUITY** or tilt-angle of the Earth occurs on a cycle of about 41,000 years. The tilt of the Earth slowly shifts between 22.1°-24.5° during this period of time, which can reduce or increase the seasonal differences accordingly. When the tilt of the Earth is greater, seasonal extremes are greater, summers are much warmer and winters are much colder. As the degree of axial tilt decreases, winters become warmer and summers cooler.

As this happens, it is believed that moisture in the atmosphere increases during a warmer winter and ice sheet growth is promoted during cooler summers. As ice sheets grow and the reflectivity of the Earth increases, more solar radiation is reflected, which promotes a cooler Earth. It does not take too much longer for the Earth to enter an ice age that lasts for thousands of years.

AXIAL PRECESSION occurs on a cycle of about 26,000 years, give or take a few hundred years, and can be described as the slow, top-like rotation that changes the Earth's orientation and what we see as the North Star. During this cycle, the Earth wobbles slowly from pointing its north axis from the present North Star, Polaris, to another star, Vega.

This change in orientation affects seasonal contrast, as well as the timing of Northern and Southern Hemisphere seasons. Today, those living in the Northern Hemisphere experience winter when the Earth is closest to the Sun. In a few thousand years, both summer and winter will be experienced when the Earth is midway between its closest and furthest point to the Sun. In about 10,000 years, winter in the Northern Hemisphere will be experienced at the furthest point from the Sun, opposite of when it is now experienced.

Currently, the Earth finds itself in the middle of all three cycles, nothing very extreme about any of them. But that will change over time and so will the global temperature, global climate systems, global greenhouse gas levels, and the Earth's steady-state of existence.

The Mayan of South America based their cosmology and calendars on the axial precession of the planet and the location of stellar objects from our vantage point. The Mayan had extraordinary knowledge about the order, movements and cycles of the cosmos. Using the 26,000 precession cycle, also called the **PRECESSION OF THE EQUINOXES**, their calendar was oriented to the movement of the Pleiades Star Cluster. The Pleiades can be seen in the night sky as a very bright 'Little Dipper' to the right of the Orion constellation. The Pleiades were not only used by the Mayan and other ancient civilizations to establish their calendric cycles, but legends from the Mayan, Egyptian, Tibeten, Cherokee, and Hopi tell the story of Pleiadien beings visiting these ancient civilizations and divulging cosmic knowledge.

Hmmmm...that wasn't taught in the classroom.

According to Mayan specialists, their calendar is divided into five 'Sun' periods of about 5125 years each. The modern Mayan period began as they entered the final Sun period in 3114 B.C.E. Dating their calendar back to the beginning of the 26,000 year cycle and projecting it through the end of this cycle, their calendar ends on December 21, 2012.

There are many ideas about what might also happen in 2012 as a result of this interstellar event, if anything.

Some believe that 26,000 years ago cosmic wisdom was divulged to the inhabitants of planet Earth and that in 2012, it will happen again.

Some believe a new order of time will reemerge in which each year will be made up of 13 months, each month made up of 28 days, each day made up of 72 hours, and each hour made up of 20 minutes, a similar timing used by the Mayan.

Some believe the Earth will go through great contortions causing magnetic pole or geographic pole shifts.

Some believe the Earth will sustain impacts with a comet or asteroid.

Some believe there will be social disintegration, global nuclear war, and the End of Times will be upon us.

Some take a bit more optimistic perspective and believe humankind will ascend to a higher awareness of reality and realize a harmonic balance, evolving into a 'new' being that will help develop a species that is globally in-tune. At this point, humankind will transition from a time of chaos and into a Golden Age.

For most, the year 2012 is fairly insignificant, because different people interpret different dates from the Mayan Calendar. These ideas are also so closely associated with New Age philosophical and spiritual movements, that it is difficult to look at them objectively.

Nevertheless, dramatic events do happen and it is fascinating that the ancient Mayan civilization made accurate observations about our planet's motion and alignment in the cosmos. It is more fascinating that the ancient Egyptians and other early civilizations possessed the same knowledge while living on opposite sides of the Earth.

Whether anything happens in 2012 will probably be more coincidence than purpose. But then again, no one can be certain.

The Earth not only moves around the Sun with a different tilt, orbit, and orientation, it wanders with the entire Solar System through the galaxy.

The Milky Way is a fantastic piece of work. It is like a big pot of melted caramel, being slowly stirred by an invisible spoon. The Milky Way contains billions of stars that spiral around a dense center which some believe is a Black Hole and others believe is a super-dense Dark Energy Star. Whatever we call it, it is a cosmic entity so dense that not even light can escape, weighing in between 2.6-4 million Sun masses, it is confined to a space about 10 times smaller than the Earth's orbit around the Sun, and rotates on its axis about every 11 minutes.

That is one heavy-duty spoon.

hello seven sisters

As our planet continues on its way, it experiences an array of stimulus. The Earth and our Solar System move through the Milky Way's long wisps of dust and stars, called its 'arms.' Although the arms appear rigid to us, they are not rigid at all, they are constantly in motion, shifting, twisting and transforming, and this creates waves of energy called COSMIC RAYS.

As cosmic rays shower the Earth, the particles they contain are believed to help generate clouds, which would increase the reflectivity of the planet and help keep it cool. However, others believe the solar wind that blankets the Earth reflects most cosmic rays, neutralizing the possible cooling effect that increased cloud cover would provide.

There has been a good deal of discussion as to whether the reduction of cosmic rays adds to the present trend of global warming. Some suggest the lack of cosmic ray influence plays a large role, perhaps central role in the present trend. Others suggest the lack of cosmic ray influence plays a minor role, perhaps an inconsequential role, because the present trend is beyond natural variability and external forcings.

There are other activities that happen beyond the boundaries of our planet and our solar system which can affect the equilibrium of global systems, the global temperature, and life on Earth.

Although most stars besides our Sun are far too distant to significantly affect the Earth, scientists believe a nearby supernova a few million years ago blew away the Earth's protective ozone layer like a bursting bubble. When this happened, the Earth and all life forms would have been suddenly bombarded with the Sun's ultraviolet radiation on a scale thousands of times more powerful than what they were used to. Exposure to ultraviolet radiation of this level would have caused severe deformities and mutations in almost every species. Extended exposure to these rays would have also caused burning and cancers as well.

Evidence from this time reveals a high number of species extinction, and as phytoplankton and forests, key regulators of greenhouse gas cycles, were scorched to death within a few years, greenhouse gas levels would have been significantly disrupted too.

Recent records show a decline in phytoplankton productivity as a result of ozone reductions of only a few percent.

°NO COSMIC RAY CLIMATE EFFECTS'
BBC NEWS ONLINE
23 JANUARY 2004

The principal cause of recent climate change is not cosmic rays but human activities, a group of scientists says. They say an article last year linking cosmic rays and changes in temperature was "scientifically ill-founded". They say the authors' methods were open to doubt and their conclusions wrong, surprising experts with their claims.

In *Eos*, the journal of the American Geophysical Union, the 11 Earth and space scientists insist that greenhouse gases remain the chief climate suspect.

They say the most important physical processes are well understood, and model calculations and data analyses both conclude the human contribution to the global warming of the 20th Century through increased emissions of carbon dioxide (CO2) and other gases was dominant. The authors of the *Eos* article - Cosmic Rays, Carbon Dioxide And Climate - are from Canada, France, Germany, Switzerland and the US.

The research by Nir Shaviv, an astrophysicist, of Hebrew University in Jerusalem, and Jan Veizer, a geologist, of the University of Ottawa and Ruhr University in Germany, was published in July 2003 in the Geological Society of America's journal *GSA Today*.

It said the Earth's climate was profoundly affected by cosmic rays, high-energy particles from outer space, which normally cool the Earth's surface by helping clouds to form. But increased solar activity lessens the cosmic rays reaching the Earth, and Shaviv and Veizer suggested this blocking effect had been the dominant cause of global warming over the past century.

They said cosmic ray changes accounted for at least 66% of the temperature variation during that period. The Eos authors, led by Stefan Rahmstorf from the Potsdam Institute for Climate Impact Research, Germany, say the paper by Shaviv and Veizer was "incorrect and based on questionable methodology"...

Hence, another piece of the evolutionary puzzle finds its place and another possible initiator for global temperature change is discovered.

Speaking of exploding stars, let us head to our Sun, since it will also explode in a few hundred million years. Our Sun is an incredibly powerful celestial being. It is more or less a giant nuclear reactor, constantly exploding with the power of billions of nuclear bombs, shifting its magnetic poles on a regular and predictable basis, and throwing massive amounts of energy through the Solar System in the form of solar wind.

The Sun has long term cycles of its brightness, occurring about every 1500 years. The Sun's magnetic field creates solar wind that travels through the Solar System at 900,000 miles per hour. The Sun's magnetic poles also flip, producing waves of sunspot activity every 11 years.

When these things happen, the Earth feels it too.

The Sun is partially associated with daily, weekly, and seasonal weather differences and can also influence local and global climate systems. Solar activity affects atmospheric and ocean systems, it helps drive ocean cycles and ocean temperatures, and helps drive evaporation and cloud cover. For example, the **LITTLE ICE AGE** in Europe during the late 17th century is attributed to a lack of sunspot activity at this time. This helped drop the average surface temperature in the Northern Hemisphere a few degrees for a few hundred years.

Some believe the Sun's current sunspot activity is playing a key role in the present trend of global warming. Others say the present trend is far outside the natural variation and predictable solar influences.

It doesn't really matter too much to the Earth.

Changes are changes.

We have to adapt no matter what the cause.

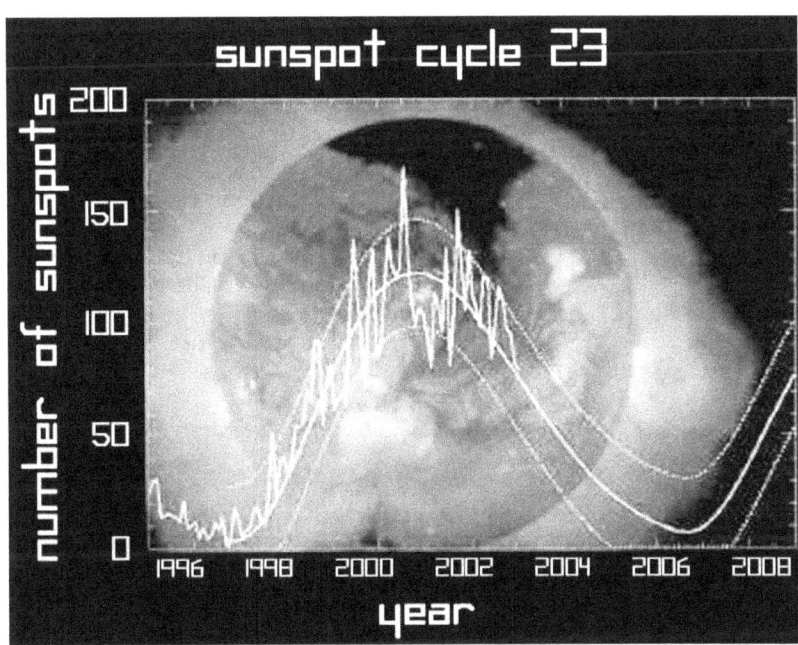

°ANCIENT SUPERNOVA MAY HAVE CAUSED ECO-CATASTROPHE
JOHNS HOPKINS UNIVERSITY 8 JANUARY 2002

An exploding star may have destroyed part of Earth's protective ozone layer two million years ago, devastating some forms of ancient marine life, according to a new theory presented at this week's meeting of the American Astronomical Society. The new theory brings together puzzling clues from several different fields of research, including paleontology, geology, and astronomy.

..."Such an extinction would have had especially pronounced effects on the plankton and the marine organisms," [Narcisso] Benitez said. [Mathilde] Canelles pointed out that evidence existed for a widespread extinction of plankton and other marine organisms about two million years ago, and noted that scientists are still debating the possible causes of the event.

He found that cosmic ray emissions from a supernova could have had a potentially devastating effect on Earth's ozone layer, an upper layer of the atmosphere that absorbs harmful ultraviolet emissions from the Sun and other sources. "This would have produced a significant reduction in phytoplankton abundance and biomass, with devastating effects on other marine populations, such as bivalves," Benitez said...

°SUN 'MINOR PLAYER' IN CLIMATE CHANGE
BBC NEWS ONLINE
3 MAY 2000

Research into the Sun's role in recent warming of the Earth's atmosphere indicates that it probably plays a relatively small part. The research, by two Danish meteorologists, suggests another factor is involved - probably human activity.

Those who challenge the consensus view that fossil fuel burning is leading to global warming have argued that increases in the intensity of the Sun is a far likelier cause. The Danish research, reported in the magazine New Scientist, means it is now harder to absolve humanity of blame for what is happening. The theory that the Sun was playing a dominant part had rested on the correlation between the sunspot cycle and temperatures in the Northern Hemisphere...

°CLIMATE FEELS THE SUN'S EFFECTS
BBC NEWS ONLINE
3 OCTOBER 2000

Recent reports that global warming is caused "mainly by the Sun" have been dismissed by leading scientists. The reports claimed that research by the European Space Agency (ESA) and others showed that computer models had severely underestimated the Sun's impact on the climate. But a conference sponsored by the ESA and the European Union has heard that the evidence is far more complex. And some participants say solar influences have diminished, while the human effects are intensifying...

°THE TRUTH ABOUT GLOBAL WARMING - IT'S THE SUN THAT'S TO BLAME
THE TELEGRAPH
18 JULY 2004

...A study by Swiss and German scientists suggests that increasing radiation from the Sun is responsible for recent global climate changes.

Dr. Sami Solanki, the director of the renowned Max Planck Institute for Solar System Research in Gottingen, Germany, who led the research, said: "The Sun has been at its strongest over the past 60 years and may now be affecting global temperatures.

"The Sun is in a changed state. It is brighter than it was a few hundred years ago and this brightening started relatively recently - in the last 100 to 150 years."

Dr. Solanki said that the brighter Sun and higher levels of "greenhouse gases", such as carbon dioxide, both contributed to the change in the Earth's temperature but it was impossible to say which had the greater impact...

°VIEWPOINT: THE SUN AND CLIMATE CHANGE
BBC NEWS ONLINE
16 NOVEMBER 2000

Natural processes involving changes in the Sun could have at least as powerful an effect on global temperature as increased emissions of carbon dioxide (CO_2). Climate scientists have already looked at changes related to Sun spot activity - a cycle of approximately 11 years - and long-term changes in the Sun's brightness, which has a cycle that lasts for centuries. They have discounted the effect of both on the temperature increase over the last century because they either happen over too short a timescale, or they are too weak.

But so far they have omitted to take two other factors into account:

°Changes in the amount of ultraviolet radiation from the Sun affect the ozone layer. This is a very important part of the atmosphere where lots of chemical reactions take place that govern the way the rest of the atmosphere works;

°The Sun's magnetic field and solar wind - mainly in the form of electrons and protons coming out of the Sun - protects the entire Solar System by acting as a sort of shield from cosmic rays (very energetic particles and radiation from outer space).

°GREENHOUSE GASES, NOT SOLAR ACTIVITY, CAUSE OF GLOBAL WARMING
ENVIRONMENT NEWS SERVICE
3 AUGUST 2004

Solar activity affects the climate but plays only a minor role in the current global warming, a German-Finnish team of scientists has found.

Since the middle of the last century, the Sun has been in a phase of unusually high activity, shown by frequent occurrences of sunspots, gas eruptions, and radiation storms.

The influence of the Sun on the Earth was believed to be one cause of the global warming observed since 1900, along with the emission of the greenhouse gas, carbon dioxide, from the combustion of coal, gas, and oil.

But Professor Sami Slanki, solar physicist and director at the Max Planck Institute for Solar System Research, is not convinced that the increased activity of the Sun is responsible for global warming

He says that based on his team's research, the Sun can be responsible for, at most, only a small part of the warming over the last 20 to 30 years.

"Just how large this role is, must still be investigated," he says, "since, according to our latest knowledge on the variations of the solar magnetic field, the significant increase in the Earths temperature since 1980 is indeed to be ascribed to the greenhouse effect caused by carbon dioxide."...

The Earth's own magnetism can affect the global temperature and prevailing climates as well. Although some people believe the interior of the Earth is hollow, with polar openings at the North and South Pole, most believe the center of the Earth is a churning and twisting liquid iron core which functions like a giant DYNAMO or electromagnet.

Scientists believe this dynamo produces the Earth's protective shield, the MAGNETOSPHERE, and also produces the Earth's magnetic fields or GEOMAGNETISM.

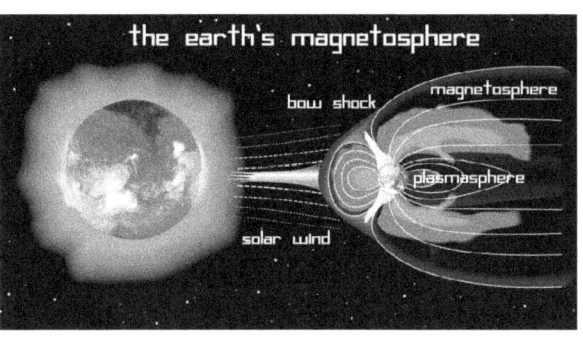

the earth's magnetosphere

The Earth's magnetic fields are strange. Not only do the magnetic poles wander about and sometimes provide simultaneous readings in different places, but they sometimes flip entirely like the Sun: magnetic North becomes magnetic South and vice versa. Research shows that the Earth's magnetic poles have flipped in a very short period of time, but the time between shifts is much longer than the decadal shifts of the Sun.

Although nobody really knows for sure how this happens, it has happened several times during the Earth's recent history; recent in geologic terms of a few million years. The Earth's magnetic field has been weakening during the past 2000 years and some believe this means the magnetic North will soon be heading South; soon in geologic terms of a couple thousand years.

The last time the Earth's magnetic poles flipped was about 780,000 years ago. At that time, magnetic North was pointing South, opposite of what we experience today.

Scientists have also found evidence that an asteroid in the range 3-7 miles across broke up and slammed into the South Pole in several places around the same time. This impact left multiple craters over an area around 1,200 miles wide and twice as long. Imagine an object striking the southern geographic and magnetic pole, that sent shockwaves through to the core and switched magnetic poles in a geological instant.

That is one big kinetic kick that could have really spun the needle.

During the transition period of a magnetic pole shift, the Earth's protective magnetosphere weakens and may entirely disappear until the Earth's magnetism reaches equilibrium and settles into a new steady-state. With the removal of the magnetosphere, the Earth is showered with solar wind, ultraviolet radiation and whatever other gamma, x, and cosmic rays may be soaking the Earth at any given time of the day. It would be similar to having the ozone layer blown off by a supernova.

During the past 2.5 million years, eight types of phytoplankton have become extinct, six of them after magnetic shifts. As these sensitive and vital regulators of global greenhouse gas cycles and the base of most food chains are significantly affected, it would make sense that global greenhouse gas cycles and global temperature regulation would be dramatically disrupted as well.

If magnetic shifts can happen over a relatively short period time and usher in global temperature changes, is it possible sudden global temperature changes could usher in magnetic pole shifts?

°SCIENTISTS EYE WHIRLPOOL IN EARTH'S CORE

SCIENCE NEWS ONLINE
13 NOVEMBER 1999

Reaching back to the compass readings of early polar explorers, geophysicists have now pieced together evidence of a slowly whirling vortex within Earth's liquid-iron interior. This result suggests that a long-nurtured theory of Earth's magnetism is on target. Einstein described geomagnetism as one of the chief unsolved puzzles of physics. Neither Earth's liquid outer core nor its solid, yet superhot, inner core has struck scientists as likely sources of its magnetic field. Heat cooks the magnetism out of magnets, and liquefaction melts it away. Physicists were flummoxed.

Over several decades, an idea has emerged: Earth's core acts like a giant dynamo. Heat causes churning in the liquid outer core, which Earth's rotation transforms into a liquid whirlpool that swirls around the planet's axis. This circulation produces a magnetic field roughly aligned with that axis. Basic physics says that electric currents give rise to magnetic fields and moving magnets generate electric currents. These two effects enable a churning outer core to magnify the small magnetic field that the planet captured from its surroundings as it formed. As opposing streams of molten iron, carrying tiny magnetic fields, sweep past one another, each induces currents in the other. This creates more magnetism, which induces more currents, and so on...

°ANTARCTIC CRATERS REVEAL ASTEROID STRIKE

THE GUARDIAN
19 AUGUST 2004

Scientists using satellites have mapped huge craters under the Antarctic ice sheet caused by an asteroid as big as the one believed to have wiped out the dinosaurs 65m years ago.

Professor Frans van der Hoeven, from Delft University in the Netherlands, told the conference that the evidence showed that an asteroid measuring between three and seven miles across had broken up in the atmosphere and five large pieces had hit the Earth, creating multiple craters over an area measuring 1,300 by 2,400 miles.

The effect would have been to melt all the ice in the path of the pieces, as well as the crust underneath. The biggest single strike caused a hole in the ice sheet roughly 200 by 200 miles, which would have melted about 1% of the ice sheet, raising water levels worldwide by 60cm (2ft).

But the climatic conditions were different at the time of the strike - about 780,000 years ago - from when the asteroid that is believed to have wiped out the dinosaurs struck Yucatan in Mexico.

...Prof Van der Hoeven said: "The extraordinary thing about this meteor strike is that it appeared to do so little damage. Unlike the dinosaur strike there is no telltale layer of dust that demonstrates the history of the event. It may have damaged things and wiped out species but there is no sign of it."

One thing that did happen at exactly the same time was the reversing of the Earth's magnetic field. There is no other explanation as to why this took place and Prof Van der Hoeven believes it was caused by the impact.

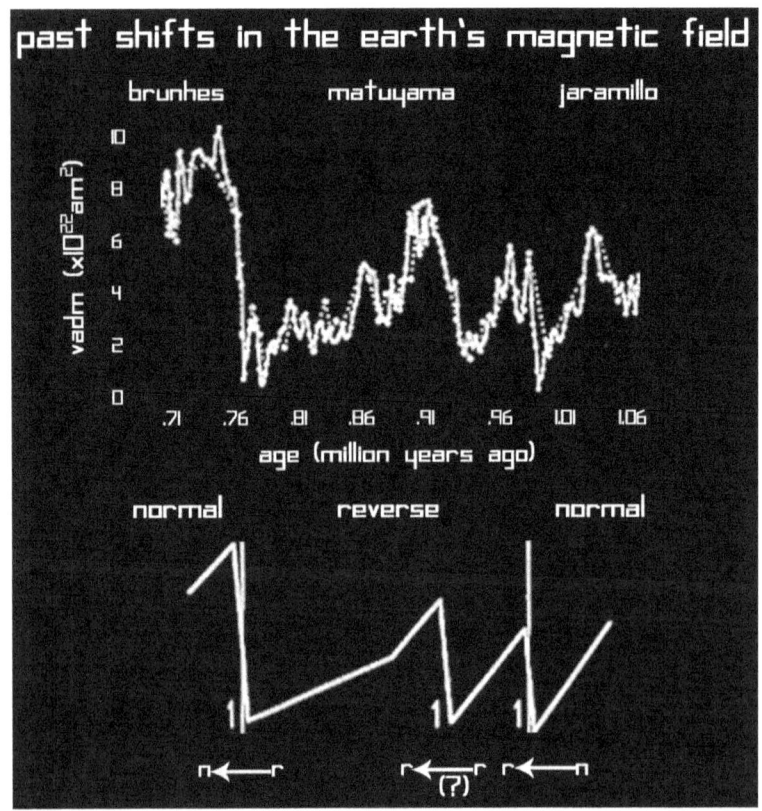

°SUN'S RAYS TO ROAST EARTH AS POLES FLIP

THE OBSERVER
10 NOVEMBER 2002

Earth's magnetic field - the force that protects us from deadly radiation bursts from outer space - is weakening dramatically. Scientists have discovered that its strength has dropped precipitously over the past two centuries and could disappear over the next 1,000 years.

The effects could be catastrophic. Powerful radiation bursts, which normally never touch the atmosphere, would heat up its upper layers, triggering climatic disruption. Navigation and communication satellites, Earth's eyes and ears, would be destroyed and migrating animals left unable to navigate.

"Earth's magnetic field has disappeared many times before - as a prelude to our magnetic poles flipping over, when north becomes south and vice versa," said Dr Alan Thomson of the British Geological Survey in Edinburgh.

"Reversals happen every 250,000 years or so, and as there has not been one for almost a million years, we are due one soon."

For more than 100 years, scientists have noted the strength of Earth's magnetic field has been declining, but have disagreed about interpretations. Some said its drop was a precursor to reversal, others argued it merely indicated some temporary variation in field strength has been occurring.

But now Gauthier Hulot of the Paris Geophysical Institute has discovered Earth's magnetic field seems to be disappearing most alarmingly near the poles, a clear sign that a flip may soon take place...

When massive amounts of solid are turned to liquid and swiftly redistributed around the planet as a result of melting ice, is it possible that the Earth becomes so unbalanced that its magnetism is thrown off?

Some people believe the Earth's geographic poles can flip as well, where the physical North Pole becomes the physical South Pole and vice versa. This is a theory called **CRUSTAL DISPLACEMENT** and is based on the premise of **PLATE TECTONICS** and **CONTINENTAL DRIFT**, the theories that propose the Earth's crust is a number of separate pieces resting on a more fluid underbelly, which allows these plates and the landmasses on them to move slowly over time. To imagine crustal displacement, think of how an orange peel covers the fruit inside. If the peel were to loosen, it would slide entirely around the fruit. Now take this idea to a planetary scale in which the Earth's upper 50 miles of crust slide around on a semi-liquid underbelly.

Those that believe in crustal displacement back up their beliefs with an event that happened about 11,500 years ago, the end of the last glacial period, in which the global temperature raised 7° Celsius in a few decades. As the world's ice quickly became water and flooded ocean systems, this would have significantly altered the weight distribution of the planet.

This temperature shift matches up with a time that crustal displacement theorists believe the entire crust of the Earth shifted about 2,000 miles. As a result of rapid ice melt, North America was dislodged from the Arctic Circle and all of Antarctica was locked into to the South Pole.

The Earth is presently changing its shape as a result of more freshwater entering the oceans from melting glaciers and polar ice caps. As the poles melt, the seas rise, and ocean circulation shifts, is it possible that the surface of the Earth could twist and churn like a meat grinder?

Conventional scientific theory states that the Earth's crust is connected to the layers of the lower mantle in a way which makes crustal displacement impossible. But it is an interesting idea to ponder in a world where right and wrong, true and not as true seem to change as quickly and abruptly as that which is being observed. People thought the idea of continental drift was absurd until a few decades ago and for a long time, the population believed the Earth was not the shape of an orange, it was the shape of a table top.

As new observations bring us new theories, our prior beliefs are taken off and tossed into the trash can of perception, replaced with shiny new glasses to look at the world around us.

As far as the observations made concerning modern-day global temperature change, the picture is almost crystal clear.

Almost.

EARTH

strong enough
to twist the
fabric of space

fragile as an
eggshell

sensitive to the
hands of man

senseless
enough to
smash them

THE BODY OF EARTH

Once upon a time, in a galaxy not so far away, a planet called the Earth began to show the vital signs of life. Around 3.5-4 billion years ago, the gases, liquids, and solids available in the middle-aged Universe gathered together in a way that created a planet with an atmosphere. The Earth's early atmosphere was composed primarily of methane and carbon dioxide and over the course of millions of years, the building blocks for life appeared. After a while, the building blocks became microscopic life forms. These life forms began changing the composition of the atmosphere by adding oxygen through photosynthesis, the process used by plants to convert sunlight into food, absorbing carbon dioxide and expelling oxygen.

After countless activities perpetrated by countless little green things, a critical threshold was reached and the atmospheric scale was tipped just enough to allow oxygen to become the dominant gas of our atmosphere.

Fortunately for us, the percentage of oxygen balanced out to be just right; just enough as not to suffocate and not enough to cause a perpetual inferno. The fraction of carbon dioxide and methane that remained was also just enough to help produce the conditions which allow life to exist.

The Earth has seen many changes throughout its life and is bound to see many more. It will experience events similar to those it has already experienced and it will experience events that are new.

Reacting to the countless activities of a species?

That is nothing new.

Reacting to the countless activities of human beings.

That is something new.

There have been times when the Earth's oceans were lava, times when the Earth's landmasses were connected, times when the poles were home to tropical climates, and times when a frozen blanket covered most of the planet.

There have been times when the Earth's tilt was different from today, times when the Earth's magnetic poles were on opposite ends, times when great balls of fire fell from the sky, and times when a distant star blew off its ozone layer as a result of a celestial sneeze.

There have been times when volcanic ash covered the sky and blocked the Sun for years, times when vast stores of greenhouse gases were expunged in one deep exhale that rapidly warmed the global temperature, times when the oceans were dozens and hundreds of meters higher than they are today, and times when the oceans shifted their currents to different modes and then back again.

Behold!

The great body of Earth!

It is not as sedentary and unchanging as we once thought.

The Earth is a big, living, changing body, filled with interrelated systems which keep it alive and breathing as opposed to a place like present-day Mars, which for all we know, is pretty much dead. But long ago, Mars was probably just as green and blue as the Earth. Some people want to make Mars like this again, and it is a fascinating idea to ponder. But most like to concentrate on the living planet that I already inhabit.

How about you? Should we spend trillions of dollars to plant turf on another planet or use the money to provide clean water to 2/3 of humankind on this planet? Should we send a few robots to another planet or remedy the conflicts between humans on this one?

Choices, choices, choices.

The Earth acts like our own body in a way: the Earth breathes; it can self-regulate its systems; it has a circulatory system which keeps its steady-state lifeblood coursing through its body; it has little organisms inside it that move around doing various activities; it has different features which play different roles within its systems; it might even 'bleed' and 'feel.'

It definitely reacts when its systems are overloaded.

The Earth may not be aware or conscious as we are, but it is alive, just like our body is alive but does not think about all of the various activities it must carry out to stay alive, it just does it. Our brain may send messages through our body relating our experience, but we do not consciously tell our body to make proteins, or tell our brain to send electrical impulses to our heart. The body just does it in an almost automatic way.

The science of understanding the Earth as a body of interrelated systems similar to our own is called GEOPHYSIOLOGY. It is a branch of science that combines both the living and non-living features of the Earth and looks at how the Earth functions in both linear and non-linear ways. It is a holistic approach in looking at how the systems of the Earth have operated in the past, how they operate today, and how they will most likely operate tomorrow in light of present activities.

geo = earth
physio = nature
logy = words

The idea of the Earth as a living being is not new, the ancient Greeks thought about the Earth in this way. What is new, is the scientific analysis and the idea/theory/hypothesis that the Earth is a living, self-regulating organism striving for an equilibrium within its systems. This concept was developed by JAMES LOVELOCK in the 1970's and he gave the name GAIA to his theory, after the Greek goddess of the Earth.

The basis of the theory is two-fold. First, the biogeochemical makeup of our living planet is affected by the planet's biosphere. Everything alive has a BIOGEOCHEMICAL IMPACT on the entirety of the Earth's interrelated systems. The rainforests' daily breath of photosynthesis, humans' perpetual belch of rush-hour traffic; it all has a significant impact on the larger systems of the Earth. Second, life on planet Earth engages with the rest of the Earth's systems, as to keep everything in general equilibrium. Although not consciously pursued, equilibrium and balance in its systems is sought, especially after significant global disruptions.

This is known as GEOPHYSIOLOGICAL HOMEOSTASIS.

bio = life
geo = earth
chemical = chemical

homeostasis

homeostasis
A relatively stable state of
equilibrium or a tendency
toward such a state
between the different but
interdependent elements
or groups of elements of an
organism or group.

In plain English, what these two tenets claim is that:

1) THE EARTH IS A LIVING BEING EXISTING AS A MULTITUDE OF DYNAMIC SYSTEMS WHICH ARE CONSTANTLY AFFECTING ONE ANOTHER.

2) THIS LIVING BEING WILL SEEK AN EQUILIBRIUM OR STEADY-STATE TO KEEP IT GOING, ESPECIALLY AFTER DRAMATIC DISRUPTIONS TO ITS SYSTEMS.

Human beings are included in the biosphere and life on planet Earth. We are a living creature within Gaia and therefore, part of her dynamic and interrelated fabric. As a part of Gaia's body, human beings and their settlements are a big part of the systems of the Earth. In fact, we now influence these systems more than any other species on the planet.

That is a big deal.

The Earth is far more stable when it is covered in ice, because there is not as much activity, as if the Earth hibernates for thousands of years. When it becomes warmer, the Earth awakens and explodes with life, in an unstable and much shorter lived experience before it goes back into hibernation.

When the systems of the Earth are disrupted, the stimulus it receives guide it into whichever steady-state follows the path of least resistance, or whatever makes it more 'comfortable.' It may take hundreds or thousands of years for the Earth to settle into a new state of equilibrium after being disturbed and will either revert back to its pre-disturbance state or shift into a state on the other end of the temperature spectrum.

If a series of events cause the Earth to warm, it will warm.

If a series of events cause the Earth to cool, it will cool.

It is pretty straightforward, but it is far from simple.

Nature does not really have a mind of its own, although sometimes it seems to have the most brilliant mind ever created.

There are structural complexities from the very small to the very large that make it seem as if a divine mind has been at work. Look at the structure of atoms and snowflakes, flowers and butterflies, seashells and galaxies.

The beauty and intricacies of all that we know, smell, see, taste, touch, and hear are astounding. We might be able to explain how something is, but we can never really explain why something is.

Nor do we really need to.

When a human mind and the natural mind coexist, a harmony is reached. Most animals are always in this state of perception, a state not governed by thinking, but of being. This is the way we all once were and we all still have it inside.

There are many ways in which humans tune in to the mind of nature. Sometimes this feeling is released for an instance, when energies coalesce and clarity emerges. Some people are able to stay in this place for days or years.

For some it takes meditation. For some it takes drugs. For some it takes physical activity. For some it takes music making. For some it comes from a life altering change. For all, it is an awakening, access to a more sensitive awareness of our surroundings.

Did you know the human body contains roughly the same proportion of water in its body as the Earth contains in proportion to its body? The human body also contains roughly the same proportion of salt in its blood as the Earth holds in its oceans. We are much closer to nature than we let ourselves believe.

God may have created man in the image of himself, but it seems more likely that God got the ball rolling and let humans develop over time into a creature which mirrors the makeup of the planet from which they came.

All ancient civilizations have worshipped and respected nature's instabilities and benevolence, as their livelihood was dependent upon favorable conditions. For our 21st century settlements, the situation is no different.

We should be very aware and careful of the blind faith we place in the stability of nature.

Just because the Sun rose today does not mean it will rise tomorrow.

Besides, the sun never really rises, that is just our perception as the world spins around.

So many wonders in the Universe, so many wonders in the Milky Way, so many wonders right here in planet Earth.

What set it all in motion? That is the greatest wonder of them all.

If we go by the measurements of modern science, life on Earth has existed for several billion years. It began as a random association of gases, culminated into some sort of chunky organic brew, took on a little more form and then just kept on going from there.

Billions and billions of species have come and gone in brilliant variety over billions of years. Sometimes they survive in cooperation, sometimes in competition, usually a little of both, and are directed with instinct and intellectual ability toward individual survival and self-preservation, as well as sacrifice for the preservation of the species or local community.

Survival of the fittest, natural selection, random luck, mutation, spontaneous change and slow emergence. Put it all together and you get **EVOLUTION**, a theory developed by **SIR CHARLES DARWIN** during the mid-1800's. Darwin was a naturalist who travelled abroad and cataloged nature as part of an early scientific expedition to observe the living planet. What he found in his travels, particularly in the isolated Galapagos Islands off the western coast of South America, was that similar, sometimes the same species, will adapt differently in different surroundings, in a way that best promotes their survival in their own unique circumstance.

Darwin believed that species struggled for existence, not necessarily in a competitive, combative way, but simply a struggle to persist. A plant on an edge of a desert struggles to persist during a lingering drought, just as an animal struggles to persist with a diminishing food source. The struggle for existence could lead to competition and combat, but his theory did not imply this.

what's up cuz

Darwin believed that as a species develops over time, **NATURAL SELECTION** and **SURVIVAL OF THE FITTEST** helps decide which members of a species persist and which fade away, taking their less adaptive traits with them. As surroundings and circumstances change over time, different traits may emerge that promote a creatures chances for survival. Some traits become obsolete.

Over the course of many generations, a member of a species may be able to survive in a way superior to others, which would give them an advantage over others, allowing them to continue passing down these beneficial traits to their descendants.

This is adaptation and evolution in action.

Darwin was harshly criticized for his theories, because at some point they meant indirectly that humans descended through evolution over billions of years from a gathering of gases, over millions of years from critters in the sea, and finally over hundreds of thousands of years from primates. In other words, human beings were not created divinely in their present form by God as the biblical story of creation claimed.

Darwin's theories were regarded by many to be totally absurd, because the more obvious explanation contained in the Bible is that all humans had descended from Adam, who God made from clay, and Eve, who was made from one of Adam's ribs.

Pooof! Just like magic. God is totally amazing like that.

If God can make the cosmos, God can surely make a man from clay and a woman from the rib of a man.

Then as the story goes, Adam and Eve made a lot of babies and all of them grew up and made babies with one another, making all of humankind the product of divinely inspired incest. But, since God is of course very clever, he gave Adam and Eve all of the genetic makeup for all of humankind, so there was no problem in making all of the variety of human beings we see today.

Adaptation and evolution? Silly monkeys. Chicks are from ribs.

The question today is not whether adaptation and evolution are scientifically 'proven' facts, because for arguments sake they are, but whether natural selection is really the dominating factor or only one of many factors in evolution. Some believe the dominating factor is the 'luck of the draw' as God rolls the dice through the Universe.

If a die lands on your head, even the best suited will perish.

The majority agree that genes, fitness and random luck all play a more or less equal role in determining what persists on planet Earth. Sometimes creatures go out of existence due to bad luck, sometimes due to bad habits, sometimes due to bad genes, sometimes due to all three.

The result of evolution, adaptation, genes, and luck over eons, culminates into the Earth's **BIOLOGICAL DIVERSITY**, also known as **BIODIVERSITY**. The variety and diversification of life is one of the greatest wonders of the Earth and of the Universe, because as far as we can find, the Earth is the only place that has it. The abundance of life that moves in, out, and around as an integrated part of the Earth's larger body processes is a truly marvelous thing.

The Earth needs biodiversity to keep its **ECOSYSTEMS** stable. There are desert ecosystems, rainforest ecosystems, river ecosystems, ocean ecosystems, prairie ecosystems, savanna ecosystems, deciduous forest ecosystems…a lot of ecosystems, containing a lot of species. All of these ecosystems play a larger role in the equilibrium of Earth's systems, and all are beginning to feel the consequences of the Earth's changing global temperature.

An ecosystem rich in biodiversity will stay healthy and relatively stable in a state of dynamic equilibrium for a long period of time. Whereas ecosystems that are lacking biodiversity quickly become unstable and cannot support the healthy functioning of individual species or the surrounding habitat.

°EVOLUTION ON FAST FORWARD: FINCHES ADAPT TO CLIMATES
NATIONAL GEOGRAPHIC TODAY 10 JANUARY 2002

In Montana, it is an evolutionary advantage for the females to be big and the males, small. In Alabama the reverse is true. That is, if you are a finch.

In less than 30 years finches have undergone a remarkable adaptation. Montana finch populations have adapted to produce large females and small males. In Alabama, by contrast, finches produce large males and small females.

Most people think of evolution as a process that takes millions of years, said evolutionary biologist Alexander Badyaev of Auburn University in Alabama, who led the study. But here is an example of real-time evolution in which two populations of finches developed characteristics to match their new environments in just a few decades, he added.

It turns out that finches are able to influence the size of their offspring by controlling the sex of their eggs according to the hatching order.

It has been well documented, particularly in the poultry industry, that the first laid egg tends to produce the biggest chick. In Montana the first-born is more often female, and thus the largest. In Alabama the first-born tends to be male…"These birds don't have more kids, they just have the right kids in the right order," said Badyaev.

…What is remarkable about this study, say scientists, is that two different populations of finches have acquired such different physical characteristics in such a short period of time—between 15 and 30 years. Badyaev believes it is this adaptability that has enabled the finches to spread rapidly throughout the United States.

…Curiously it was the finch populations on the Galapagos Islands that focused Charles Darwin's studies of evolution. He noted that each of the 13 species of finch had a characteristic beak shape that was tailored to a specific habitat and food source…

°BIODIVERSITY: BUFFER AGAINST CLIMATE CHANGE
ENVIRONMENTAL NEWS NETWORK
10 MAY 2001

...A research team based at the University of Minnesota concludes that biodiversity is an important factor regulating how ecosystems will respond to increasing atmospheric carbon dioxide, a component in global warming.

The team of investigators, led by Peter Reich at the University of Minnesota, found that more diverse plant ecosystems are better able to absorb carbon dioxide and nitrogen, both of which are on the rise due to human activities and industrial processes. The more of a mix of species, the better job they do at absorbing these greenhouse gases...

°CLIMATE CHANGE WILL UNBALANCE ECOSYSTEMS, STUDY SAYS
REUTERS
10 APRIL 2002

Climate change over the next 50 years will throw delicate ecosystems off balance, reduce the geographical range of many species and bring new predators and prey together, scientists said on Wednesday...

With biologically diverse ecosystems, many **SYMBIOSES** between species and habitat occur, keeping populations in balance and keeping the surrounding environment vibrant and robust. Some symbioses are so intimate that species will **COEVOLVE**, in which features of both species are tailored to one another. Such is the case with many flowers and the insects that pollinate them.

All in all, within a dynamic and robust ecosystem, species will come in and out of existence, but the ecosystem will sustain itself and its larger purpose within the body of the Earth, including the regulation of the Earth's greenhouse gas cycles. However, when a significant event disrupts the balance to such a degree, then the ecosystem can unravel quickly.

Ecosystems unravel all of the time, be it from natural causes, like a volcanic eruption, or from human activity, like rainforest removal. But, when it comes to global greenhouse gas cycle regulation and the pressures of global warming, people are not only concerned about ecosystems being lost, but are concerned about the central role that all of the Earth's ecosystems play together within the fabric of the Earth. When the entirety of the Earth's living system, the **BIOSPHERE**, begins to unravel, greenhouse gas cycles and global temperature regulation are disrupted too.

There have been countless types of algae, reptiles, mammals, reptile-mammals, flowers, plants, fish, vines, trees, bugs, birds, rodents; everything imaginable has probably touched the planet at some point. Oodles and oodles of life forms have graced planet Earth, oodles more are gracing it now, and oodles more will grace it in the future.

That is a lot of oodles. And a lot of grace.

Some are here for thousands of years, some for many millions of years, some for just a few centuries, years, days, or hours. We see the bones of ancient monsters in museums, we use carbonized ancient plants to propel our vehicles, and we use ancient microorganisms to fuel nuclear reactors.

There are fish that look like coral, bugs that look like leaves, moths that look like butterflies. We find out that some of the ancient monsters became birds of flight, some big cats became house pets, some hairy elephants became bald elephants, and some monkeys became people. For some species, we can say for certain they are a direct descendant of a particular being from the past. For other species it is not as easy, because so many variables go into the adaptation and evolution of life.

There are thousands of species that share the bounty of the Earth in this place in time and there are thousands more that we do not even know exist. All of the past, present, and future actors play a starring role in what did, does, and will help to maintain the planet's dynamic equilibrium.

The world's scientists have calculated that there are between 10 -100 million different living species on planet Earth today.

That is a pretty big gap.

The Earth has seen many species come and go and has experienced five mass extinctions since life began on the planet. Mass extinctions are called 'mass' extinctions, because more than 65% of all living things cease to exist in a relatively short period of time.

At the beginning of the 21st century, the world's scientists believe the Earth is going through another extinction of mass proportion. It is the first mass extinction to be initiated by a single species: human beings. It is predicted that by 2050, one-fifth of all living species could disappear due to the pressures applied by expanding human settlements, combined with other pressures like global warming, climate change and ozone layer depletion.

As a result of human interference, global temperature change, and climate changes today, pressures on all ecosystems around the world are mounting. Sure, humans are part of the Earth, this is our home too, but our collective activity is having such an impact on the larger systems of the Earth, that we are undermining the stability of the very systems we all depend on for survival.

Talk about shooting ourselves in the foot.

More like shooting ourselves in the head.

The wonders of the natural world are so awesome.

So beautiful. So dynamic. So brilliant.

And we can be so blindly careless in our own struggle to survive.

°BIODIVERSITY CRUCIAL TO EARTH'S ECOSYSTEMS
UNIVERSITY OF CHICAGO PRESS RELEASE
24 APRIL 2004

For more than half a century, ecologists have been aware of the devastating effects of species loss within an ecosystem. University of Chicago researchers have found that not only the number of species lost within the system, but also the identity of the species lost plays a vital role.

"How diverse the ecosystem is and how a particular species interacts with the rest of the system is perhaps more important than the actual number of species," said Mathew Leibold, an associate professor in the department of ecology and evolution, and co-investigator of the study, which is published in the April 25 issue of *Nature*.

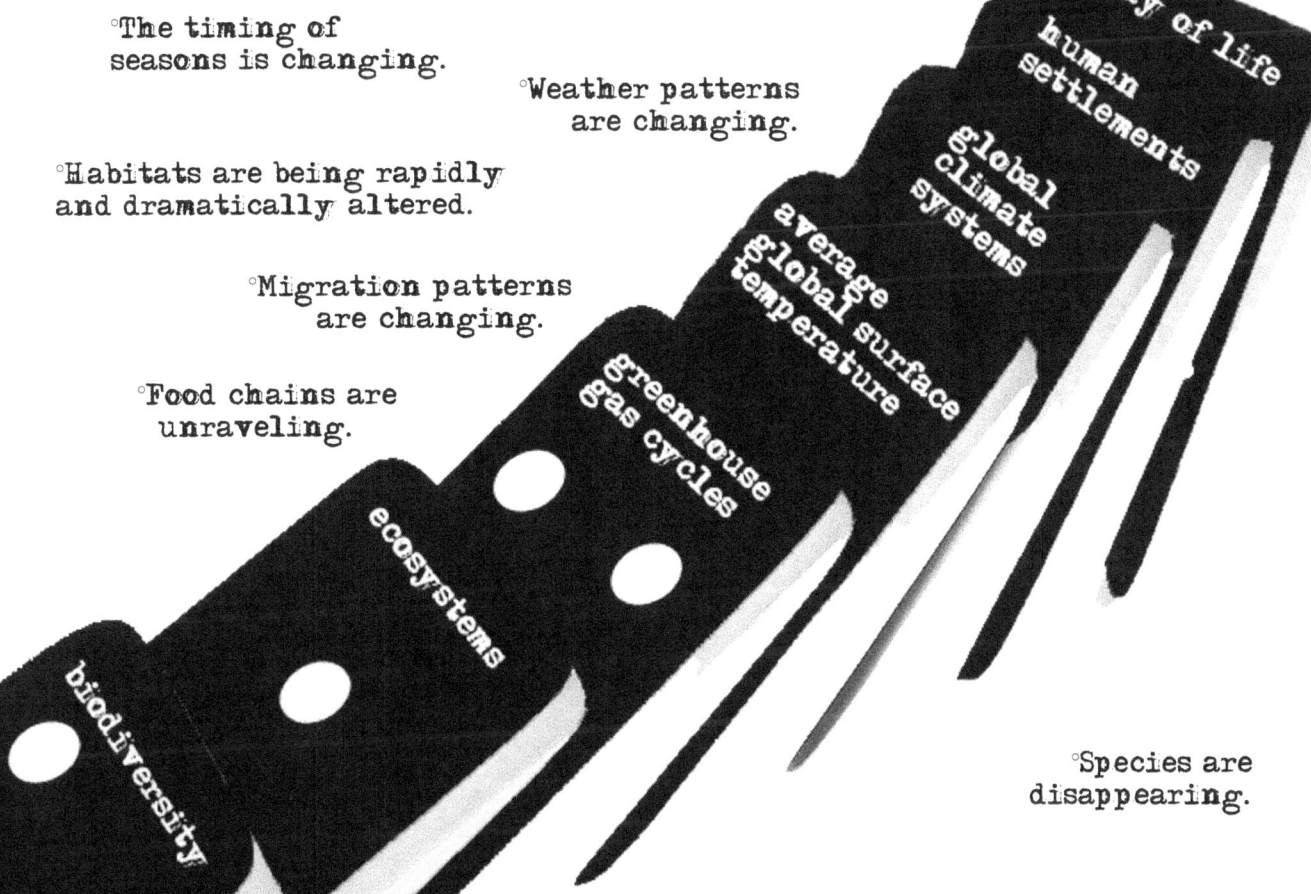

°The timing of seasons is changing.

°Weather patterns are changing.

°Habitats are being rapidly and dramatically altered.

°Migration patterns are changing.

°Food chains are unraveling.

way of life

human settlements

global climate systems

average global surface temperature

greenhouse gas cycles

ecosystems

biodiversity

°Species are disappearing.

°HIGHER ANIMALS SUFFERING
MAJOR DECLINES
ENVIRONMENTAL NEWS NETWORK
27 MAY 1998

°WORLD'S REPTILE POPULATIONS
RUNNING THIN
ENVIRONMENTAL NEWS NETWORK
11 AUGUST 2000

°ORANGUTANS EDGING CLOSER TO
BRINK OF EXTINCTION
NATIONAL GEOGRAPHIC NEWS
24 OCTOBER 2000

°BRITISH BUTTERFLIES
'FACE EXTINCTION'
BBC NEWS ONLINE
2 MARCH 2001

°WORLD PLANTS NEAR EXTINCTION
CLOSE TO 50 PERCENT-STUDY
REUTERS
31 OCTOBER 2002

°LIONS 'CLOSE TO EXTINCTION'
BBC NEWS ONLINE
18 SEPTEMBER 2003

°AMPHIBIANS FACE
A BLEAK FUTURE
NATURE NEWS SERVICE
14 OCTOBER 2004

°HUMAN IMPACT TRIGGERS
MASSIVE EXTINCTIONS
ENVIRONMENT NEWS SERVICE
2 AUGUST 1999

°CORAL REEFS WILL BE GONE IN
20 YEARS, SCIENTISTS SAY
ASSOCIATED PRESS
23 OCTOBER 2000

°HUMANS MOVING CLOSER TO
EXTINCTION, STUDY SAYS
SEATTLE POST-INTELLIGENCER
5 JANUARY 2001

°THIRD OF PRIMATES
'RISK EXTINCTION'
BBC NEWS ONLINE
7 OCTOBER 2002

°STUDY: ONLY 10 PERCENT OF
BIG OCEAN FISH REMAIN
CNN NEWS ONLINE
14 MAY 2003

°DECLINE IN OCEANS'
PHYTOPLANKTON ALARMS
SCIENTISTS
SAN FRANCISCO CHRONICLE
6 OCTOBER 2003

°CLIMATE WARMING SPELLS
SPECIES WIPEOUT
REUTERS
2 FEBRUARY 2005

POSITIVE FEEDBACK

global warming and climate changes
swift habitat changes
ecosystem and biodiversity disruption
reduced greenhouse gas sinks

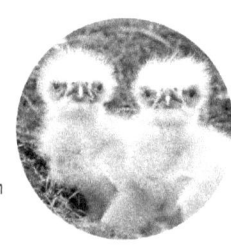

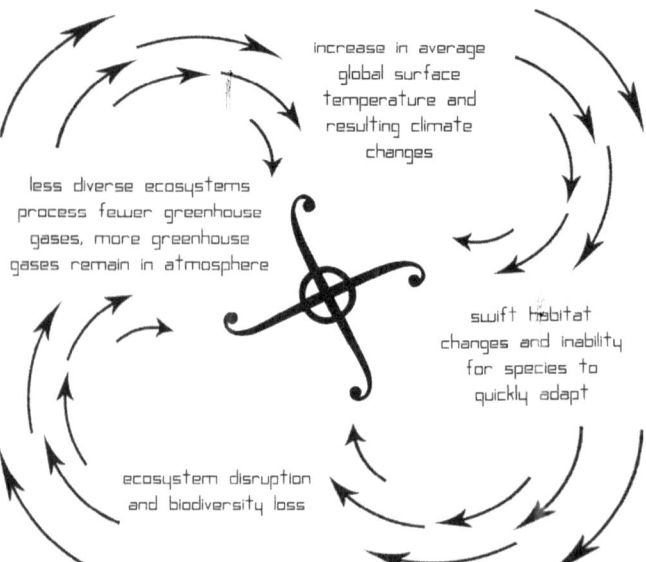

increase in average
global surface
temperature and
resulting climate
changes

less diverse ecosystems
process fewer greenhouse
gases, more greenhouse
gases remain in atmosphere

swift habitat
changes and inability
for species to
quickly adapt

ecosystem disruption
and biodiversity loss

°EARTH FACES SIXTH
MASS EXTINCTION
NEW SCIENTIST
18 MARCH 2004

BE WARY OF YOUR WINGS
YOUNG CONOLEY-SAN

FOR WHEN YOU FLUTTER
THEM IN JAPAN

A HURRICANE MAY LAND
THOUSANDS OF MILES AWAY

MANY DAYS LATER

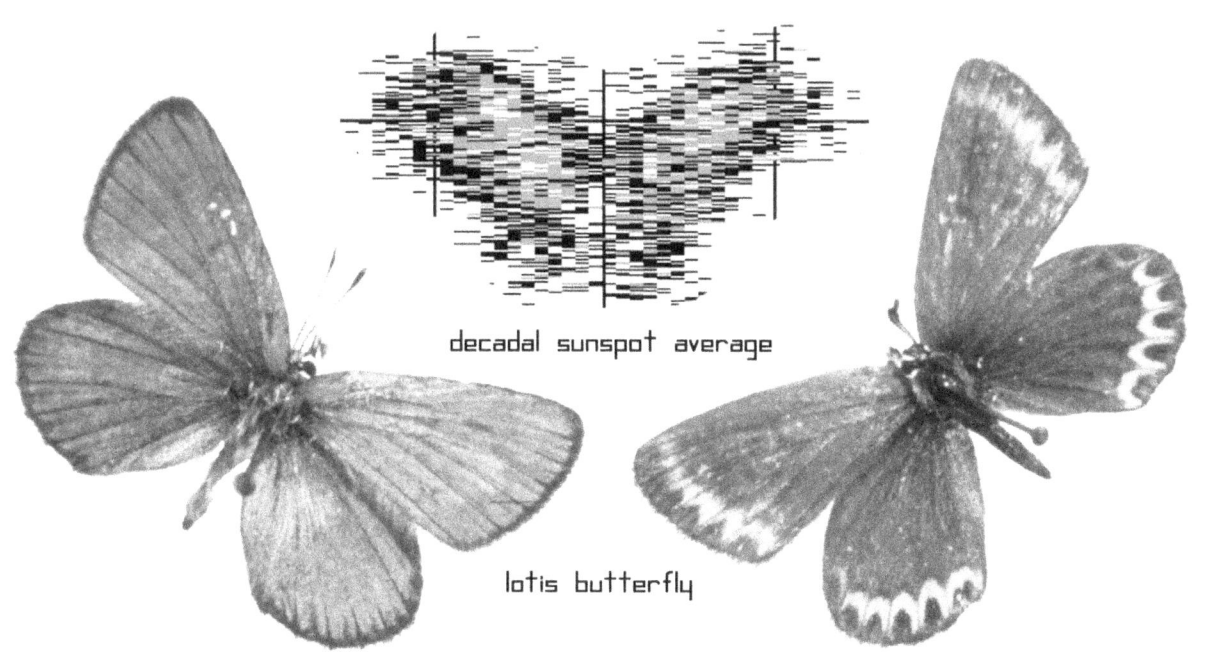

decadal sunspot average

lotis butterfly

3
HUM OF HUMANITY

What was once stardust will become stardust again.

But before that happens, there is a long, long, long journey through particles, places, and things the stardust will travel. It will travel through gases and galaxies, diatoms and dinosaurs, planets and people. As we move through time and life on planet Earth, through a few billion years and through a few billion species, we come to a point around 8-12 million years ago where stardust formed an organism different from any other organism ever created.

Different, because it became aware and believed it was different.

The warm-blooded mammals had become the dominant group of beings on planet Earth and a particular group of mammals, the primates, were living, among other places, in the African savannah and forest, roaming the grasslands and hanging around in trees. These creatures had a body like ours, two arms and two legs, hands, feet, and a head. Their brains were smaller, their bodies harrier and they scuttled around with their hands near the ground. Each successive generation becoming better and better equipped to survive.

The time span that separates modern human beings from their primate ancestors is long and the Grand Tree of our genesis is full of bending and twisting branches of HOMINID creatures: species that are not totally ape-like and not totally human-like, they are somewhere in between.

They go by the names of *Australopithecus anamensis, Australopithecus africanus, Australopithiecus robustes, Homo habilis, Homo ergaster, Homo heidelbergensis* and *Homo sapiens neanderthalensis*. Although we are very different from the primates on our grand tree, we all share 96-98% of our genetic makeup with other primates.

We are called *Homo sapiens sapiens*.

Dozens of bones and artifacts have been found here and there and tell the story from ancient apes that became modern humans. Fossils from our ancestral branches have been found all over the planet, but the oldest ones have been excavated near Lake Victoria in Africa, along the Great Rift Valley located in modern-day Ethiopia.

Most paleoanthropologists believe that our prehistoric ancestors developed on a path of CONVERGENT EVOLUTION, the case when different species develop similar features. Another theory is that the branches of our tree were twisted

°U.N. SAYS GREAT APES IN DANGER WORLDWIDE
ASSOCIATED PRESS
28 NOVEMBER 2003

...The great apes - gorillas, chimpanzees, bonobos and orangutans - are in great danger, the United Nations said as its environment and cultural agencies ended a meeting Friday on how to protect the closest relative to humans in the animal world...The great apes share more than 96 percent of their DNA with humans...

hey, hey, hey, just wait a minute??

Are you saying that me and all the other humans on the planet have 98.4 percent identical DNA to Chimpanzees and Bonobos and 96 percent identical DNA makeup to Gorillas??

That's more shared genetic identity than Indian elephants have with African elephants.

Well...I guess we are monkey's uncles dude.

among one another so that traits were passed between different species, for better or for worse. The tree of apelike mammals is varied indeed and the path from primates to humans is not necessarily explainable in a linear way.

Whether all of modern humankind came from one ancestral family or several, all modern humans share family members that appeared on the planet a couple hundred thousand years ago, were fine tuned, developed different features for different circumstances, and took a modern form about 20,000-25,000 years ago.

There are, of course, many interesting stories that diverge from the conventional understanding about the genesis of humans. Some question if modern humans evolved at all. The Creationists believe the Earth is only 6500 years old and humankind is even younger. Intelligent Design followers believe that humans, all life on Earth, and the rest of the cosmos were not only created over billions of years from a mixture of chance, selection, luck, and natural laws, but that an intelligent force also played a role. Others believe humans descended from a lineage of extraterrestrial beings at some point not too long ago. Although this theory is not generally accepted, it is generally accepted that all life on this planet originated from the game of Galactic Marbles, with the building blocks for life arriving on meteors or asteroids.

There are also anomalies concerning the lineage of our species, evidence that throw a curveball of uncertainty and mystery into the origins and development of humankind. Some of these anomalies have turned out to be hoaxes, some have been proven beyond a shadow of a doubt, and some still lie under a thick haze of confusion and curiosity.

People have found footprints, bones, tools, paintings, temples, walls, dams, and entire civilizations which either substantiate conventional wisdom about the origins and evolution of humankind, or make us rewrite our history books.

In Texas during the 1930's, locals claimed to have found tracks of a dinosaur side-by-side the 16-inch footprints of a hominid in a layer of Earth that dated back 100,000,000 years ago. Human-like tracks 21-inches long, eight inches wide with a gait of seven feet were found in 1973 in a layer of rock dating back many millions of years before such ape-like creatures were thought to exist. In 1947, mummies of men 8-9 feet tall were found in the Death Valley desert of Nevada and dated back 80,000 years. Stories of Bigfoot and the Yeti continue to arise from the forests of Northwest Canada to the Himalayan mountains.

A recent expedition on an isolated island in Indonesia has turned up of a race of tiny humans that lived with tiny elephants and giant komodo dragons. *Homo floresiensis* were one meter humans with little brains living as recently as 18,000 years ago, most likely alongside more modern humans.

There are stories of people finding artifacts buried in layers of coal, that were far too modern to have been created at the same time the coal was formed. There are reports of cave paintings dating back thousands of years that detail advanced technologies like aviation and surgical procedures.

°HUMANS DID COME OUT OF AFRICA, SAYS DNA
NATURE NEWS SERVICE
7 DECEMBER 2000

Archaeologists are still not sure when and where modern humans first appeared. Some believe that Homo sapiens evolved independently in several places around the globe. But research revealed in this week's *Nature* lends support to the idea that we appeared in one location in sub-Saharan Africa and spread from there, replacing Neanderthals and other early humans as we went.

Researchers led by Ulf Gyllensten of the University of Uppsala in Sweden have found evidence that we are all descended from a single ancestral group that lived in Africa about 170,000 years ago. And they suggest that modern humans spread across the globe from Africa in an exodus that took place only around 50,000 years ago.

Gyllensten's team didn't scrutinize fossils to come up with these results -- instead the group examined DNA from living people around the world. The genetic material in our chromosomes is a combination of genes from our parents. But each cell also contains structures called mitochondria, and these house DNA that is independent of that found in chromosomes.

...They discovered that the most recent common ancestor of everyone in the sample group lived in Africa $171,500 \pm 50,000$ years ago. They also found a significant branch in the tree that separates most Africans from non-Africans. This genetic divide probably represents an exodus of people from Africa that took place $52,000 \pm 27,500$ years ago...

°SCIENTISTS FIND FOSSILS OF MAN'S EARLIEST RELATIVE
REUTERS
11 JULY 2001

An international team of scientists said Wednesday they had dug up bones and teeth of an early pre-human that could help fill the gaps in mankind's evolutionary tree. The fossils of a new subspecies of an early relative of humans were found 140 miles northeast of Addis Ababa in Ethiopia. They are about 5.4 to 5.7 million years old, about a million years older the previous oldest known hominid, according to the researchers.

"It is the earliest hominid. We are pushing back the hominid record by more than a million years," Yohannes Haile-Selassie, a researcher at the University of California at Berkeley who discovered the fossils, told Reuters...

°SKULL MAY SHIFT EVOLUTION THEORIES
ASSOCIATED PRESS
10 JULY 2002

In what may be the most startling fossil find in decades, scientists in central Africa say they have unearthed the oldest trace of a pre-human ancestor - a remarkably intact skull of an apelike species that walked upright as far back as 7 million years ago. The thick-browed, flat-faced skull was found in Chad, 1,500 miles west of pre-human discoveries in east Africa.

Exactly where the skull fits into man's family tree is not clear. But the skull's age, shape and location challenge basic beliefs about the evolution of man's earliest ancestors. Among other things, the find could push back the date at which humans are believed to have diverged from apes. And it suggests that upright-walking human ancestors evolved not just on the grasslands, as anthropologists have long thought, but in forests as well...

Everyone knows there are pyramids in Egypt and Mexico, but how about those found under water off the coast of Japan. What civilization built those? Did the civilizations of ancient Egypt, Mexico, and Japan visit one another and share their wisdom of the cosmos. Did these civilizations all originate from the same place? Were they acting in tandem?

Artifacts and bones from ancient civilizations have recently been found in caves in northern Spain and Israel, both dating back 780,000 years and both showing a high level of sophistication, much higher than was expected for this time period. If you recall, the Earth received a shock from a giant space rock and reversed its magnetic poles 780,000 years ago. As the planet lost its magnetic field for a while, radiation would have blasted all life on the planet for many years. Around this same time, anthropologists believe our ancestors went through a possible 'brain boom' of expanded cranial capacity and intellectual ability, which would substantiate the findings in Spain and Israel.

It is certainly possible that generational mutations occurred, bringing both positive and negative attributes to our ancient ancestors. An acceleration of intellect is certainly imaginable, as is gigantism and miniaturism. Perhaps even different species fused together and genetic codes were intertwined, producing entirely new creatures every generation.

Millions of years have passed and events have happened beyond our imagination which certainly could have mutated, changed, or accelerated evolutionary processes. Look at the dinosaurs of antiquity, some are the size of a train car. It is not so unbelievable that our own evolution may have also had bizarre episodes.

As the Earth has churned and shifted for millions of years, moving at great speeds through the cosmos, it is certainly possible that events occur that do not fit into our conventional understanding of the past.

Whether we all came from the Garden of Eden in the African Steppe or were plopped here by something else, the fact remains that we are all human and are all members of the same global species.

We may look different, speak different, and think a bit different than those we share the planet with, but we are all the same.

We all need water.

We all need food.

We all need shelter.

We all need energy.

We all need love.

And we will all have to face the circumstance provided by a changing planet during the 21st century and beyond.

The possibilities for human's Genesis are interesting to think about.

But the possibilities for our adaptation are necessary to survive.

brown, black, white, orange, pink, yellow, tan, cream, beige...

african, asian, european, middle-eastern, indo-american...

hindu, christian, muslim, catholic, jewish, protestant,
athiest, agnostic, adventist, buddhist...

congolese, nepalese, lebenese, chinese, japanese...
isreali, iraqi, thai, french, greek, english, dutch, danish...
mexican, iranian, canadian, australian, brasilian, syrian...

no matter how we look, where we are from, or what we believe...

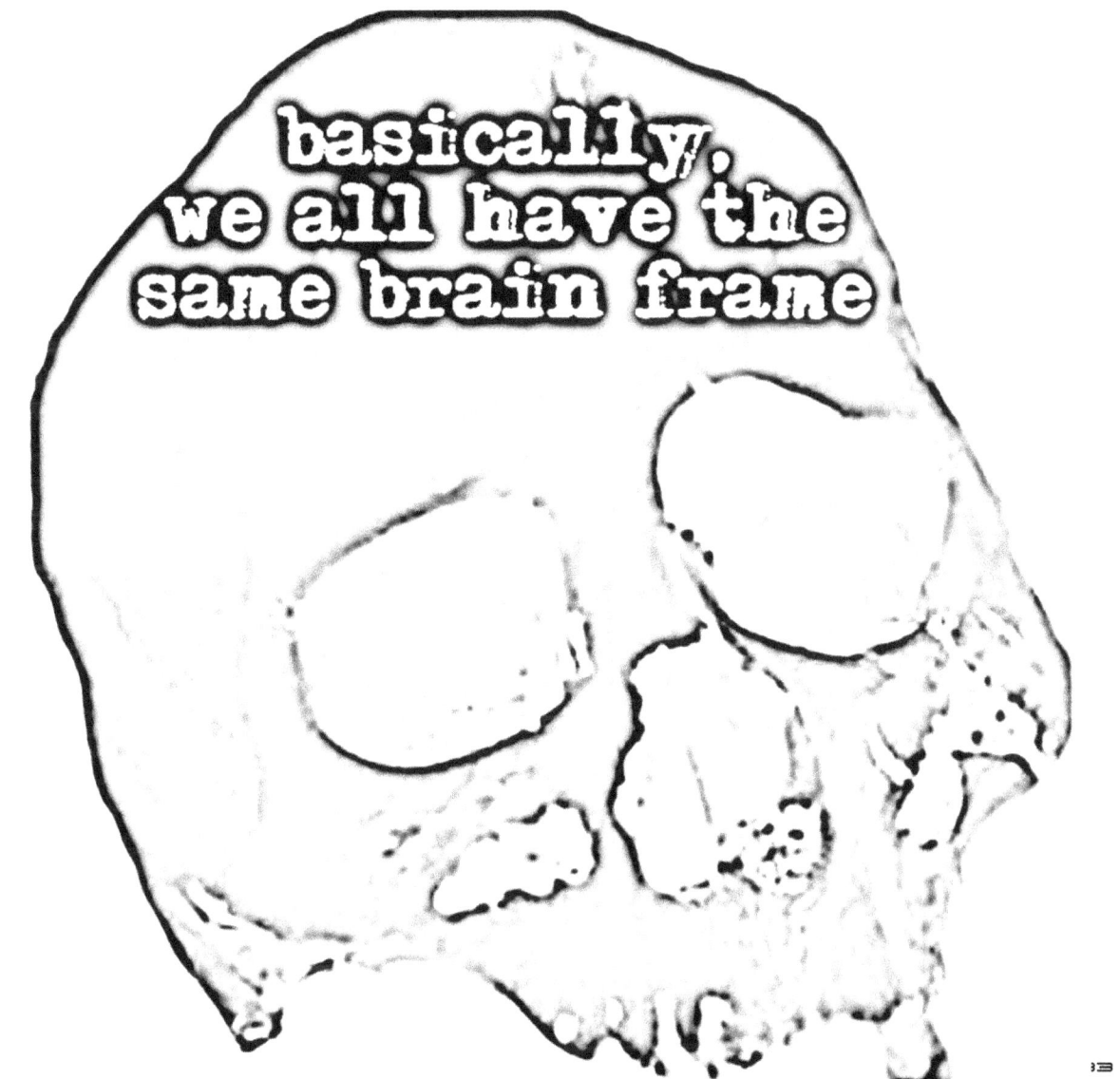

GREAT DESCENT, GREAT MIGRATIONS
GREAT EVENTS, GREAT PERTURBATIONS

Over a long period of time, our bipedal ancestors spread themselves around the Earth. Different individuals in different places developed different features which reflected their adaptation to different surroundings.

Some stayed dark skinned, which helped protect them against intense sunlight, some lost their pigment, because their skin was exposed to little sunlight and covered in clothing to protect them from the cold.

Some became tall and skinny to efficiently cool in a hot climate, some became short and stout to efficiently store heat in a cold climate. But many similar traits emerged and were retained by them all and passed on from generation to generation to generation. Everyone was still related, everyone still brothers and sisters in a diverse global family.

Our ancestors evolved and struggled in a haze of animal-like survival for a long time. Then we discovered fire. The gift of Prometheus lifted the animal out of darkness and into the light of being human. The world could now be lit day and night, the Sun had come to Earth and could be used at will.

The mind began to move toward logic and away from instinct. With the mind's development also came a new sense of bewilderment with the world and a need to understand the Universe in a way different than the instinctual acceptance of our animal ancestors.

Our ability to cultivate increased, our need to be nomadic decreased. We became secure and comfortable in one place and as a result, settlements developed, cultures were founded, and civilizations established. Diseases and illness were reduced, clothing was tailored, tools were made, food became more abundant, and intellect continued to grow. Our time and energy focused on improving the security of the community, which was a much safer haven than the wilderness our ancestors inhabited.

As intellect expanded, it allowed us to create and manipulate with deliberate purpose and reason. To explain what we did or what we saw, we created symbolic art and language, which enabled us to pass along stories from to our descendants. The ability for future generations to learn from the achievements and mistakes of the past was a very beneficial trait for those individuals and settlements who chose to listen and act accordingly.

There have been many civilizations that have come and gone during the history of humankind. There are many civilizations we know a great deal about and there are those which seemed to have disappeared without a trace.

There are stories from the ancients about fantastic events and when the time period is matched with modern scientific findings, we see that these fantastic stories may actually be the recording of eye-witness accounts.

There are stories of fire falling from the sky, stories of lost civilizations, and stories of great social demise. There is one story of a fantastic event that stands out more than any other: the story of a Great Flood.

One historian has tallied the reference of a Great Flood from the stories of our common ancestry and found it was mentioned in over 80,000 works in 72 languages.

The Hawaiians tell of a flood.

The Irish tell of a flood.

The Native Americans tell of a flood.

The Greeks and Romans tell of a flood.

The Babylonians tell of a flood.

They all tell the story of a flood which bathed the face of the Earth and washed away everything that stood in its path. It is believed that the flood story was used in ancient literature to illustrate the virtue of humility before Nature and the Gods, as those perpetuating evil and arrogance upon the Earth suffered a torturous demise.

Perhaps the most popular flood story is that of Noah's Ark. It is a good story about a monumental flood that cleansed the wickedness from the world and washed away the evil of humankind, leaving only one righteous man and his family to care for the rest of God's creatures.

Noah was told to prepare for this flood from a source of divine intervention. He was told that God's creations - Man, Earth, and all of its creatures- had become polluted with evil and needed to be cleansed, so that a new beginning could ensue.

And God saw that the wickedness of man,
Great in the Earth, and every imagination of the thoughts of his heart.
Only evil continually.
And it repented the Lord that he had made man on the Earth, and it grieved him
at his heart.
And the Lord said, I will destroy man whom I have created from the face of the
Earth, both man, and beast, and the creeping thing, and the fowls of the air, for it
repenteth me that I have made them.
But Noah found grace in the eyes of the lord.
These, the generations of Noah.
Noah was just a man. Perfect in his generation,

Noah walked with God.

The Earth also was corrupt before God,
and the Earth was filled with violence.
And God looked upon the Earth, and, behold, it was corrupt,
for all flesh had corrupted his way upon the Earth.
And God said unto Noah, The end of all flesh is come before me,
for the Earth is filled with violence through them,
and, behold, I will destroy them with the Earth.
Make thee an ark of gopher wood, rooms shalt thou make in the ark,
and shalt pitch it within and without with pitch.

And, behold, I, even I, do bring a flood of waters upon the Earth,
to destroy all flesh, wherein the breath of life,
from under heaven, every thing that is in the Earth shall die.

And the rain was upon the Earth forty days and forty nights.

Noah was told to build a boat large enough to harbor a male and female for all of God's creatures and strong enough to withstand a flood of the ages. He was told to prepare in anticipation of the event and Noah took these words seriously, doing what he was told.

The flood came and washed away the evil. Noah survived with all of God's creatures that we now have on Earth, or so the story goes.

There is also a Native American story about a flood from the Cherokee Indians in which a Dog comes and tells his Master of a pending flood:

–YOU MUST BUILD A BOAT, THE DOG SAID, AND PUT
IN IT ALL THAT YOU WOULD SAVE; FOR A GREAT RAIN
IS COMING THAT WILL FLOOD THE LAND.–

A God tells his Creation, a Dog tells his Master. The story is more or less the same and the message passed down tells us to perpetuate goodness during our time on the planet and be humble before nature.

Catastrophic and sudden flood events have indeed occurred on planet Earth and most likely our ancestors witnessed and recorded them, either through verse or through scribe. Although the story of Noah dates back to about 3,600 years ago, a real flood believed to be the basis of the story, happened about 7,600 years ago. During this time, water from the melting ice age blankets in the northern hemisphere had already filled the oceans with water, raising them many dozens of feet.

As the water in the Mediterranean Sea swelled, it pushed northward through what is now the country of Turkey. The water then funneled through the Bosporus River into the Black Sea and reached a critical threshold. The small canal could no longer hold back the immense pressure and in a single burst, over 10 cubic miles of water began racing through the canal at over 50 miles per hour each day for almost a year. This inundated 60,000 square miles of land and raised the water level in the Black Sea hundreds of feet, changing it from a land-locked freshwater lake to a salt water lake connected to the world's oceans.

Imagine if your village was located along this small river, it would have been swept away faster than you could open your mouth to scream.

A flood like this would have caused an immediate catastrophe for the people and villages affected, and would have also had long term environmental and social consequences.

Now, that is a flood worth telling your descendants about.

This particular flood is probably not the flood referred to by the Hawaiians or the Native North and South Americans. But there have been other occasions when massive floods have happened and these were probably witnessed, recorded, and passed down as well.

The global thaw that ended the last ice age around 11,500 years ago would have dramatically altered the face of the planet as the oceans raised over 130 feet in a matter of decades. A period of significant Antarctic ice melt happened about 14,200 years ago that raised oceans around 70 feet during a couple of centuries. These are dramatic changes in sea levels that happened over the course of a couple generations. Anyone living on a coast or island would have noticed and probably would have told their descendants about it, assuming they survived.

What of the civilizations swallowed by such floods?

Perhaps the most famous story of a lost civilization is that of Atlantis.

The Greek philosopher Plato was the first to refer to this lost civilization, a civilization that apparently held incredible military power, exceptional wealth, and was swallowed whole by floodwaters. In two of his dialogues he wrote of a huge landmass in the Atlantic Ocean, home to a civilization that may have been global in breadth with outposts throughout the world.

In his writings, Plato tells of Atlantis' attempt to stretch their dominance over much of Europe, he tells of its war with Athens, and he tells of its underwater burial. Plato wrote that the Great Flood which buried Atlantis happened about 9,000 years before his time. This happens to coincide to about 11,500 years ago, the end of the last ice age, a time of rapid and significant flooding of the Atlantic Ocean from glacial retreat in the Northern Hemisphere.

Quite a coincidence.

There are other ideas regarding the location of Atlantis. Some believe Atlantis was one of the Greek islands in the Mediterranean Sea, home to the Minoan civilization and was submerged by the global deluge at the end of the last ice age.

Some believe Atlantis was an ancient Incan civilization located in the Bolivian highlands under Lake Titicaca.

°**FINDING NOAH'S FLOOD: EVIDENCE OF ANCIENT DISASTER IS LINKED TO BIBLICAL LEGEND** COLUMBIA UNIVERSITY PRESS RELEASE NOVEMBER 1999

Scientists have retrieved sonar images of an ancient coastline 550 feet below the surface of the Black Sea that are strong evidence that a sudden violent flood destroyed a fresh water oasis and inspired the story of Noah's Ark, a theory advanced by two oceanographers at Columbia.

Explorer Robert Ballard, who found the remains of the Titanic, led a research team this summer that discovered the ancient beach dated 7,500 years ago. That date supports the ideas of Walter Pitman and Bill Ryan, senior scientists at the Lamont-Doherty Earth Observatory, who believe that a flood 7,500 years ago forced the Diaspora of an advanced civilization.

Their research has generated new discussions about the role climate change has played in human history...

°METEOR SHOWERS
BLOTTED OUT MAN'S
FIRST CIVILISATIONS
THE TIMES ONLINE
14 DECEMBER 1997

A cataclysmic shower of giant meteors destroyed the great Bronze Age civilisations in Egypt, Mesopotamia and Greece by provoking a series of natural disasters. New archeological and astronomical evidence indicates that a huge number of extraterrestrial bodies caused famine, flooding and bushfires thousands of miles wide that led to the collapse of the world's first sophisticated civilisations.

...Dr Benny Peiser, an anthropologist from Liverpool John Moores University, has analysed 500 excavation reports and climatological studies from the sites of ancient civilisations and found they all suffered huge changes in climate at exactly the same time. Previous explanations for the collapse of the ancient civilisations have pointed to warfare, volcanoes and earthquakes.

But Peiser's findings show that the worldwide devastation could only have been provoked by an external cosmic event. "There is very strong evidence to suggest that massive meteor storms are the real scientific reason why these ancient societies collapsed," he said last week.

...A new finding by Victor Clube, an astrophysicist at Oxford University, appears to confirm Peiser's theory that meteorites were responsible for the Bronze Age catastrophe. Clube claims to have identified a meteor cluster in an orbit around Jupiter which has collided with the Earth about every 3,000 years...

Others think Atlantis was one of many coastal civilizations submerged by tsunamis caused by cometary impacts.

It might remain a mystery whether a place called Atlantis really did exist, but it will always be a symbol of how delicate our relationship with nature is and how susceptible even the most advanced civilizations are to her whims. Especially a whim which buries your settlement in a day.

As recently as 5,000 years before present, our ancestors experienced some other very disruptive events, events that brought their settlements to their knees as well. Until 2300-2200 B.C.E., there were several highly advanced civilizations scattered about the planet: the fifth dynasty of Egypt's Old Kingdom, the Akkad culture of central Iraq, the Hongsan culture in China, and others in the Indian subcontinent, Persia, Iberia, South America, the British Isles, and Greece. The people in these cultures lived in agrarian-based, self-reliant and self-sufficient settlements that were totally dependent on the predictable weather patterns nature provided them. They all had common attributes, they were all experiencing similar levels of development, and they all witnessed great balls of fire raining from the sky.

Back then, the area near the Tigris River and the Euphrates River was home to the Sumerien and Akkadian cultures, two different groups of people who lived peacefully together and created perhaps the first high-level civilization in modern history. By the end of the fourth millennium B.C.E., the capital Uruk became the largest city-state in Mesopotamia, if not the world. It was here that the first evidence of a written language was found, the first wheeled vehicles were found, and a successful urban way of life was established.

No wonder it is called the cradle of civilization.

Then one day, the sky began to fall and the cradle fell with it.

Into the cosmos we go and travel to a birthplace of asteroids and comets, into a comet system called Comet Encke and the Taurid Complex. The Comet Encke passes by the Earth on a regular basis and is believed to have passed by around 1,800 and 5,000 years ago and around 2,100 and 4,700 years ago. During these fly-bys, scientists believe the comet was breaking apart, draping its debris over the Earth like a dusty veil, scarring the land and sending the civilizations of the time into a frenzy.

Scientists have found a giant pock-mark on the surface of the Earth in a lakebed at the southeastern tip of the Fertile Crescent, close to the point where the Tigris and Euphrates River converge. It is located a few miles away from yesterday's Uruk, in today's maimed Iraq.

The heart of Mesopotamia.

Close your eyes and send your body and mind back a few thousand years to the place that many believe was the biblical Garden of Eden. All is well with the world, the weather is pleasant, the ground is fertile, harvests are good, food and water are plentiful, and the Gods are at peace with humankind and the Earth. Your life revolves around a center square of commerce and cosmic spirituality, great temples and idols of worship are being constructed and everything exists in harmony.

You are enjoying an evening with friends, wine, music and frivolity, discussing the wonders of the cosmos and human's prime place in it all.

syria

iraq

euphrates river

tigris river

giant crater found here ●

saudi arabia

Everything moves around the Earth and we are all simply visitors, here for a time to enjoy the bounty it provides. Someone is pointing out Orion's Belt, explaining the temples built by the region's God-Kings were positioned on Earth to replicate the star's positioning in the heavens.

"You know what they say: As above, so below."

You are discussing the motion of the stars and the motion of the constellation, the night is dark, lit only by the Milky Way.

"Do you see that dark patch?" Someone asks.

"That is where we all came from, the celestial birth canal."

"As above, so below."

Then someone points to a bright light in the sky, a light no one has seen before. It is a very intense light, brighter than Venus, brighter than Saturn, brighter than Sirius. The star is shaking and then suddenly convulses.

You see another and another. The lights appear from nowhere.

The lights spark, bursting and splitting into several streaks across the night sky. Then you feel a burning flash and your hair is on fire. This is followed by an ear-splitting crash, you are surrounded by explosions, your eyes are blinded, your ears are numb, your body is burning and suddenly you and your friends are sent hurling hundreds of feet through the air by a shock wave. Cauterized boulders the size of a Sphinx fall around you, smashing houses, temples, idols, and people.

The village is in panic and everyone is running, screaming, burning. Everything is in slow motion.

ANCIENT LEGENDS TELL THE TALES OF DRAMATIC EVENTS

AS WRITTEN IN THE EPIC OF GILGAMESH

...and the Seven judges of hell, raised their torches, lighting the land with their livid flame.
A stupor of despair went up to heaven, when the god of the storm turned daylight into darkness, when he smashed the land like a cup...

AND RECORDED IN THE ADMONITIONS OF IPUWER AT THE END OF ANCIENT EGYPT'S OLD KINGDOM

...The Sun is occluded and will not shine, that men may see. None may know that it is midday, and the Sun will cast no shadow...

°METEOR CLUE TO END OF MIDDLE EAST CIVILISATIONS
THE TELEGRAPH
4 NOVEMBER 2001

Scientists have found the first evidence that a devastating meteor impact in the Middle East might have triggered the mysterious collapse of civilisations more than 4,000 years ago.

Studies of satellite images of southern Iraq have revealed a two-mile-wide circular depression which scientists say bears all the hallmarks of an impact crater. If confirmed, it would point to the Middle East being struck by a meteor with the violence equivalent to hundreds of nuclear bombs.

Today's crater lies on what would have been shallow sea 4,000 years ago, and any impact would have caused devastating fires and flooding. The catastrophic effect of these could explain the mystery of why so many early cultures went into sudden decline around 2300 BC.

...A date of around 2300 BC for the impact may also cast new light on the legend of Gilgamesh, dating from the same period. The legend talks of "the Seven Judges of Hell", who raised their torches, lighting the land with flame, and a storm that turned day into night, "smashed the land like a cup", and flooded the area. The discovery of the crater has sparked great interest among scientists...

GREAT CHANGES IN THE EARTH CAUSED GREAT CHANGES IN ANCIENT EGYPT'S OLD KINGDOM

...The fruitful water of the Nile is flooding.

The fields are not cultivated.

Robbers and tramps wander about and foreign people invade the country from everywhere.

Diseases rage and women are barren.

All social order has ceased, taxes are not paid and temples and palaces are being insulted.

Those who once were veiled by splendid garments, are now ragged.

Noble women wander around the country and lament, if only we would have something to eat.

Men throw themselves in the jaws of crocodiles—
So out of one's senses are people in their horror.

Laughter has ceased everywhere.

Mourning and lament are in its place.

Both old and young wish they are dead...

More lights appear and strike the village with equal or greater intensity. Utter destruction is all around you, searing boulders continue to rain down and liquid glass is sprayed around like water. You can no longer hear the sky falling, you can only feel the hot thuds.

You are powerless.

Everyone is powerless.

You have just witnessed the beginning of the end.

And then it is quiet.

Your ears are bleeding, your body is scorched.

Clouds of colored vapor pass by.

Everything is enshrouded in a gaseous haze.

Everything is destroyed.

Everyone is losing their mind.

You are awake for what seems to be days, but the Sun is gone.

You will not see it again for months, if you are still alive.

What could you have done to make the Gods so angry?

The history of meteors pummeling ancient civilizations can be read in many legends dating back to this time. Stories detailing cometary impact and societal collapse can be found in the Sumerien legends of Gilgamesh, the first King of Uruk. The Mayan culture entered their final Sun period at this time, 3114 B.C.E., with stories of fire raining from the sky. There are similar stories from China, as well as from South America. The first Stonehenge monument on the British Isles was also built around this time and believed to be a type of celestial forecasting device to predict the return of these meteor showers.

The legends from our ancestors have not only been substantiated with the finding of a crater in the Fertile Crescent, but other craters have been found in Argentina that date back to the same time, as well as evidence from tree rings in Ireland to dust samplings in the Gulf of Oman. It was not only a time when numerous meteors fell to the Earth, but it is also believed that seismic and volcanic activity increased as the planet readjusted its crust after absorbing the collisions.

The stable weather patterns of our ancestors would have also changed and as a result, economic and social systems would have been weakened to the point of collapse. Famine, hunger, disease and illness would have taken hold, and a general state of demise would have quickly engulfed all levels of society.

Significant, rapid and persistent changes to one's prevailing local climate can bring an advanced civilization to its knees. Some events that usher in such changes, like a meteor impact, cannot be avoided. If a comet wants to shower the Earth with meteors, let the stones land where they will. However, some events that usher in such changes can be avoided. Unfortunately, a society usually finds out only too late that the cumulative result of their activities is the cause for their undoing.

A classic example of this comes from the Classic Period of Mayan culture. The people of this period lived during the first centuries C.E. and thrived for six centuries; then their civilization collapsed between 800-900 C.E. The remnant of this grand culture can be found on the Mexico-Guatemala border, in the forest buried ruins of Tikal.

The people of Tikal lived in a tropical rainforest, an ecosystem which self-regulates itself with a wet climate upon which the Mayan civilization was based and dependent.

As their population grew, they began clearing the trees for farmland, which is not so problematic, unless too much of the forest is removed. If a critical threshold is breached, the forest can no longer produce its microclimate and things become very dry, very fast.

As the Maya expanded their settlement, at one point reaching a density of 500-2,000 people per square mile (equivalent to modern-day Los Angeles County), they removed more and more forest cover for farmland.

As the surrounding area became drier and more trees were removed, erosion became more common and the fertile topsoil was easily washed or blown away. Without the shade of tree-cover, the land would have warmed, leading to further drying and making the soil even less suitable for farming. At some point, the threshold was breached and the Classic Mayan civilization experienced several decades, perhaps a century of dryness and were no longer able to provide their large settlement with the necessities to survive.

When they cut down the trees, the climate dried up.

When the climate dried up, the land dried up.

When the land dried up, their civilization withered away.

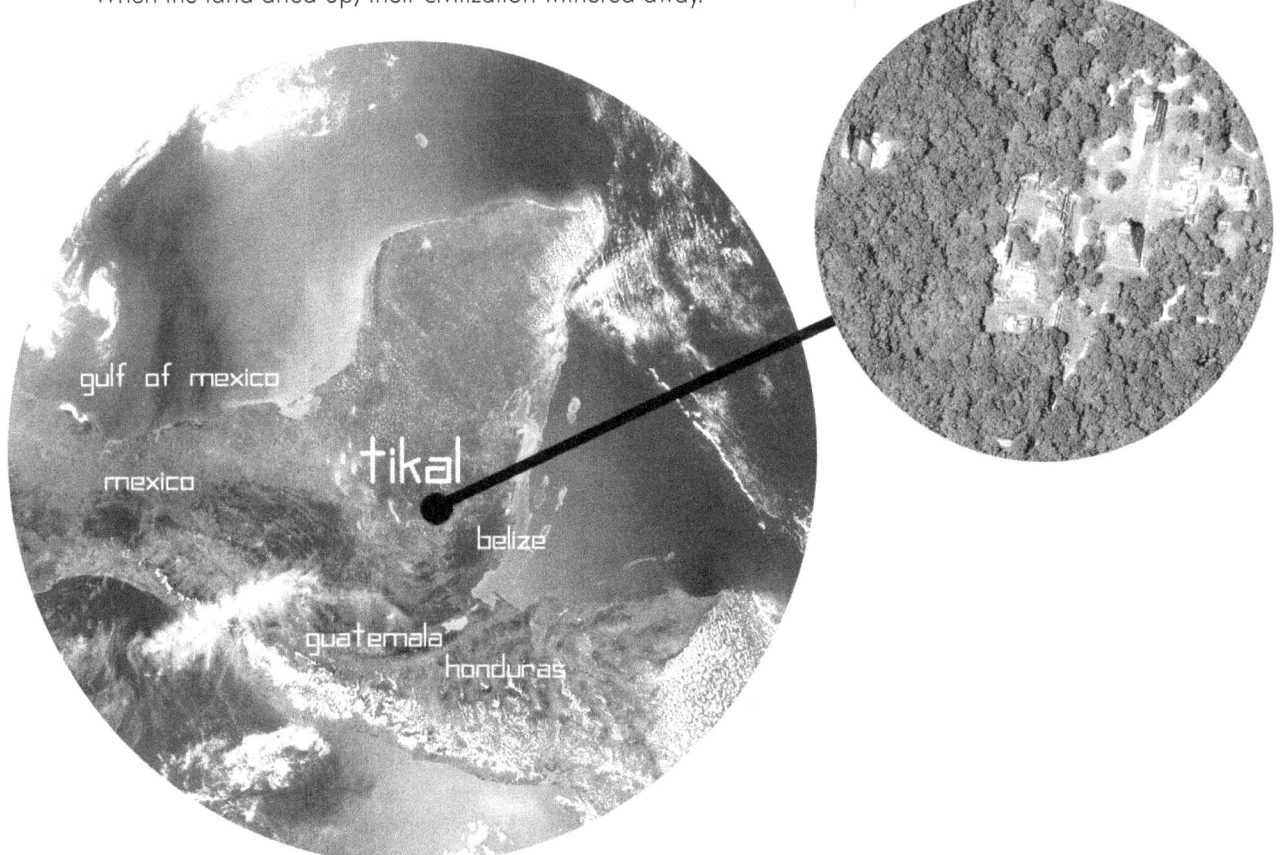

gulf of mexico

mexico

tikal

belize

guatemala

honduras

a message from our ancestors............

The Mayan of Tikal were not our only ancestors to cause their own undoing by unravelling the natural systems they depended on for survival. The Polynesian culture that inhabited Easter Island, also called Rapa Nui, in the South Pacific Ocean did more or less the same thing, right around the same time.

This culture also built up a great civilization that depended on the forest for their way of life. They used the trees for housing, fuel, and fishing vessels.

Slowly over decades and centuries, the population grew and the trees were removed at a pace faster than nature could replenish them. The soil eroded and became infertile for farming, the wildlife that inhabited the forests disappeared, as did their source for fuel, housing, and fishing transport.

As the population outpaced the resource base, famine took hold and they later resorted to cannibalism. The society finally dissolved into disorder, conflict, and chaos.

Individual benefits from the trees' harvest were realized immediately, while community consequences were realized only too late. The elders, who may have realized the problem of continued deforestation, were likely ignored as the leaders needed a larger supply of wood to satisfy their growing settlements.

The Earth and our global population is often compared to Easter Island. Our way of life feeds our growing population and growing settlements in a way which affects the Earth in many complex ways. Whether it is overfishing, deforestation, or fossil fuel use, at some point, the disruption we have caused to the systems of the Earth will come back and affect the vital systems of our settlements in various ways.

Fortunately we have episodes in the past that can teach us about how to behave in the present and future.

But only those who listen and act accordingly will be fit enough to survive.

hmmm...

...i wonder if they'll get it?

The fate of the Sumerian, Egyptian, Mayan, and Easter Islanders has met hundreds of human settlements, in hundreds of places, for the past hundreds of thousands of years. Our settlements are no different than those of our ancestors, they are vulnerable to changes, especially changes to the stable weather patterns produced by local and global climate systems.

If the conditions change slightly, it can be inconvenient.

If the conditions change greatly, it can be disastrous.

The events experienced by our ancestors caused decades of despair and tragedy. But the survivors of these events were able to pass down the history of civilization-altering events, so that successive generations would not repeat the mistakes of the past. The survivors also helped to usher in advances in technology and understanding, raising the bar of cultural progress to levels that had never been experienced before.

Crisis and struggle will force ingenuity into the mind of any creature.

At the beginning of the 21st century, our global civilization faces a situation similar to that of our ancestors with regard to the consequences of global temperature change. But there is one important difference with our situation: we know what is coming with a high degree of certainty and can adapt in anticipation many years before anything happens.

We have knowledge of the impending flood.

We can judge where the meteor will strike.

We can see what will happen if we cut down all of the trees.

We can adapt before the Earth forces us to conform.

This understanding is a great gift.

It should not be taken for granted.

The changes predicted for our settlements are dramatic, like those of the past. It may once again take dramatic events to give our species the brain boom needed to provide us with the technologies and understanding needed to live in a more advanced and benign way in planet Earth.

Some will see these future events as simply another piece in the puzzle of who is the luckiest and who is the fittest.

Some will see them as a way to lead humankind to a greater level of cooperation within our global civilization.

Some will see them as an opportunity.

Some will see them as an inconvenience.

Some will refuse to see anything at all until it crashes into their face.

Great apes, brace yourself for the rocky coast.

Get ready for a brain boom.

Get ready to adapt.

Get ready to evolve.

Just get ready.

The history of our ancestors will repeat itself time and time again.

If we let it.

FEELING LUCKY?

FEELING FIT?

ARE YOU FEELING ANYTHING AT ALL?

4
THE MODERN DARK AGES

WATER COMES
FROM A FAUCET

FOOD COMES
FROM A BOX

ENERGY COMES
FROM A SWITCH

FUEL COMES
FROM A PUMP

WELCOME TO THE MODERN DARK AGES

HUMANTS

The humans. They are a grand species. Truly grand.

Humankind has done relatively well for themselves up to the 21st century. We have magnificent settlements all over the world, buildings that resemble mountains, airplanes that resemble birds, tools to take pictures from the beginning of the Universe, machines that unravel the genetic code of our own being, and processes to clone God's creatures.

We have built up a great global civilization and are a button away from destroying it all 10,000 times over.

We are truly awesome.

We are ingenious and creative problem solvers that have jumped leaps and bounds in clarity and awareness.

But, sometimes we are just plain stupid.

We have developed a perspective of separation from the Earth and a dominance over it, which has led us to believe we can control its processes related to global temperature change.

Unfortunately, whether we want to admit it or not, we are not in control of the Earth's processes. We may significantly influence these processes, but we are not the master of nature and will never be. Nature will always direct the course of humankind, just like it does with all other life on Earth. Nature makes the rules and humankind has to live with them, whether they are good or bad, just or unjust.

We have been led to believe we are no longer animals, because *Homo sapiens* are very advanced creatures, both mentally and manipulatively. We are very unique in a very unique way, but we are still just one creation of billions which have been allowed to bask here on planet Earth.

To some we may be the Crown of Creation, but sometimes we act more like the Asshole of Apes.

In the process of becoming something new, we have denied what we are and where we come from.

We have forgotten what we need to survive.

In our person, we have forgotten how to live healthily, both mentally and physically. In our settlements, we have forgotten how to be self-sufficient and rely on a great web of cooperation.

We are a young species, having been in our present form only a few dozen millennia, if that. It may take some time for us to figure out how to live on the planet without jeopardizing our own existence. Perhaps new societies will emerge after our modern settlements endure the consequences of global temperature changes. Maybe then humankind will remember the precarious and precious relationship it has with nature and one another for a healthy survival.

Maybe then cooperation will prevail and the global human species will be reunited in the great global family that it is.

But why wait until we are forced?

HOW DO YOU VIEW?

One's WELTANSCHAUUNG or WORLDVIEW, is created by a number of different factors that are internalized to give each person their own unique understanding of the world around them. Worldviews are shaped by many things: religion, science, cultural norms, language, economic standing, political systems, natural surroundings, level of development, and so on. Our worldview affects our relationship to one another, our relationship to all living and non-living things, and our understanding of place and time in the cosmos.

There are differences in worldview depending on where and how one is brought up and the circumstances of their experience. The way I look at the world is a bit different than the way you look at the world. Neither is totally right or totally wrong, and neither are totally better or totally worse. There are things we perceive differently and things we perceive similarly.

Throughout the history of civilization, major shifts in worldview have taken place as a result of creative inventions, intellectual revelations, and amazing discoveries. The way humankind's worldview has changed during the past few hundred years is of particular interest when framing our present situation in light of global temperature change.

Until a few hundred years ago, the prevailing worldview of our place in the planet was one of an involved and integrated actor. Rocks and rivers, mountains and seas, fish and birds, trees and grass; everything was alive, everything possessed a spirit, everything was connected, and humans were part of it all.

By definition, ANIMISM is 'the belief that the vital and principle element of an organic being is an immaterial spirit.' The term animism is also used when we give objects the attributes of a life-force. This is the way our ancestors looked at the world and how many indigenous and aboriginal cultures still understand the world around them.

As far as our time and place in the cosmos was concerned, western civilization believed that time moved in cycles over and over and over again, forever. The Earth was located at the center of a closed system Universe that consisted of concentric, crystalline spheres of stars and planets with God inhabiting the very last ring. Everything was always in a state of becoming, except God, who was being.

Everybody knew 'why' something happened: because God deemed it so and that answer was final. That answer was enough.

'How' something happened was totally irrelevant.

Gravity? What is that?

By the end of the 16th century, some people were growing tired with this worldview and the SCIENTIFIC REVOLUTION was born.

Everyone already knew why something happened, God made it so. And that was boring.

There were new tools, new thoughts, new discoveries.

–A human being is part of the totality, which we call the Universe, a part which is limited in time and space. Mankind experiences himself, his thoughts and feelings as something which is separated from the rest, a kind of optical delusion of his consciousness.

This delusion creates a prison for ourselves, it limits us to our own personal desires and to affection for those few people who are the nearest to us. It is our task to free ourselves from this prison, by enlarging the circle of our compassion and to embrace all living creatures and the beauty of all nature.–

–Albert Einstein

There was a new way to look at the world.

People were discovering 'how' things happened.

And that was exciting!

The Earth became objectified, it became something full of secrets just waiting to be heard. Those secrets could be discovered by observing what happened when one took the Earth and dissected it, manipulated it, broke it down, and pieced it back together again.

The Universe was still a big place, in fact it was infinite, it went on forever and was not bound by a final ring of God. The heavens did not revolve around the Earth and the Earth had no particular role, it was just one of many glowing objects that revolved around the Sun. Just a speck of dust in the cosmos, it was not very special at all.

The ability to explain how things happened was established with sets of principles and 'provable' theories. Things happened and could be explained without God deeming them so. Time moved in a straight line, from point A to point B, further to point C, and into infinity. No cycles and no end.

Nature? Well, nature was dead, it was inanimate, it worked like cogs in a grand machine or the gears in a pocket watch and could be manipulated in whichever way we liked.

Nature? Spirits? Ha!

This shift in worldview occurred with vehement protest from those clinging to the way things were. Those manipulating the Earth, the Alchemists, who later became the Scientists, were viewed as Heretics, because they were degrading the Earth and defiling its spirits through manipulation and experimentation.

The Catholic Church protested severely, because they were no longer privileged to the answers of Universal questions. Any ordinary person could discover them too, as long as they understood certain principles and concepts. This was a serious threat to the legitimacy, control, and social stability of the prevailing order at the time.

There are four key people who helped shape this new worldview that removed humans from an active role in the spirited body of the Earth and made them simply an observer conducting tests on dead matter. Those four people are the English philosopher Sir Francis Bacon, the Italian astronomer Galileo Galilei, the French philosopher and logician René Descartes, and the English mathematician Sir Isaac Newton. They are not the only individuals responsible for bringing about this century long shift, but they are the primary movers that have led us to the way we presently perceive our relationship to one another and the planet.

SIR FRANCIS BACON was called 'The great secretary of Nature and all learning' by the renowned English outdoorsman Izaak Walton. Bacon was adamant about the need to disregard all knowledge up to this point and begin anew, basing all understanding of how the world works with the new means of data collection and experimentation emerging at the time.

Disregard all knowledge up to this point? That is pretty severe.

Bacon saw the Universe as a great riddle to be examined, meditated upon, and solved, not just a heavenly blanket that was laid upon us. He perceived the Earth in the same way. The Earth was not just a 'fixed stage' that humans walked upon. Nature was something that needed to be tested in order to yield definitive 'yes' or 'no' answers. Nature was not to be left alone to simply exist, it needed to be probed, broken, and burned to see how it reacted.

Nature was mechanical and the mind had to work this way too.

While Bacon was talking about the need for experimentation, the Italian astronomer GALILEO GALILEI was finding evidence contrary to the Catholic Church's belief that the Earth was located at the center of the Universe. He bolstered Nikolaus Copernicus' idea that the planets were moving around the Sun, the theory of heliocentrism. Although Galileo believed it was the Sun that lay at the center of the Universe, which is not the case, Galileo's observations and reports at the time shattered the idea that the Earth held a special place in the cosmos.

Imagine, one day we are living at the center of the Universe, the next day, we are not. That is pretty severe too, it is more or less similar to believing one day we are the only sentient species in the Universe and the next day, we are not.

Because Galileo's theories were contrary to the powers that be, Galileo was put under house arrest for the majority of his adult life until his death. Although it is not always in one's self-interest to question the powers that be, sometimes it is the only way to discover the truth, even though the 'truth' changes over time.

RENÉ DESCARTES is regarded by many as the 'Father of Modern Philosophy' and agreed with Bacon's idea of rejecting traditional knowledge and conventional methods of learning. He was a great mathematician and believed the only way for scientific findings could be ascertained for certain, was to base them on mathematics: numbers and equations.

He also saw the Universe as being mechanical, able to be broken down into individual parts, manipulated at leisure, and put back together again. His name is frequently mentioned in the schools of logic, as his Cartesian Method is something still taught and relied upon today for problem solving.

The Cartesian Method basically proposes to:

1) ACCEPT NOTHING AS TRUE WHICH HAS NOT BEEN PRESENTED CLEARLY AND DISTINCTLY OR IS SELF-EVIDENT.

2) BREAK DOWN COMPLEX IDEAS INTO THEIR SIMPLER ELEMENTS THAT CAN BE INTUITIVELY UNDERSTOOD.

3) RECONSTRUCT SIMPLE PROBLEMS FIRST AND PROGRESSIVELY MOVE TOWARD RECONSTRUCTING MORE COMPLEX PROBLEMS.

4) EXERCISE THOROUGHNESS OF THE DATA RELATING TO THE PROBLEM AND USE BOTH INDUCTIVE AND DEDUCTIVE REASONING TO REACH YOUR ANSWER.

−It is good to know something of the customs of various peoples, in order to judge our own more soundly and not to think that everything that is contrary to our way of doing things is worthy of scorn and against reason, as those who have seen nothing commonly think.−

−It is not enough to have a good mind. The main thing is to use it well.−

−René Descartes

The idea being that by following this method for investigation and analysis, one will be able to produce the most thorough and overwhelming evidence to make one's case.

Descartes wrote many treatises in his time and is well known for the statement "cogito ergo sum" - "I think, therefore I am" which refers to his theory that even when one can doubt the existence of everything in the Universe, the one thing which is undoubtedly true, is the existence of one's own mental self.

Descartes was in the final stages of completing his first Treatise on Physics when he received word that Galileo was imprisoned. Having a hunch that his findings would also put him in opposition with the powers that be, he held off, and this particular treatise was printed only in part after his death.

SIR ISAAC NEWTON is considered to be the first modern scientist and considered to be the last of the alchemists. Newton believed he was chosen by God as the next in a lineage of great thinkers placed upon the Earth to unwrap the riddles of the cosmos; riddles that had been handed down since the Babylonians. He believed this for many reasons, but particularly because he survived a difficult birth on Christmas Day in 1642 (under the Julian Calendar used at the time).

During his adult life, Newton bordered on psychological mania, partially due to his belief that divine purpose placed him on the planet. But he was not chastised for his beliefs, in fact, he was chosen to be the first scientist ever knighted by the English court and became president of the Royal Society of London. He also developed the theory for the Three Laws of Thermodynamics and the Theory of Gravity, which are hugely significant events in the history of science. The scientific methods he established for examination and experimentation have also carried through to the 21st century in almost every aspect of scientific discovery.

Whether he was chosen by God or created his own destiny by achieving a self-fulfilling prophecy is irrelevant. The relevance lies in his helping to seal the shift in worldview from human's living in the center of the Universe to human's living on just another place in the cosmos; from human's living 'in' a planet to human's living 'on' a planet; from human's respecting the spirit of all things to human's manipulating without thought of consequence.

Before this shift, Humankind, Nature, Earth, Universe, and God were all united and bound to one another in a big bean-bag of cosmic coziness. Everything was alive, everything had a spirit, everything was understood, everything just was and humans were a part of it all.

Then the scientific thinkers came tinkering about and removed Humankind from the warm hearth of Earth, putting them in the sterile laboratory of observation, microscopes and dissection: an environment devoid of feeling. But also an environment full of new discoveries and a new way to look at the world.

Another hundred years passed and an economist came on the scene to perpetuate the worldview of manipulation without consequence and added economic self-interest to the mix. That economist's name was ADAM SMITH and the INDUSTRIAL REVOLUTION was born.

Adam Smith published a book called THE WEALTH OF NATIONS in the same year the United States of America became a country, 1776 C.E. *The Wealth of Nations* was very broad in scope and a very involved piece of economic philosophy. *The Wealth of Nations* described an economic system based on 'the private interests and passions of men' that lead toward a result 'which is the most agreeable to the interest of the whole society.' The 'invisible hand,' guided a 'free-market' economy which allowed both people and nations to prosper in a free society with the free flow of goods, services, resources, and knowledge.

Although often forgotten or ignored in the Age of Global Consumer Capitalism, there was a strong moral strain in Smith's theories. This involved the pursuit of social harmony through checks and balances of the free-market system. These checks and balances are often disregarded in pursuit of ever-increasing profit margins, producing ever-cheaper products, acquiring an ever-cheaper labor force, and using an 'endless' supply of resources. This faulty mentality never considers the full social, environmental, or economic costs of such activities which Smith's doctrine held as a central tenet.

Smith also believed the 'private interests and passions of men' should be pursued with a deep understanding of the following tenet:

THAT WHICH MAKES THE SOCIETY STABLE AND PROSPEROUS FOR YOUR NEIGHBORS, MAKES IT STABLE AND PROSPEROUS FOR YOU, INDIVIDUALLY.

This tenet, although often ignored as well, may be more important than ever as we are now living in the context of a global civilization and the definition of 'neighbor' has been considerably expanded.

The thinkers of the scientific and industrial revolutions unknowingly created the backbone of a worldview that ignores the possible consequences of actions. By successfully separating humankind from nature, they removed the understanding of connectedness and replaced it with an understanding of mere mechanical manipulation.

This UTILITARIAN perception of the Earth and its resources allowed our ancestors to collectively transform the Earth without much thought of what might happen as a result.

Here is a truer representation of Smith's economic proposal:

In most cases today, people in developed countries receive their electricity from a few sources: coal, oil, nuclear, and natural gas. They are supplied by a few large companies and transmitted through centalized networks.

This method has been established because such energy sources and their infrastructure are expensive to acquire and maintain, which makes it is easier for well funded companies to provide a stable flow of energy to the population. These electricity systems are large and vulnerable to any disruptions down the line, from energy source acquisition to transmission to reception.

A check in the system for Smith would be to promote alternative energy sources, like home solar electricity systems and community networked wind electricity systems that could all be interconnected to the main network, thus easing the population off conventional sources while continuing energy security.

As new products and businesses are created, self-interests and the society benefit from new economic impulses stimulating the free market and new employment opportunities. Individuals and society benefit further as other energy sources are integrated into the existing energy network, diversifying it and creating a more secure and stable electricity.

Sound good? It's already happening.

°THE EARTH MOVES MOST FOR HUMANS
NATURE NEWS SERVICE
7 MARCH 2005

Agriculture and excavations shape the landscape more than rivers and glaciers. Human activities shift ten times as much material on the Earth's surface as all natural geological processes put together.

That's the conclusion of geologist Bruce Wilkinson at the University of Michigan, who has used the geological record to estimate the earth-moving capacity of natural processes over the past half a billion years. He publishes his findings in *Geology*.

Wilkinson was inspired to calculate a natural 'baseline' for the movement of soil and sediment after reading a paper published five years ago by geomorphologist Roger Hooke of the University of Maine in Orono.

...Hooke found that this impact has been increasing exponentially over the course of human civilization, and suggested that "we have now become arguably the premier geomorphic agent sculpting the landscape."

Wilkinson's calculations now show that there's no argument about it: we are ten times more active at land-shifting than nature. "It doesn't surprise me," Hooke says of the finding -he already suspected that the human impact on the landscape is alarming. "We're headed for disaster," he says...

utilitarian
a) Using the Earth simply for its usefulness.

b) Doing so to supply the greatest amount of good to the greatest amount of people.

°STUDY: EARTH CAN'T MEET HUMAN DEMAND FOR RESOURCES
REUTERS
24 JUNE 2002

The consumption of forests, energy and land by humans is exceeding the rate at which Earth can replenish itself, according to research published on Monday in the Proceedings of the National Academy of Sciences.

The study, conducted by California-based Redefining Progress, a non-profit group concerned with environmental conservation and its economics, warned that a failure to rein in humanity's overuse of natural resources could send the planet into "ecological bankruptcy."

Earth's resources "are like a pile of money anyone can grab while they all close their eyes, but then it's gone," said Mathis Wackernagel, lead author of the study and a program director at Redefining Progress.

Scientists said humanity's demand for resources had soared during the past 40 years to a level where it would take the planet 1.2 years to regenerate what people remove each year.

The impact by humans on the environment had inched higher since 1961 when public demand was 70 percent of the planet's regenerative capacity, the study showed.

"If we don't live within the budget of nature, sustainability becomes futile," Wackernagel said...

For over 350 years, nations have pursued economic and geopolitical superiority by manipulating and using the resources needed for a thriving and technological civilization: coal, oil, water, agricultural lands, and humans. During this time, humankind has acted as a collective global force, greatly transforming the larger systems of the Earth.

350 years to noticeably alter the global systems of the planet.

That is a big deal.

But we really did not realize this was happening until the last few decades. However, at the beginning of the 21st century, there is no denying that our anthropocentric, mechanical, and industrial-scale consumptive practices over the past few centuries have left us with one serious global predicament: we have a great number of people, living in great settlements which are based on complex and sensitive infrastructures. All of the above are at significant risk of the changes brought on by global temperature change.

That is a big deal too.

The worldview that produced our present predicament, the worldview that is still adhered to today, is a remnant of a bygone era. Believing that we are separate from the Earth and acting as if our activities will have no consequences is passé, it is a relic of a time that is no more.

It has had its day.

The belief that we can use the resources of the Earth forever? That the Earth's resources are without limit? Well, that is physically impossible. The Earth simply cannot meet our growing demands at the rate we are using them and continue to provide stability to its systems.

THE PROBLEM IS NOT THE NUMBER OF PEOPLE ON THE PLANET, BUT RATHER THE WAY IN WHICH THE NUMBER OF PEOPLE ON THE PLANET LIVE.

We are recognizing that the stable systems of the Earth are not as stable as we once believed.

We are noticing that the cumulative activities of a global species have indeed affected the global systems of the planet.

We are realizing that the Earth has certain limits to its dynamic equilibrium.

We are remembering that our actions do have consequences in the short and long term.

The Mind of Humankind is waking up from The Dream.

'Limitless' transformation of the Earth and careless utilization of the planet and the human species does have consequences.

Man, waking life sure looks different than the dream.

I guess this is what a 350 year hangover feels like.

Anyone have some adaptive aspirin?

a message from our descendants..........

WHEN PEOPLE LOOK BACK UPON OUR 21ST
CENTURY GLOBAL CIVILIZATION,
THEY PROBABLY WILL NOT SEE IT AS WE DO

JUST AS WE PERCEIVE THINGS DIFFERENTLY
TODAY THAN THOSE WHO LIVED 1500 YEARS
AGO OR 15,000 YEARS AGO OR 150,000 YEARS
AGO OR 1.5 MILLION YEARS AGO

OUR DESCENDANTS IN THE YEAR 2300 MAY
THINK TO THEMSELVES

Hmmm. Quite peculiar ancestors back then. Those 21st century folk believed they were separate from the Earth? Strange indeed. What made them different from the humans thousands of years before them?

I can't believe this is true, but I heard they tried to control the Earth's greenhouse gas cycles? Although that sounds mad, I don't think they were crazy. But it would seem to suggest a delusion of grandeur.

Hmmmm.

I heard they killed each other by the millions over the course of centuries. It must have been a type of primitive population control. Yes, that must have been the reason. Why else would one kill off his own species? It seems like quite a contradiction to the principle of cooperation and preservation.

Have you heard of the inequalities of that time? Billions struggled for basic necessities and a few others hoarded the riches of the Earth.

What a shame indeed.

No wonder violent outbreaks were common. Quite unfortunate. It would seem to imply a severe case of intraspecific schizophrenia and a self-destructive nature.

What more could one ask for than secure shelter, well made clothes, free energy, all the fresh food and water one could want and peace with your fellow humans and all that exists upon this grand planet?

Peculiar ancestors they were. To think they called themselves 'civilized.' It must have meant something different back then.

Of course, I wouldn't be where I am today without their trials, tribulations, successes and failures.

They must have made their share of good decisions as well.

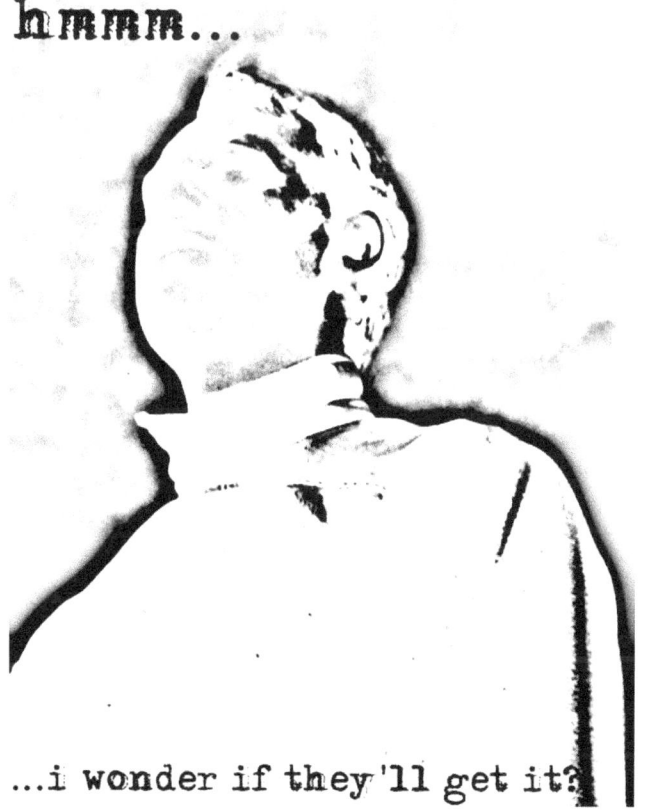

hmm...

...i wonder if they'll get it?

six senses of sickness
six senses of sickness senses of sickness
six senses of sickness
six senses of sickness
six senses of sickness
six senses of sickness
six senses of sickness
six senses of sickness
six senses of sickness sickness
six senses of sickness
six senses of sickness
six senses of sickness
six senses of sickness

six senses of sickness

Do we see the death inflicted upon ourselves as we destroy others?
Do we see the mushroom cloud rising high or the atomic generator in the sky?
Do we see the dead-zone oceans, the cancered skin, the laboratory grain?
Do we see the congestion, the walls, the earth imprisoned by our fences?
Do we see the devil disguised in a ballcap and bomber jacket?
Or have we lost the ability to see?

Do we smell the decay of humans huddled in shanty towns?
Do we smell the sewer running in front of their welcome mat?
Do we smell the garbage piled in their back yard?
Do we smell the genetically altered flowers placed upon their graves?
Do we smell the sweat oozing from the pores of the overworked and the underpaid?
Or have we lost the ability to smell?

Do we touch the tears of children who died of starvation today?
Do we touch the blood of children on our beloved's diamond ring?
Do we touch the faces branded by the ozone hole?
Do we touch the trees we've cleared from the forest to wipe away our waste?
Do we touch the 100,000 innocent bones littering the freeways of `freedom`?
Or have we lost the ability to touch?

Do we hear the whining of the chainsaw, over the whining for a cheeseburger?
Do we hear the 8-lane highway grumbling beneath our feet as we make it 10?
Do we hear the hate espoused by the righteously wrong?
Do we hear the last breath of a thousand species?
Do we hear the icebergs crack?
Or have we lost the ability to hear?

Do we taste the pesticide in our food?
Do we taste the fertilizer in our water?
Do we taste the soot in our air?
Do we taste the cancer in our blood?
Do we taste the bitter pill we leave for our children to swallow?
Or have we lost the ability to taste?

Do we sense the agony we create by no fault of our own?
Do we sense what is really right and what is really wrong?
Do we sense the pressure and the pain?
Do we sense the strain?
Do we sense the winds of change?
Or have we already lost our senses?

BLINDED BY OUR MIGHT

Let's face it.

We humans have a superiority complex, which can make us a little arrogant sometimes.

We kind of have a right to be arrogant, we have created the most technologically advanced settlements ever known to the planet.

We don't see squirrels running things do we?

But as a result of our perceived primary position on planet Earth, present day humankind has had a shadow cast over their foresight. It is a veil of arrogance developed from our sense of greatness that seems to have shortened our sight and ability to think thoroughly ahead to see the full consequences of our individual and collective actions.

For a humankind that did not know any better in the past, this error of judgment can be excused.

For a humankind at the beginning of the 21st century, with all of the knowledge it has about its past, present, and future, for us to all know better and still not do better?

Well, there is no excuse really.

Our descendants will not forgive us for this either.

There is a great paradox that has occurred during the past few centuries which relates to our present level of technological development and the predicament of global temperature change. Had humankind not experienced the shift in worldview following the scientific and industrial revolutions, the predicament of collective human activities causing global changes may not have happened. But without the collective activities of humankind during the past few centuries, we would not have created the level of technology we have today that allows us to see what the collective impact of our species upon the planet's systems has been.

Oh how the Gods play tricks on us mighty humants.

Sure, the temperature of the Earth has changed over time, climates and weather patterns have changed over time, and those alive during the time of these changes simply adapted or disappeared.

This is nothing new.

The Earth is not punishing us because of the way we have treated it and the Gods are not punishing us because we have become evil and betrayed the sacred gift. Although, one need not look any further than our alienation to the Earth, the way we exploit its bounty and the evilness with which we treat our fellow humans to see that there is certainly a degree of wickedness perpetuated by humankind on planet Earth.

But we are told this has always been the case, there are winners and losers, good and bad. The trick as an 'evolved,' 'civilized,' and 'humane' species is to make more winners and to make more good.

Although our manipulations and technologies may seem to be far more advanced and enlightened than those of the past, our present situation is not much different than those of Easter Island or Tikal.

hubris
Exaggerated pride or self-confidence often resulting in retribution.

ate
A blindness which comes from arrogance, also the Greek goddess personifying foolhardy and ruinous impulse.

nemesis
An act or effect of retribution, also the Greek goddess of retributive justice.

This is where useless rainforests are transformed into valuable grazing lands

The Hamburger Corporations Partnering with the Third World

Hier wird nutzloser Regenwald in kostbares Weideland verwandelt

Hamburger Konzerne
Partner für die Dritte Welt

Foresight is foresight, whether we are living within a dynamic and sensitive closed island system or within the dynamic and sensitive closed system of planet Earth.

The reason history repeats itself is because we let it.

It is not that the wisdom of past experiences fails to be transmitted into the present, but rather the present fails to listen, fails to learn, and fails to act.

As humankind continues to stretch their resources and infrastructure thinner and thinner, we do not pay attention to the fact that our systems are becoming increasingly susceptible to disruptions. Humankind does not think about the day when it has no more oil, when wars are fought for water, when weather no longer permits growing seasons, when disruptions cause a society to fall into disarray.

Arrogance, blindness, demise.

The classic tragedy played on a stage where everyone knows the outcome.

Not to downplay the importance that the scientific and industrial revolutions and the benefits they have created for humankind. There have been great local improvements in many areas, like cleaner water in some places and cleaner air in other places. We are becoming more vigilant and concerned about the places we live and the environment we inhabit. We tend to live longer than we have in the past or at least can artificially extend our lives.

But on a global scale, the planet has been transformed so much that the stability of its systems is coming into question, which brings into question the stability of our own systems which we depend upon for survival. We live in a world and in a way that is perceived to be unusually stable, but in reality, it is unusually precarious.

Humankind has tinkered and is tinkering with forces beyond its control and those forces are beginning to tinker back.

It is important to hold on to some of the scientific and economic principles of the past, they are certainly due some respect and have brought us great advancements. But, holding on too tightly without incorporating a new understanding in light of a new situation will lead our global mind into a cavern so dim, that we will not be able to see our hands in front of our face...or the rocky coast shrouded in fog ahead.

We could not see the rocky coast 50 years ago, the fog was still too thick. We were finally seeing some progress after cleaning up the mess World War II left behind and we had new pursuits to focus on, like atomic weapons development, space travel, and plastic wrap. Great advances were made in the areas of medicine, telecommunications, and suburbs.

And our global family continued to grow.

The fog started dissipating about 30 years ago, when we saw the Earth from space and began to internalize the limits and boundaries of living here. We saw a fine film of atmosphere that protected us from the Sun's radiation, we saw the blue oceans, the green forests, and the brown deserts. From outer space we could not see war, or starvation, or inequalities. All we saw was a lonely, beautiful living planet floating in space and time.

We were part of this great glowing body, all of us together.

And our global family continued to grow.

About 10 years ago, the rocky shore became quite visible. The global population learned of the disappearing ozone layer, biodiversity loss, greenhouse gas emissions, global warming, and the effects of global warming upon climate stability. We had begun to see the rocks, but we still believed we could turn our global boat around.

And our global family continued to grow.

The year is 2005 and our boat is still on course for a rough ride.

We did not see ourselves as a globally disruptive force when we began our journey into the unknown, but we see it now. We realize we have started something rolling which we cannot control and cannot stop.

We are being dragged in by the swell and see no escape.

The rocks are in plain sight now and it is too late to turn around.

But, it is not too late to brace ourselves for the impact.

Realize it, deal with it, accept it, get on living with it.

So, out with the old worldview and in with the new.

Time for a new way of thinking.

Time for a new way of seeing.

Time for a new human in the Age of Global Responsibility.

HOMO SAPIENS CONSCIENTIAS, welcome aboard.

Let us give ourselves a fresh start.

Let us start by taking some time to get to know our planet.

Again.

hello...i haven't seen you for a while

WATER
WHERE ARE WE

5

GAIA'S PROTECTIVE EMBRACE

CLIMATE OF CONFUSION

°GREENHOUSE EFFECT
SAID TO BE PROVED
ASSOCIATED PRESS
14 MARCH 2001

A comparison of satellite data from 1970 and 1997 has yielded what scientists say is the first direct evidence that so-called greenhouse gases are building up in Earth's atmosphere and allowing less heat to escape into space.

The study contains no evidence on whether Earth's surface temperature is actually increasing. In fact, whether this greenhouse effect will lead to global warming or global cooling is unclear, the scientists said.

That is because the greenhouse effect could start a cycle in which more clouds are formed, stopping the Sun's energy from reaching Earth's surface in the first place, said John Harries, who led the study.

Scientists have long theorized that carbon dioxide and other waste gases are increasing the trapping of heat close to Earth in what is called a greenhouse effect.

..."We're absolutely sure, there's no ambiguity: This shows the greenhouse effect is operating and what we are seeing can only be due to the increase in the gases," Harries said.

..."One of the main things that cause people to be skeptical of global warming is the lack of that real, definite connection between greenhouse gases and the planet getting warmer," Shindell said. "This really gives concrete evidence for the first time that greenhouse gases are changing the energy balance of the planet."...

What is global temperature change? What is global warming? What is the greenhouse effect? What is global climate warming? What is global climate change? What is global change?

They all mean the same thing, right?

Not exactly.

Television, radio, newspapers, popular magazines, and the Internet are great purveyors of the research pursued by the world's scientific community. However, they sometimes accidentally mislead the public by misusing terms, exaggerating the findings, or exaggerating the lack of findings.

Misleading and confusing the public is not always the intention of the media, although some media sources are certainly guided by such objectives, reporters are not climate change experts and they are not policy analysts.

They are reporters and they sometimes make unintentional mistakes.

GLOBAL CHANGES include changes in things like global reductions in ice cover, global reductions in forest cover, ozone layer depletion, and ocean disruption. Global changes also include global temperature changes and climate changes.

The **GREENHOUSE EFFECT** refers to the natural phenomenon of trace 'heat-trapping' **GREENHOUSE GASES** like methane, carbon dioxide and water vapor that reside in the upper atmosphere and 'reradiate' heat energy back to the Earth. **GLOBAL WARMING** refers to the measured and projected increase in the average global surface temperature: the 'surface' being from the surface of the Earth up to 8 kilometers above it. Global warming is a result of an **ENHANCED GREENHOUSE EFFECT** that is caused by an increase in greenhouse gas levels in the upper atmosphere.

GLOBAL COOLING is the opposite of global warming that occurs as a result of a **DIMINISHED GREENHOUSE EFFECT**, caused by a decrease of greenhouse gas levels in the upper atmosphere.

GLOBAL TEMPERATURE CHANGE simply refers to the change in the Earth's average global surface temperature; the peaks and dips, the warming and cooling of the Earth over time, be they slight or significant.

Some reporters and scientists use the term **CLIMATE WARMING** to describe global warming, which is incorrect, because climate warming only exists in places where the climate in general becomes warmer on average. Climate warming is not a term that can be blanketed over the entire Earth, because as the Earth warms, its average surface temperature increases, but not all of its climates warm, the climates simply change. Some places become warmer on average, some cooler, some wetter, some drier.

Changes in the prevailing local weather patterns that result from changes in global climate systems are known as LOCAL CLIMATE CHANGES. The general change in climate systems around the world is called GLOBAL CLIMATE CHANGE. The term global climate change bundles all of the changes from everywhere on the planet under three big words. Both global climate changes and local climate changes can be dramatic or slight, long lasting or short lived.

Under the umbrella of global climate change there are rates of speed at which the climates may change. A GRADUAL CLIMATE CHANGE is what is usually understood when climate change is discussed. Gradual climate changes are the changes in weather patterns around the world that are predicted to happen as a result of a gradual increase in atmospheric greenhouse gas levels and a gradual increase in the average global surface temperature. Gradual climate changes include a global average increase in the frequency and severity of extreme weather events, a shift in the onset of seasons, and an average increase of mountain precipitation falling as rain rather than snow.

Then there is the big daddy of climate change: ABRUPT CLIMATE CHANGE, which is caused by an abrupt hemispheric or global temperature shift. This phenomenon has recently received a lot of attention from not only the scientists, but from politicians, the business community, military strategists, and Hollywood.

Abrupt climate changes have occurred many times during the history of the Earth and have not always been global in scale. Most often abrupt climate changes have been isolated to the North or South Pole, coinciding with an abrupt change in the average surface temperature of these areas.

On a scale of 1-10 in severity, abrupt climate change is 1000. We are talking about a change of 5-15° Celsius in 20 years or 7° Celsius change in a decade. Were a hemispheric dip in temperature to occur today, it would put Scandinavia under a layer of ice…within the time span of two presidential terms.

That is abrupt.

A lot of terms, a little bit of confusion, that is all it takes for someone to be stuck in their tracks.

And then be buried by an avalanche.

°NO DOUBTS GLOBAL WARMING IS REAL, U.S. EXPERTS SAY
REUTERS
3 DECEMBER 2003

There can be no doubt that global warming is real and is being caused by people, two top U.S. government climate experts said.

Industrial emissions are a leading cause, they say -- contradicting critics, already in the minority, who argue that climate change could be caused by mostly natural forces.

"There is no doubt that the composition of the atmosphere is changing because of human activities, and today greenhouse gases are the largest human influence on global climate," wrote Thomas Karl, director of the National Oceanic and Atmospheric Administration's National Climatic Data Center, and Kevin Trenberth, head of the Climate Analysis Section at the National Center for Atmospheric Research.

"The likely result is more frequent heat waves, droughts, extreme precipitation events, and related impacts, e.g., wildfires, heat stress, vegetation changes, and sea-level rise," they added in a commentary to be published in Friday's issue of the journal *Science*.

…"Climate change is truly a global issue, one that may prove to be humanity's greatest challenge," they wrote. "It is very unlikely to be adequately addressed without greatly improved international cooperation and action."

°ABRUPT CLIMATE CHANGE LIKELY
NATURE NEWS SERVICE
13 DECEMBER 2001

Abrupt changes in global climate, common in the past, will become more so in future, thanks to man's impact on the environment, according to a new report presented at the American Geophysical Union meeting in San Francisco, California.

The report from the US National Research Council, to be published next year, calls for more research into rapid climate change, and for policy to equip societies to deal with it better.

"Realization has been growing over the past decade that climate change is not always gradual," says Richard Alley of Pennsylvania State University in University Park, chair of the committee that produced the report.

For example, data increasingly suggest that about 11,500 years ago, during a period called the Younger Dryas, global temperatures fell by up to 16 degrees within a decade and rainfall halved. Things stayed that way for more that 1,000 years…

WHO IS IN THE HOTHOUSE

The research on the relationship of global temperature to atmospheric gas concentrations began about 150 years ago. Back then, measurements were not as thorough or as numerous as they are today, and technologies were not as advanced as they are today, but a few clever chaps in Europe had a sneaking suspicion that atmospheric gases, particularly carbon dioxide and water vapor, played a key role in regulating the global temperature.

The first to fiddle with the idea was a French scientist and mathematician named **JEAN-BAPTISTE FOURIER** who lived during the reign of Napoleon and worked for him as well. In 1827, Fourier was the first person to describe the phenomenon keeping the Earth temperate as something similar to a garden room greenhouse and noted that human activities could possibly modify the natural climate.

In the 1860's, a British scientist named **JOHN TYNDALL** became involved with the research on atmospheric gases and global temperature. He is most notably known for explaining why the sky is blue, a phenomenon known as the Tyndall Effect: during the daytime, molecules in the air scatter sunlight more from the blue color spectrum, rather than red. But he also noted that water vapor might affect the temperature of the planet. He was the first to suggest that ice ages were perhaps caused by lower levels of 'heat-trapping' gases present in the atmosphere.

Then along came a Swede, **SVANTE ARRHENIUS**, at the end of the 19th century and followed the dynamics of global temperature as a professional hobby. He predicted correctly that a doubling of 'heat-trapping' gases, particularly carbon dioxide, would cause the average global temperature to rise 5-6° C and that the warming would be the most intense at higher latitudes. Even with fairly remedial scientific measuring devices Arrhenius' predictions were as accurate as those made today.

Arrhenius also developed the idea of the positive feedback model, predicting that as the Earth warms, increased water vapor resulting from evaporation of warming oceans would enhance the heat budget of the Earth, which would warm the oceans more, which would increase water vapor from evaporation, which would enhance the heat budget of the Earth... This discovery sometimes goes unnoticed, but is perhaps the most important discovery of them all, as positive feedbacks are an important key to understanding how sharp shifts in global temperature can happen so quickly.

There were several others who followed Arrhenius' study of 'heat-trapping' gases and their effect on global temperature and climate systems, but the next significant impulse of important new discoveries came in the 1950's with **CHARLES KEELING** and **ROGER REVELLE**.

In 1956, Keeling was assigned to take measurements of atmospheric carbon dioxide in the pristine conditions atop the world's largest volcano,

Mauna Loa on the island of Hawaii. Since this place is far from any industry, the measurements are a very accurate representation of the global atmospheric carbon dioxide level.

The next year, Revelle published a ground breaking article which associated the combustion of fossil fuels with a rise in atmospheric carbon dioxide. The subject of human's influence on global warming had been catapulted into the public sphere.

In 1958, the measurement of atmospheric carbon dioxide levels from Mauna Loa commenced and Keeling's initial measurement was 315 parts per million (ppm). Today we have reached 379 ppm and take readings from another 60 sites around the world.

As the world continues to combust more and more fossil fuels as a result of increasing energy uses, the level of carbon dioxide in the atmosphere continues to increase like a step-ladder every year.

Step by step by......

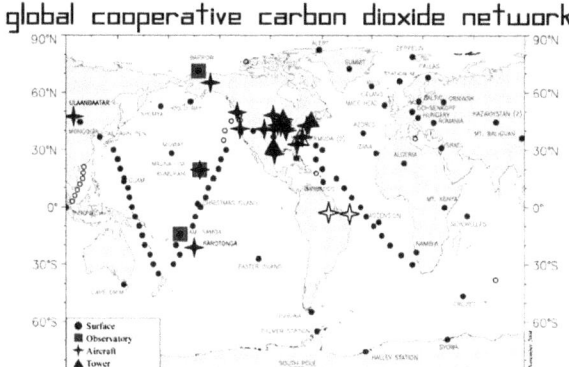

global cooperative carbon dioxide network

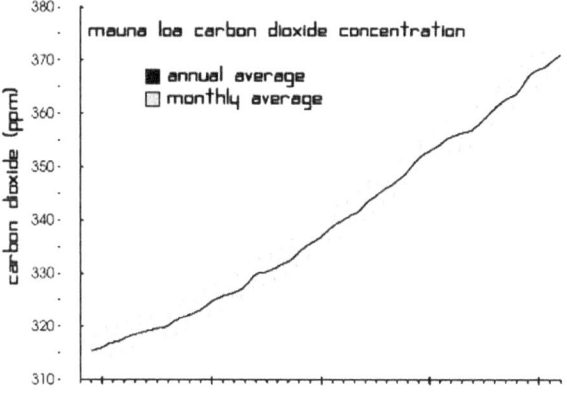

mauna loa carbon dioxide concentration

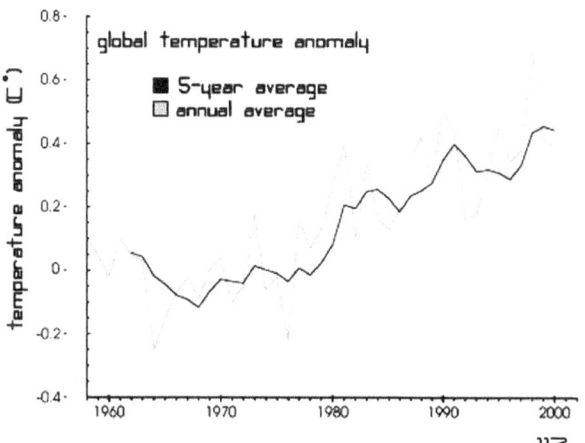

global temperature anomaly

THERMOTAXIS

THERMOTAXIS is defined as the 'regulation of body temperature,' which is more or less what the Earth does on a global scale.

The Earth seems to have a broad temperature band in which it seems comfortable. Sometimes the temperature of its body is very cold, as is the case during longer lasting ice ages. Sometimes, the temperature of its body is warmer, like it is today during a shorter lasting interglacial, or even much warmer, like it was many millions of years ago.

The Earth seems to be able to regulate the prevailing temperature for long periods of time in a state of dynamic equilibrium depending on what is alive at the time and what it experiences from terrestrial and extraterrestrial activities.

Sometimes the temperature of the Earth allows for a flourishing of life and robust exchanges of gases, heat, and energy. Sometimes there is less life to play a role and the Earth is comfortable being covered in a frozen blanket. But regardless of the dynamic steady-state it is in, when something shocks its systems enough, the Earth transitions and adjusts, either settling back into the previous state or into a different steady-state. This transition period and new state will alter the average global surface temperature, alter climate regimes, and alter weather patterns.

This is what we see when we look at the Earth's global temperature shifts that go up and down and up and down over time; the Earth is in a steady-state, then something happens to make it shift into a different steady-state. Back and forth and back and forth, all the while bringing species in and out of existence, changing global systems, changing the levels of greenhouse gases present in its cycles and changing the intensity of the greenhouse effect. Sometimes the shifts take 10,000 years, sometimes 1000 years, sometimes 100, sometimes 10.

There are many phenomena that influence the average temperature of the Earth's surface. We know the tilt of the Earth's axis, the shape of its orbit around the Sun, and the degree of rotation act as orbital forcings for the global temperature. There are also many RADIATIVE FORCINGS, like the amount of greenhouse gases that enhance or diminish the amount of heat energy reradiated back to the Earth.

Besides the radiative forcings of greenhouse gases, there are many other radiative influences to consider, that have either a positive or negative effect on the greenhouse effect. These influences come from sources like SOLAR ACTIVITY AND SUNSPOT CREATION; CLOUD DYNAMICS AND COVERAGE; OCEAN HEAT VENTING; THE REFLECTIVITY OF OCEAN WHITE CAPS; AEROSOLS AND DUST; ICE AND SNOW COVER; FOREST FIRES; VOLCANISM; POLLUTION; FOREST COVER; URBAN EXPANSION; OZONE LAYER DYNAMICS; and BIODIVERSITY INTEGRITY to name a few.

albedo
The degree of reflection of light or radiation by a surface, often expressed as a percentage or a fraction of 1. Snow or cloud covered areas have a high albedo (0.9 or 90%) due to their white color. Vegetation has a low albedo (0.1 or 10%) due to the light absorbed for photosynthesis, as do dark rock surfaces.

All of these things and more influence how the Earth regulates its average global surface temperature.

One of the most difficult radiative forcings to account for when making future climate predictions is the significance of cloud cover. As the Earth warms, more water evaporates and enters the atmosphere, which should increase the amount of cloud production. Because clouds are reflective, the albedo of the Earth increases, reflecting more solar energy away from the surface before it hits, and acting as a negative forcing. However, when there is more water vapor in the atmosphere, the greenhouse effect is enhanced. Also, when more clouds are present during the evening, more warmth is trapped near the surface, reducing the disparity between daily minimum and maximum temperatures, increasing the average surface temperature and acting as a positive forcing.

Another difficult-to-measure radiative forcing comes from aerosols and dust from soot and volcanism. When more particles are present in the atmosphere, a general cooling effect is usually produced through an increase in cloud formation these aerosols help produce, hence more solar energy is reflected. But if the aerosol particles are dark, like soot, then solar energy is absorbed, which enhances warming.

Aerosols from pollution also affect the climate on a more regional scale. Immediate changes in rainfall patterns have been observed as a short-term result of severe pollution in places like China, Bangladesh, Pakistan, and India.

The radiative forcing from forest fires is also quite significant.

Forest fires release huge amounts of greenhouse gases into the atmosphere when the trees and biomass burn. Furthermore, as the greenhouse effect intensifies and the Earth warms in general, forest fires are predicted to become more common, more fierce, and longer lasting, because the areas in which fires occur become drier and more prone to burning. Also, when fires remove forest cover, there are fewer places for carbon dioxide to be absorbed through photosynthesis.

Ocean phenomena affect the global temperature as well. Recent studies have shown that as the temperature of the oceans rise, the oceans are capable of releasing large amounts of heat energy, a type of venting, which could offset some global warming. There are also ocean whitecaps which have a reflective property, and would also help cool the planet.

There are many forcings that are affected by changes initiated inside and outside of the Earth and all of them affect greenhouse gas cycles, which affects the strength of the greenhouse effect, which affects the average global surface temperature, which affects global climate systems, which affects regional weather patterns, which affects our systems.

So how does the greenhouse effect work?

I thought you would never ask.

It is, after all, the only reason that any life on Earth exists.

Which is a good thing if you are part of that life.

°WILDFIRES ADD CARBON TO THE ATMOSPHERE
ENVIRONMENT NEWS SERVICE
9 DECEMBER 2002

Wildfires contribute tons of carbon dioxide to the atmosphere, increasing global warming as part of an accelerating cycle, U.S. researchers said this weekend.

The fires take carbon out of storage in vegetation and soils, and feed it into the atmosphere as carbon dioxide, a greenhouse gas. Drought makes the problem worse by stunting tree growth and turning forests into dry tinderboxes, and environmental disturbances caused by human activities further reduce the ability of forests to store carbon.

These are some of the preliminary findings from computer modeling studies of the 2002 Colorado wildfires, conducted by a team of researchers from Colorado State University, the U.S. Geological Survey (USGS), and the National Center for Atmospheric Research (NCAR). Their results were at the annual meeting of the American Geophysical Union (AGU) in San Francisco.

"We're using the western U.S. as a case study area where climate and land use are interacting in several interesting ways," said NCAR senior scientist David Schimel. Western lands, particularly evergreen forests, represent about half of all U.S. carbon storage, he said.

The researchers developed a new computer model of a complex forest ecosystem to simulate the release of carbon during the 2002 fire season in Colorado. The findings estimate how much carbon would be stored in a normal year compared to a drought year, such as 2002.

More carbon is freed from storage during droughts, not only because more dry vegetation burns, but also because plants deprived of water grow slower, absorbing and storing less carbon in their tissues...

THE RESULT PRODUCED WHICH IS SIMILAR TO A GLASSED ENCLOSURE USED FOR THE CULTIVATION OR PROTECTION OF TENDER PLANTS

The greenhouse effect.

It is the heat-trapping phenomenon of planet Earth.

Without it, the Earth would resemble Mars. With too much of it, the Earth would resemble Venus. With any significant change in its intensity, be it enhanced or diminished, climate systems and weather patterns change, which changes the foundation and the living conditions for everything on planet Earth.

This means the foundation for all human settlements change too.

Many scientists will tell you the term 'greenhouse effect' is not the best metaphor for how the Earth produces the conditions for life to persist.

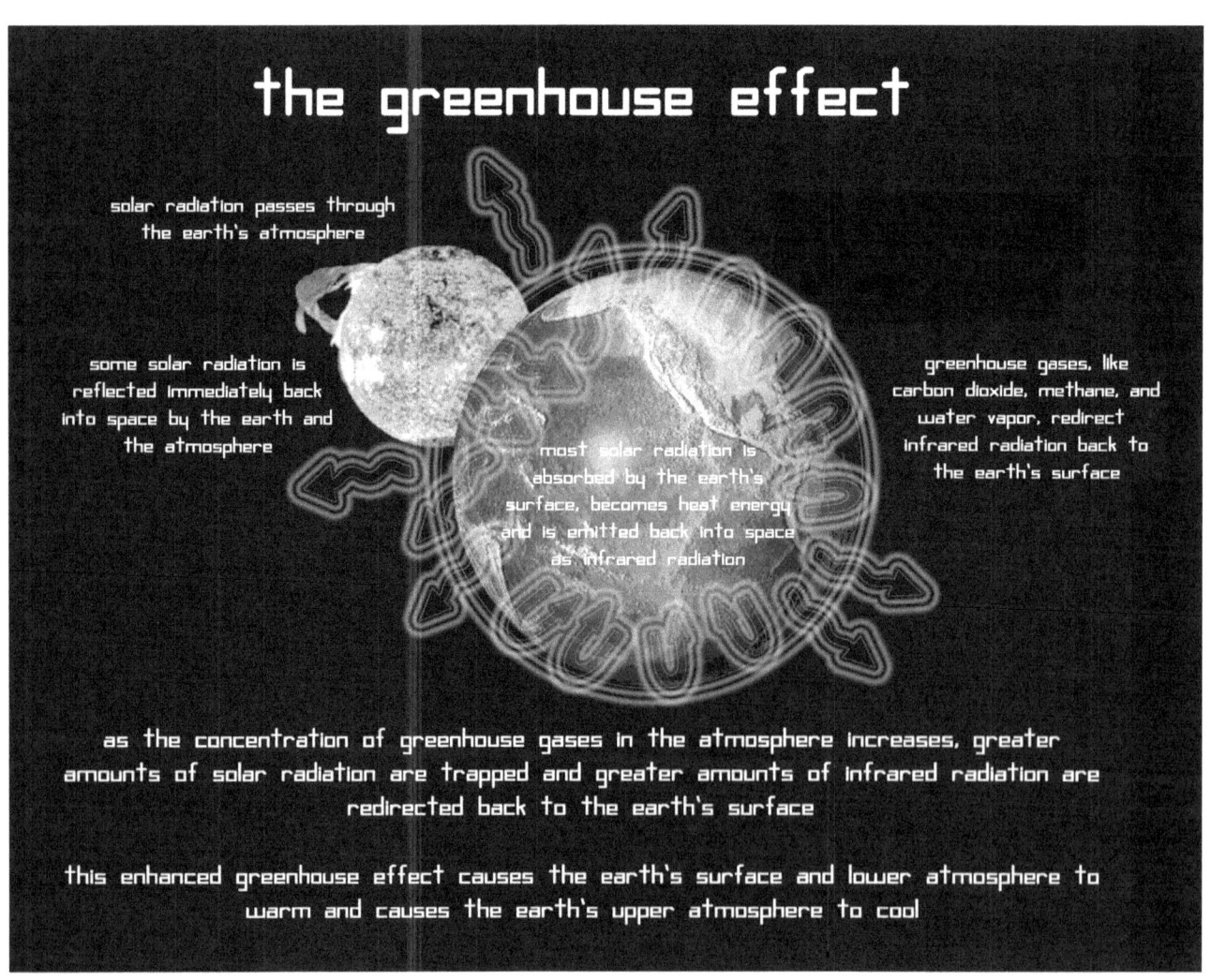

the greenhouse effect

solar radiation passes through the earth's atmosphere

some solar radiation is reflected immediately back into space by the earth and the atmosphere

most solar radiation is absorbed by the earth's surface, becomes heat energy and is emitted back into space as infrared radiation

greenhouse gases, like carbon dioxide, methane, and water vapor, redirect infrared radiation back to the earth's surface

as the concentration of greenhouse gases in the atmosphere increases, greater amounts of solar radiation are trapped and greater amounts of infrared radiation are redirected back to the earth's surface

this enhanced greenhouse effect causes the earth's surface and lower atmosphere to warm and causes the earth's upper atmosphere to cool

Many scientists will also tell you the way the popular media communicates how the greenhouse effect works and how atmospheric gases create a greenhouse effect are a bit misleading. Phrases like 'heat-trapping gases,' 'trap warmth,' 'act as a blanket,' and 'reradiate heat,' do not describe exactly how the greenhouse effect works. Although this may be the case, these are the most descriptive and conceptually non-complex phrases that relay the general idea of the greenhouse effect.

So, to the scientists, sorry for the simplicity.

A common definition of the greenhouse effect is a 'warming of the surface and lower atmosphere of a planet caused by a conversion of solar radiation into heat in a process involving selective transmission of short wave solar radiation by the atmosphere, its absorption by the planet's surface, and reradiation as infrared which is absorbed and partly reradiated back to the surface.'

Greenhouse gases, like water vapor, carbon dioxide, nitrous oxide and methane, have been circulating through the Earth's systems for a long time in varying levels, producing a greenhouse effect of varying strength.

These gases eventually collect in the upper atmospheric regions, in the **TROPOSPHERE** and **STRATOSPHERE**. In the upper atmosphere, these gases allow about 70% of the Sun's light energy to pass through and strike the Earth, while the remaining 30% is reflected back into space. As solar energy from the Sun hits the Earth, it is transformed into heat energy, some is absorbed and some is reflected back toward space. Greenhouse gases 'trap' some of this heat energy and 'reradiate' it back to the Earth. As the percentage of greenhouse gases in the atmosphere increases, the amount of heat energy which is trapped and reradiated increases. The warmer the planet becomes, the more heat there is to radiate and have reradiated.

Greenhouse gases make up less than 1% of the global atmospheric composition and are measured in parts per million or parts per billion. That there is a measurable and discernible difference in the global temperature correlated to an observable change in global greenhouse gas levels shows us just how powerful these gases are.

Greenhouse gases have different strengths and stay in the atmosphere for different lengths of time. For example, methane is a more powerful greenhouse gas than carbon dioxide, but resides in the atmosphere only a quarter of the time. In looking at the radiative forcing capacity that individual greenhouse gases have, two things are considered: **LONGEVITY** and **GLOBAL WARMING POTENTIAL (GWP)**.

Carbon dioxide is used as the base value 1 GWP.

Over the course of the Earth's history, the greenhouse effect has been enhanced and diminished and the average global surface temperature has risen or fallen in tandem with the rise and fall of greenhouse gases. They are intertwined like a fair maiden's braided hair.

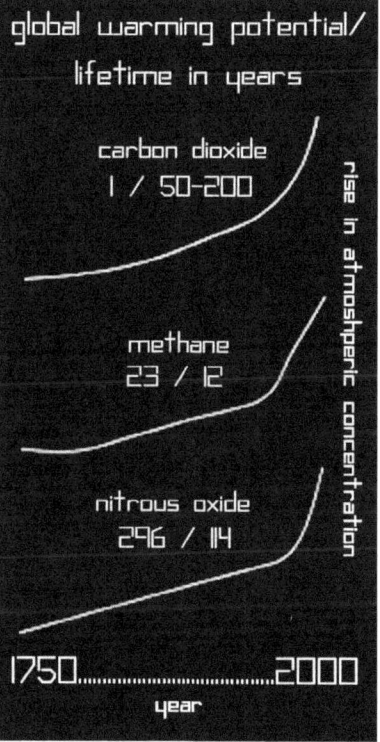

atmospheric altitudes
(mi)(km)

stratosphere
ozone layer
troposphere

global warming potential/ lifetime in years

carbon dioxide
1 / 50-200

methane
23 / 12

nitrous oxide
296 / 114

rise in atmospheric concentration

1750.............2000
year

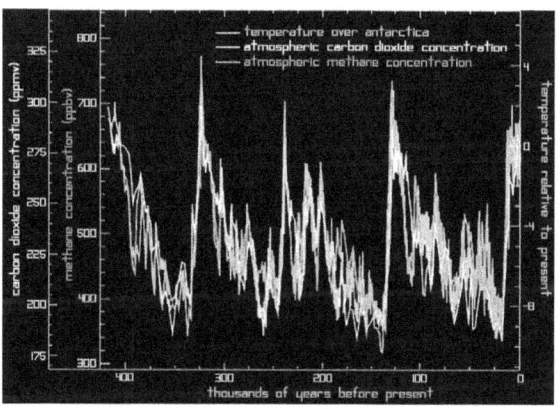

A vast belch of gas from beneath the North Atlantic 55 million years ago may have warmed the planet and hold clues to threats from an even faster modern surge in greenhouse gases, scientists reported in the journal *Nature*.

...Until now, scientists have been at a loss to explain the trigger for a 10-20 Fahrenheit global warming over about 10,000 years in the Eocene -- a blink in geological time.

...The scientists said the annual rate of modern human emissions of greenhouse gases to the atmosphere in the 1990s -- from fossil fuels burnt in cars, factories and power plants -- was more than 35 times as fast as the pace of the Eocene gas buildup.

"We can cause the same amount of global warming ourselves in a few hundred years at current rates," Svensen said...

POSITIVE FEEDBACK

rainforest reduction

increased greenhouse gas release

enhanced greenhouse effect

global temperature increase

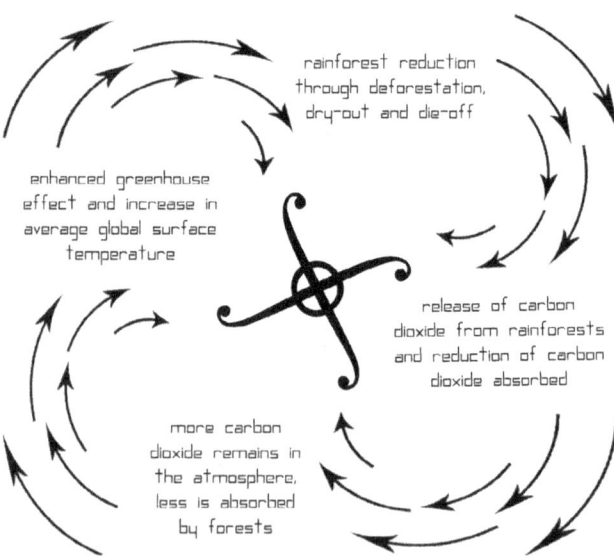

rainforest reduction
through deforestation,
dry-out and die-off

enhanced greenhouse
effect and increase in
average global surface
temperature

release of carbon
dioxide from rainforests
and reduction of carbon
dioxide absorbed

more carbon
dioxide remains in
the atmosphere,
less is absorbed
by forests

Humankind's emissions of greenhouse gases to the Earth's global greenhouse gas cycle are considered an **ANTHROPOGENIC** source.

In other words, a result of human activity.

Humankind adds nitrous oxide to the atmospheric mix through fuel combustion, industrial-scale agricultural practices, manure management, human sewage, waste combustion, natural gas leakage, and fertilizer production.

Humankind adds methane to the atmospheric mix with landfill emissions, manure from industrial-scale animal farms, and emissions from industrial-scale rice cultivation.

Humankind adds carbon dioxide to the atmospheric mix through fossil fuel combustion when we burn oil and coal, which we do a lot of, through industrial processes like cement manufacturing and lime manufacturing used to make steel, pulp and paper, water and sewage treatment.

Humankind also releases carbon dioxide to the atmospheric mix when we clear cut or burn large areas of forest. At the same time, we diminish the ability of the Earth to absorb carbon dioxide with these areas.

There is an unsettling trend in Borneo right now, in which the people are clear-cutting and burning the forest for large-scale industrialized rice production. Not a very wise choice in light of rapidly rising greenhouse gases and the specter of accelerated global warming.

But when one is struggling for survival and a better way of life today, long-term consequences are not always considered.

°PLANTING NEW FORESTS CAN'T
MATCH SAVING OLD ONES IN
CUTTING GREENHOUSE GASES,
STUDY FINDS
NEW YORK TIMES
22 MARCH 2000

A new study has cast doubts on an important element of a proposed treaty to fight global warming: the planting of new forests in an effort to sop up carbon dioxide, a heat-trapping gas.

The research concludes that old, wild forests are far better than plantations of young trees at ridding the air of carbon dioxide, which is released when coal, oil and other fossil fuels are burned.

The United States and other countries with large land masses want to use forest plantations to meet the goals of the proposed treaty. The study's authors say that any treaty also needs to protect old forests and that, so far there is no sign that such protections are being considered.

Without such protections, the scientists conclude, some countries could be tempted to cut down old forests now and then plant new trees on the deforested land later, getting credit for reducing carbon dioxide when they have actually made matters worse.

The analysis, published in the journal *Science* today, was done by Dr. Ernst-Detlef Schulze, the director of the Max Planck Institute for Biogeochemistry in Jena, Germany, and two other scientists at the institute...

The activities of humankind during its recent industrialized history have also created a large number of artificial greenhouse gases that stick around a lot longer than the natural ones. These chemicals were made by us largely for industrial processes and used as cooling agents. Many of them also deplete the ozone layer. Some of these artificial chemicals have lifetimes of 50,000 years and a global warming potential of over 32,000. That is quite a strong chemical that circulates for quite a long time.

The planet breathes, it is alive, inhaling and exhaling with its oceans and forests, its new life and its decaying death. The natural greenhouse gases water vapor, carbon dioxide, methane, and nitrous oxide are intimately intertwined with the natural processes of a life-bearing Earth in a state of dynamic equilibrium.

Greenhouse gas SOURCES and SINKS are where greenhouse gases come from and where they go. For example, the activities of humankind are more of a source of greenhouse gases. The planet's forests are more of a sink, large old-growth forests are better than small new forests at soaking up carbon dioxide through photosynthesis. The planet's oceans are also more of a sink for carbon dioxide through photosynthesis in phytoplankton.

However, should the oceans warm to a point where phytoplankton is no longer productive, a positive feedback cycle could be unleashed that would make the oceans become a source of carbon dioxide. The same is true with old growth forests. When deforestation reaches a level where a forest is destabilized, there is not only less forest to absorb and process carbon dioxide, but the forest begins emitting carbon dioxide as it dries up and dies off.

There are also huge amounts of methane and carbon dioxide locked in the permafrost of northern latitudes and in Antarctica. When the average global surface temperature rises to a point in which a positive feedback begins, the tundra thaw, greenhouse gas release, and surface warming will reinforce one another. The amount of greenhouse gases stored in the permafrost of the North Pole make our annual emissions of several billion tons look like a drop in the ocean.

Most scientists agree the past and present activities of global scale transformation and rise in the level of greenhouse gases is tipping the scale, disrupting the ability of the Earth to regulate itself and producing a situation in which the Earth's vast greenhouse gas sinks are now becoming vast sources.

If this is the case, the flood gates of global warming have been opened and evidence from the past show that when this happens, things can change very quickly.

It has happened before under similar circumstances.

Why would it not happen again?

°GLOBAL WARMING CREATES GRIM FUTURE FOR FORESTS
REUTERS
5 MARCH 2002

Global warming is becoming an increasing threat to forests in much of the world, paving the way for fires, droughts and pest infestations, officials told an environmental conference on Tuesday.

Ola Ullsten, former Swedish prime minister and co-chairman of the World Commission on Forests and Sustainable Development said the latest evidence indicates that over half the world's boreal forest could disappear due to the effect of climate change as conditions shift.

"It's a very severe problem," Ullsten told Reuters ahead of a conference in Winnipeg attended by timber industry representatives and environmentalists.

"If you want to illustrate the environmental dilemma the world is in, I think what has happened to the boreal forest in Manitoba, or you can probably take some other provinces as well, is a very good or tragic example of what might happen if actions aren't taken."...

°DECLINE IN OCEANS' PHYTOPLANKTON ALARMS SCIENTISTS
SAN FRANCISCO CHRONICLE
6 OCTOBER 2003

Plant life covering the surface of the world's oceans, a vital resource that helps absorb the worst of the "greenhouse gases" involved in global warming, is disappearing at a dangerous rate, scientists have discovered.

Satellites and seagoing ships have confirmed the diminishing productivity of the microscopic plants, which oceanographers say is most striking in the waters of the North Pacific -- ranging as far up as the high Arctic.

...The significant decline in plankton productivity has a direct effect on the world's carbon cycle, Gregg said. Normally, he noted, the ocean plants take up about half of all the carbon dioxide in the world's environment because they use the carbon, along with sunlight, for growth, and release oxygen into the atmosphere in a process known as photosynthesis.

Primary production of plankton in the North Pacific decreased by more than 9 percent during the past 20 years, and by nearly 7 percent in the North Atlantic, Gregg and his colleagues determined from their satellite observations and shipboard surveys. Combining all the major ocean basins of the world, Gregg and his colleagues found the decline in plankton productivity more than 6 percent...

THIN OF A SKIN

OZONE, it smells like burning electrical wires. When it is on the surface, we develop breathing problems due to its ground-level pollutive qualities. Without it in the stratosphere, we and everything on Earth would be scorched by intense ultraviolet (UV) radiation.

Ozone is a molecule made up of three oxygen atoms and is an abundant gas in the stratosphere, constantly being created and destroyed in relative equilibrium through natural chemical processes involving oxygen, hydrogen, nitrogen, and other elements. The OZONE LAYER is a diffuse collection of ozone molecules located between 10-60 kilometers above sea level with varying degrees of concentration. 90% of the Earth's ozone resides at this height and guards the Earth from 97-99% of the Sun's harmful UV rays, reflecting them back into space. Although the ozone layer appears to be very thick, if the entire ozone layer were compressed to sea level pressure, it would only be a few centimeters thin.

Ozone layer dynamics, general ozone layer depletion, and the ozone hole are interesting atmospheric-land coupled phenomena, involving atmospheric vortex currents, interactions with the surface of the Earth, water vapor collection, cloud formation in the upper atmosphere, solar storms produced by sunspots, and the activities of humankind.

Ozone in the upper atmosphere is constantly interacting with other chemicals and swirling around the Earth on atmospheric currents. Since the invention of OZONE DEPLETING CHEMICALS, ozone in the upper atmosphere is being destroyed faster than the Earth can produce it. These chemicals, like CHLOROFLUOROCARBONS (CFCs) and HALONS, were created in the laboratories of man during recent decades for use in styrofoams, refrigerants, propellants, and industrial processes. These chemicals were created because they are very, very stable. They are so stable, that some hang around for many thousands of years before breaking down completely.

Ozone depleting chemicals drift around the planet and eventually reach the stratosphere. There, the complex molecules are broken down by the Sun's UV rays and release chemicals like chlorine and bromine which bond to ozone and break it down.

Although there are some natural processes which produce chlorine, over 85% of the chlorine in the stratosphere is a direct product from the artificial compounds created by humans. One free atom of chlorine in the stratosphere has the ability to destroy over 100,000 molecules of ozone.

That is a lot of damage from a pretty small particle.

Ozone depleting chemicals were not detected until the 1970's, when the scientist James Lovelock developed a tool to detect CFCs. Although he was unaware of the CFC impact to the ozone layer at the time, he

°OZONE-DEPLETING
GASES ARE NOT
NATURAL
ENVIRONMENTAL NEWS
NETWORK
2 JULY 1999

Most of the gases responsible for stratospheric ozone depletion are produced by human activities and are not naturally occurring in the atmosphere, according to measurements of air trapped in the polar snowpack...

detected CFCs in every air sample he took on a voyage from England to Antarctica which led to the understanding that CFCs are carried on atmospheric currents.

Several scientists followed up on Lovelock's work, some of them receiving Nobel prizes in chemistry, and a decade later the leaders of the world came together in Canada to discuss the potential consequences of ozone depletion.

The **MONTREAL PROTOCOL** was drafted and signed by the leaders of the world and was the first international agreement of its kind to deal with a global environmental issue. The Montreal Protocol effectively banned the production and use of CFCs and other ozone depleting chemicals.

Although most countries have signed on and followed the agreement in good faith, ozone depleting chemicals continue to persist and new ones continue to be created.

All of them continue to circle the planet, eventually reaching the stratosphere and breaking down the ozone layer.

Here is a list of some of the ozone depleting chemicals banned by the Montreal Protocol: CFC-11 (CCl3F) Trichlorofluoromethane, CFC-12 (CCl2F2) Dichlorodifluoromethane, CFC-113 (C2F3Cl3) Trichlorotri-fluoroethane, CFC-114 (C2F4Cl2) Dichlorotetrafluoroethane, CFC-115 (C2F5Cl) Monochloropentafluoroethane, Halon 1211, (CF2ClBr) Bromochlorodifluoromethane, Halon 1301 (CF3Br) Bromotrifluoro-methane, Halon 2402 (C2F4Br2) Dibromotetrafluoroethane, CFC-13 (CF3Cl) Chlorotrifluoromethane, CFC-111 (C2FCl5) Pentachlorofluo-roethane, CFC-112 (C2F2Cl4) Tetrachlorodifluoroethane, CFC-211 (C3FCl7) Heptachlorofluoropropane, CFC-212 (C3F2Cl6) Hexa-chlorodifluoropropane, CFC-213 (C3F3Cl5) Pentachlorotrifluoropro-pane, CFC-214 (C3F4Cl4) Tetrachlorotetrafluoropropane, CFC-215 (C3F5Cl3) Trichloropentafluoropropane, CFC-216 (C3F6Cl2) Dichlo-rohexafluoropropane, CFC-217 (C3F7Cl) Chloroheptafluoropropane, CCl4, Carbon tetrachloride, Methyl Chloroform (C2H3Cl3), Methyl Bromide (CH3Br), CHFBr2, HBFC-12B1 (CHF2Br), CH2FBr, C2HFBr4, C2HF2Br3, C2HF3Br2, C2HF4Br, C2H2FBr3, C2H2F2Br2, C2H2F3Br , C2H3FBr2, C2H3F2Br, C2H4FBr, C3HFBr6, C3HF2Br5 , C3HF3Br4 , C3HF4Br3 , C3HF5Br2, C3HF6Br, C3H2FBr5, C3H2F2Br4, C3H2F3Br3, C3H2F4Br2, C3H2F5Br, C3H3FBr4, C3H3F2Br3, C3H3F3Br2, C3H3F4Br, C3H4FBr3, C3H4F2Br2, C3H4F3Br, C3H5FBr2, C3H5F2Br, C3H6FBr, CH2BrCl Chlorobromomethane, HCFC-21 (CHFCl2) Dichlo-rofluoromethane, HCFC-22 (CHF2Cl) Monochlorodifluoromethane, HCFC-31 (CH2FCl) Monochlorofluoromethane, HCFC-121 (C2HFCl4) Tetrachlorofluoroethane, HCFC-122 (C2HF2Cl3) Trichlorodifluoro-ethane, HCFC-123 (C2HF3Cl2) Dichlorotrifluoroethane, HCFC-124 (C2HF4Cl) Monochlorotetrafluoroethane, HCFC-131 (C2H2FCl3) Tri-chlorofluoroethane, HCFC-132b (C2H2F2Cl2) Dichlorodifluoroethane, HCFC-133a (C2H2F3Cl) Monochlorotrifluoroethane, HCFC-141b (C2H3FCl2) Dichlorofluoroethane, HCFC-142b (C2H3F2Cl) Mono-chlorodifluoroethane, HCFC-221 (C3HFCl6) Hexachlorofluoropro-pane, HCFC-222 (C3HF2Cl5) Pentachlorodifluoropropane, HCFC-

°DISCOVERY OF NEW OZONE-DESTROYING CHEMICAL
CSIRO AUSTRALIA
31 AUGUST 1998

CSIRO scientists have discovered a new ozone-destroying chemical in the atmosphere, as positive signs emerge that damage to the ozone layer should decline in the next decade.

Halon-1202, which has an ozone depletion potential approximately half that of the common CFCs, has increased five-fold in the atmosphere since the late 1970s. During the past two years the atmospheric concentration of halon-1202 has been growing at 17 per cent per year...

°SCIENTISTS FIND OZONE-DESTROYING MOLECULE
NASA PRESS RELEASE
9 FEBRUARY 2004

Using measurements from a NASA aircraft flying over the Arctic, Harvard University scientists have made the first observations of a molecule that researchers have long theorized plays a key role in destroying stratospheric ozone, chlorine peroxide...

...Chlorine monoxide and its dimmer originate primarily from halocarbons, molecules created by humans for industrial uses like refrigeration. Use of halocarbons has been banned by the Montreal Protocol, but they persist in the atmosphere for decades...

223 (C3HF3Cl4) Tetrachlorotrifluoropropane, HCFC-224 (C3HF4Cl3) Trichlorotetrafluoropropane, HCFC-225ca (C3HF5Cl2) Dichloropentafluoropropane, HCFC-225cb (C3HF5Cl2) Dichloropentafluoropropane, HCFC-226 (C3HF6Cl) Monochlorohexafluoropropane, HCFC-231 (C3H2FCl5) Pentachlorofluoropropane, HCFC-232 (C3H2F2Cl4) Tetrachlorodifluoropropane, HCFC-233 (C3H2F3Cl3) Trichlorotrifluoropropane, HCFC-234 (C3H2F4Cl2) Dichlorotetrafluoropropane, HCFC-235 (C3H2F5Cl) Monochloropentafluoropropane, HCFC-241 (C3H3FCl4) Tetrachlorofluoropropane, HCFC-242 (C3H3F2Cl3) Trichlorodifluoropropane, HCFC-243 (C3H3F3Cl2) Dichlorotrifluoropropane, HCFC-244 (C3H3F4Cl) Monochlorotetrafluoropropane, HCFC-251 (C3H4FCl3) Trichlorofluoropropane, HCFC-252 (C3H4F2Cl2) Dichlorodifluoropropane, HCFC-253 (C3H4F3Cl) Monochlorotrifluoropropane, HCFC-261 (C3H5FCl2) Dichlorofluoropropane, HCFC-262 (C3H5F2Cl) Monochlorodifluoropropane, HCFC-271 (C3H6FCl) Monochlorofluoropropane.

Dizzy yet?

If not, here are some chemicals currently used as replacements: NF_3, HFC-23 (CHF_3), HFC-32 (CH_2F_2), HFC-41 (CH_3F), HFC-43-10mee ($C_5H_2F_{10}$), HFC-125 (C_2HF_5), HFC-134 ($C_2H_2F_4$), HFC-134a (CH_2FCF_3), HFC-143 ($C_2H_3F_3$), HFC-143a ($C_2H_3F_3$), HFC-152a ($C_2H_4F_2$), HFC-227ea (C_3HF_7), HFC-236fa ($C_3H_2F_6$), HFC-236ea ($C_3H_2F_6$), HFC-245ca ($C_3H_3F_5$), HFC-245fa ($C_3H_3F_5$), HFC-365mfc ($C_4H_5F_5$), Perfluoropropane (C_3F_8), Perfluorobutane (C_4F_{10}), Perfluorocyclobutane (c-C_4F_8), Perfluoropentane (C_5F_{12}), Perfluorohexane (C_6F_{14}), HFE-7100 ($C_4F_9OCH_3$), HFE-7200 ($C_4F_9OC_2H_5$).

POSITIVE FEEDBACK

increased UV exposure
reduced phytoplankton productivity
enhanced greenhouse effect
increased ozone layer depletion

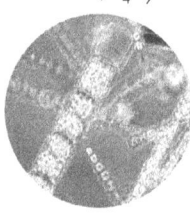

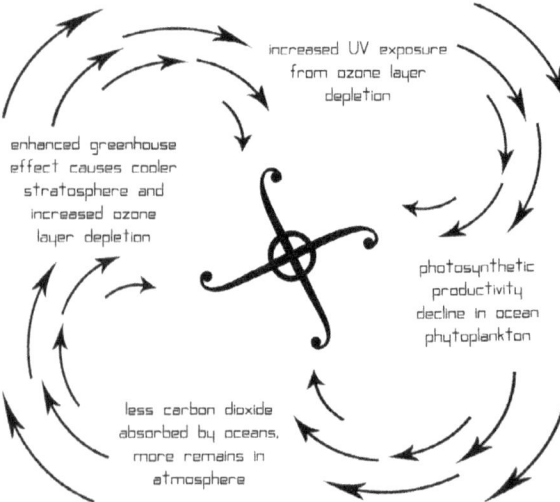

increased UV exposure from ozone layer depletion

photosynthetic productivity decline in ocean phytoplankton

enhanced greenhouse effect causes cooler stratosphere and increased ozone layer depletion

less carbon dioxide absorbed by oceans, more remains in atmosphere

While these replacements are apparently safer for the ozone layer, some of them have incredible longevity and very significant global warming potentials. Sulfur hexafluoride (SF_6) has a lifetime of 3200 years and is over 22,000 times more powerful as a warming agent than carbon dioxide. Perfluoroethane (CF_4) lasts over 10,000 years and has a global warming potential of more than 10,000. Perfluoromethane (C_2F_6) lasts for 50,000 years and has a global warming potential more than 6,000 times that of carbon dioxide.

So what is **OZONE LAYER DEPLETION**?

General ozone layer depletion is something that happens to the entire ozone layer as a result of chemical reactions with ozone depleting chemicals. It describes a general thinning of the ozone layer in which more of the Sun's UV radiation penetrates and reach's the Earth's surface.

Increased exposure to UV radiation causes many maladies for life on Earth. As higher levels of UV radiation reach the surface, there is an increased potential for skin cancers, eye cataracts, and a suppressed immune system. For every 1% decrease in ozone, there is a corresponding 2% increase in the amount of UV radiation bathing our skin and a 4% increase in certain skin cancers. As a result of a thinning ozone layer during the past few decades, rates of skin cancer have risen 40% around the world.

Increased radiation exposure also causes maladies for the rest of the life on Earth. It causes mutations in sensitive life forms, burns the leaves of trees, and reduces photosynthetic productivity in the ocean's food chain base and key regulator of the global carbon dioxide cycle: phytoplankton.

Those important little plants sure take a beating.

The term OZONE HOLE is used to describe an area in the ozone layer that has been depleted by more than 60% from normal levels. This phenomenon of severe ozone layer depletion has been occurring over the South Pole for the past few decades and presently occurs over the North Pole as well, where far more people are at risk.

°GLOBAL WARMING MAY INCREASE OZONE HOLE
ENVIRONMENTAL NEWS NETWORK
29 MARCH 1999

Despite international measures to reduce atmospheric concentrations of ozone destroying halogens to protect the ozone layer, global warming may lead to a weakening of the ozone layer, according to international scientists.

A team of German, Swiss and British scientists reported in the March 26 issue of the journal *Science* that while future climate change is expected to heat up the lower atmosphere it is likely to cool the air at the altitude of the ozone layer in the stratosphere...

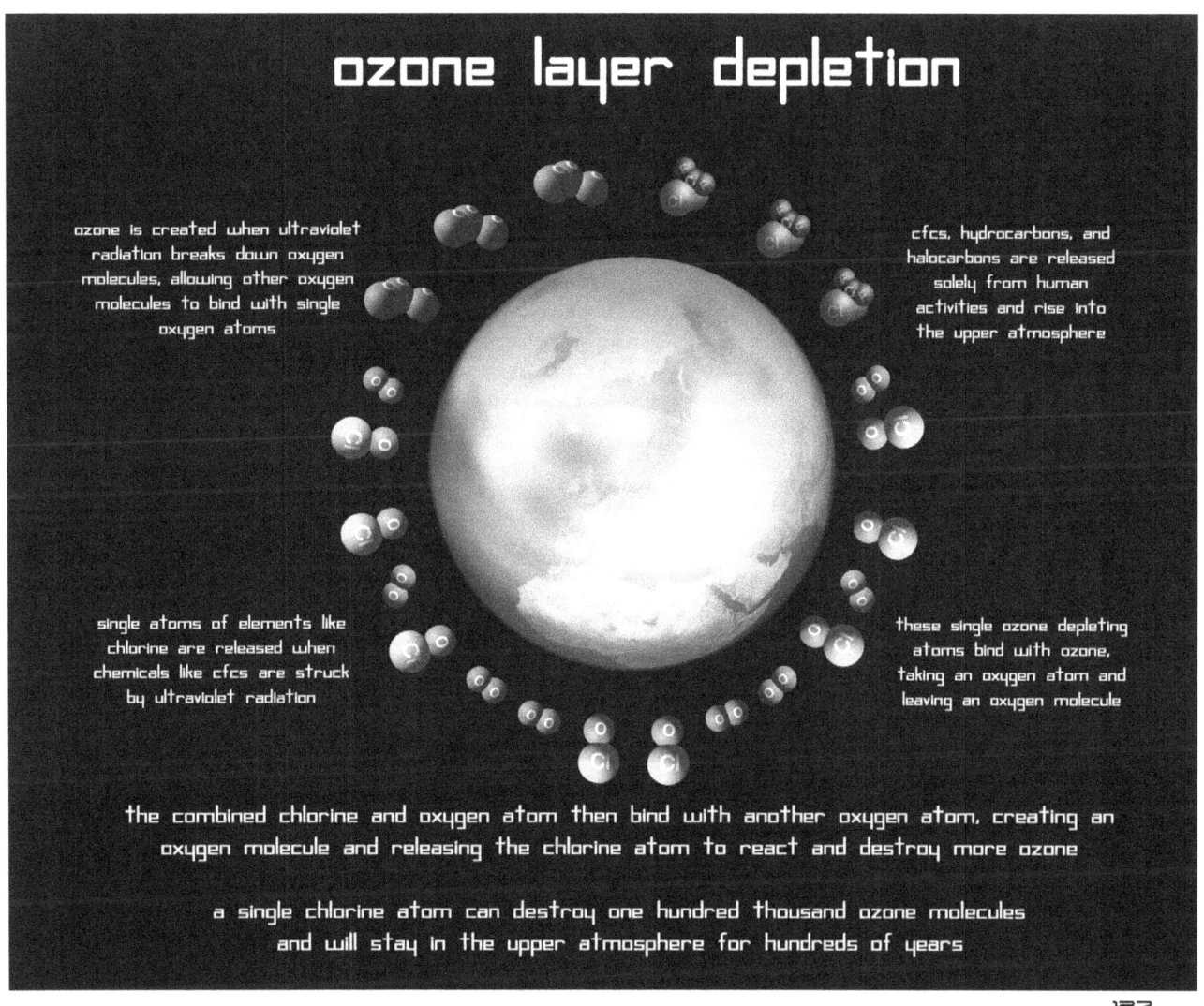

ozone layer depletion

ozone is created when ultraviolet radiation breaks down oxygen molecules, allowing other oxygen molecules to bind with single oxygen atoms

cfcs, hydrocarbons, and halocarbons are released solely from human activities and rise into the upper atmosphere

single atoms of elements like chlorine are released when chemicals like cfcs are struck by ultraviolet radiation

these single ozone depleting atoms bind with ozone, taking an oxygen atom and leaving an oxygen molecule

the combined chlorine and oxygen atom then bind with another oxygen atom, creating an oxygen molecule and releasing the chlorine atom to react and destroy more ozone

a single chlorine atom can destroy one hundred thousand ozone molecules and will stay in the upper atmosphere for hundreds of years

°ARCTIC OZONE LOSS MORE SENSITIVE TO CLIMATE CHANGE THAN THOUGHT
NASA - JPL
23 APRIL 2004

A cooperative study involving NASA scientists quantifies, for the first time, the relationship between Arctic ozone loss and changes in the temperature of Earth's stratosphere.

The results indicate the loss of Arctic ozone due to the presence of industrial chlorine and bromine in Earth's atmosphere may well be sensitive to subtle changes in stratospheric climate. Such ozone depletion leads to increased exposure to harmful, ultraviolet solar radiation at Earth's surface.

According to the study, the sensitivity of Arctic ozone to temperature is three times greater than predicted by atmospheric chemistry models. This leads to the possibility decreases in stratospheric temperatures may have significantly larger impacts on future Arctic ozone concentrations than have been expected in the past.

Dr. Markus Rex of the Alfred Wegener Institute for Polar and Marine Research, Potsdam, Germany, led the study. It also included scientists from NASA's Jet Propulsion Laboratory (JPL), Pasadena, Calif.

The researchers analyzed more than 2,000 balloon measurements collected over the past 12 years. They found the amount of ozone loss occurring in any given Arctic winter is closely related to the amount of air exposed to temperatures low enough to support the formation of polar stratospheric clouds. Reactions occurring on the surface of these clouds convert chlorine from unreactive forms to other forms that quickly deplete ozone.

Based on the relation between ozone loss and polar stratospheric cloud existence, the researchers found every degree Kelvin (equal to one Celsius degree) cooling of the Arctic results in an additional ozone destruction of five percent. This sensitivity is a factor of three larger than previously predicted by state-of-the-art, coupled climate- chemistry computer models.

...Researchers are trying to understand why the Arctic stratosphere cools. It may be due to a number of factors: rising levels of greenhouse gases such as carbon dioxide; a feedback between ozone depletion and stratospheric temperature; and natural variability. Higher amounts of greenhouse gases trap heat near Earth's surface, warming the surface and preventing the heat from reaching the stratosphere, thus cooling the upper atmosphere. However, climate models vary widely in their estimates of how much stratospheric cooling has occurred due to rising greenhouse gases over the past 40 years...

Each year the ozone hole over the South Pole reaches an average of 20 million square kilometers from August through October, sometimes reaching the tip of Chile, Australia, and New Zealand. This is three times the size of the entire North American continent

That is one big hole and a lot of UV streaming through.

Imagine a seven minute sunburn on the tip of Chile.

Ozone holes occur at the poles due to the extreme cold temperatures and the shroud of darkness they experience for half of the year. The colder stratospheric conditions at the poles offer ideal conditions for the formation of POLAR STRATOSPHERIC CLOUDS. These clouds provide both a surface for the chemical reactions to occur and a place in which benign chlorine atoms become ozone-destroying chlorine atoms.

When sunlight returns to the pole, ultraviolet radiation drives the chemical reactions that break down the ozone layer. The hole drifts about in the upper atmosphere above the poles, reaching a maximum size and then returning to normal ozone levels when the Sun stops shining for the other half of the year.

As a result of ozone depletion and the ozone holes, the most intense radiation strikes the poles and high altitudes. High latitudes and altitudes are also the places most strongly affected by global warming.

Ozone depletion and an enhanced greenhouse effect have been shown to be a positive feedback on one another. As the greenhouse effect intensifies, more heat energy is reradiated back to the Earth, which means less heat passes through to warm the stratosphere, keeping the stratosphere cooler and accelerating ozone layer depletion.

Furthermore, as the Earth warms, more water vapor is created through evaporation, increasing the likelihood that polar stratospheric clouds will form. As more clouds are formed, there are more surfaces where ozone breakdown can take place.

And the positive feedback continues.

It might be time to think about sunscreen.

ANYONE HAVE SOME SUNSCREEN FOR PLANET EARTH?

IT HAS A BIT OF A BURN ON ITS BUM

°OZONE HOLE OVER CHILE PUTS SOUTHERN CITIES ON SUNBLOCK ALERT
THE INDEPENDENT
11 OCTOBER 2000

Ultraviolet radiation warnings have been keeping 120,000 residents of Punta Arenas, the most southerly city in Chile, virtual prisoners inside their homes this week while the Sun burns high overhead.

A hole in the ozone layer, which shields the earth from harmful rays that can cause skin cancers and cataracts, now extends from Antarctica over the tip of South America into populated areas.

Second-stage health alerts, cautioning that unprotected skin would burn after just seven minutes exposure, were also issued in Ushuaia, Argentina, on the nearby island Tierra del Fuego...

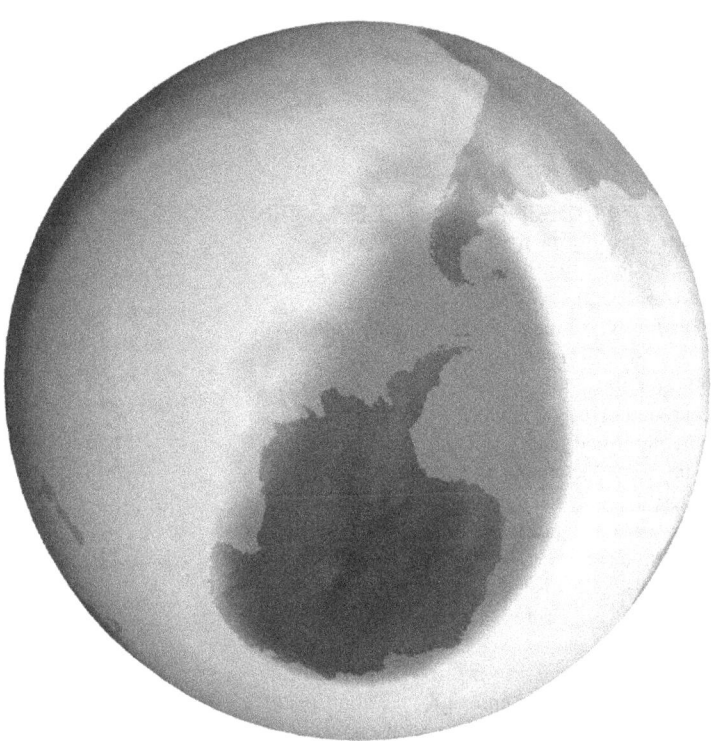

september 2004 ozone hole over antarctica

POSITIVE FEEDBACK

ozone layer depletion
increased heat absorption
enhanced greenhouse effect
cooler stratosphere

thinner ozone layer to reflect solar energy

enhanced greenhouse effect causes cooler stratosphere and increased ozone layer depletion

more solar energy reaches earth's surface

more heat energy absorbed and reradiated

°HUGE 2004 STRATOSPHERIC OZONE LOSS TIED TO SOLAR STORMS, ARCTIC WINDS
UNIVERSITY OF COLORADO
PRESS RELEASE
1 MARCH 2005

A new study led by the University of Colorado at Boulder indicates that two natural atmospheric processes in 2004 caused the largest decline in upper stratospheric ozone ever recorded over the far Northern Hemisphere.

According to Research Associate Cora Randall of CU-Boulder's Laboratory for Atmospheric and Space Physics, nitrogen oxide and nitrogen dioxide gases in the upper stratosphere climbed to the highest levels in at least two decades in spring 2004. The increases led to ozone reductions of up to 60 percent roughly 25 miles in altitude above Earth's high northern latitudes, said Randall.

"This decline was completely unexpected," she said. "The findings point out a critical need to better understand the processes occurring in the ozone layer." Randall is chief author of a paper on the subject appearing in the March 2 online issue of *Geophysical Research Letters*, published by the American Geophysical Union...

BREATHE DEEPLY

°EARTH LOSING AIR-CLEANSING ABILITY, STUDY SAYS
LOS ANGELES TIMES
4 MAY 2001

The Earth's atmosphere is beginning to lose its natural ability to remove air pollutants, a condition that could spread smog and accelerate the accumulation of greenhouse gases, according to a study published today in the journal *Science*.

The study documents for the first time the modest, two-decade long worldwide decline of a key molecule that cleanses the air. Without enough of the molecules, emissions that contribute to the greenhouse effect, smog and the hole in the ozone layer do not get destroyed as fast as humans release them.

"The one molecule is very, very important. It is the critical cleaning chemical for the atmosphere," said Ronald G. Prinn, the Massachusetts Institute of Technology scientist who led a 13-member research team responsible for the study. "If this free-radical (molecule) is deceasing, it could add to global warming."

But the losses of the chemical, called a hydroxyl radical, are slight so far and are not currently cause for alarm, experts say

...The latest findings are further evidence that the planet's fragile atmosphere is undergoing profound change and may have crossed a threshold that threatens its self-cleansing ability...

°METHANE EXPLOSION WARMED THE PREHISTORIC EARTH, POSSIBLE AGAIN
NASA/GSFC PRESS RELEASE
10 DECEMBER 2001

...In the last 200 years, atmospheric methane has more than doubled due to decomposing organic materials in wetlands and swamps and human aided emissions from gas pipelines, coal mining, increases in irrigation and livestock flatulence.

...When methane (CH_4) enters the atmosphere, it reacts with molecules of oxygen (O) and hydrogen (H), called OH radicals. The OH radicals combine with methane and break it up, creating carbon dioxide (CO_2) and water vapor (H_2O), both of which are greenhouse gases...

There sure are a lot of gases circulating in the atmosphere for a long time to come with significant ozone depleting and global warming potentials. Fortunately, the Earth has a mechanism to cleanse itself of excess chemicals in the atmosphere. However, as can be seen by the annual polar ozone holes and the enhancement of the greenhouse effect, the Earth has not been able to cleanse itself as well as it used to and its systems are becoming overwhelmed as a result.

Whether during an ice age or interglacial, without significant internal or external disruptions, the Earth keeps itself in relative equilibrium concerning atmospheric gas concentrations, able to cleanse itself through a process called **OXIDATION**.

Oxidation occurs when chemicals called **HYDROXYL RADICALS** mix with greenhouse and other gases, breaking them down and neutralizing them. Hydroxyl radicals break down many gases including carbon monoxide, nitrogen dioxide, nitrous oxide, and methane. Hydroxyl radicals also neutralize ozone depleting chemicals.

When the Earth is overwhelmed by too many gases and not enough hydroxyl radicals, it can not cleanse its systems properly and can fall out of equilibrium, experiencing what is called **REDUCED OXIDATION**.

As the ozone layer repairs itself, hydroxyl radicals are needed in the upper atmosphere. When the majority of hydroxyl radicals are required in the upper atmosphere, they are removed from the lower atmosphere. As these levels drop, the lower atmosphere remains uncleansed and the concentrations of greenhouse gases increase.

But there are a couple catches. Hydroxyl radicals are best formed when there is a large amount of water vapor in the atmosphere, which is one of the results of a warming planet. But as water vapor increases, the greenhouse effect is intensified and the ozone layer thins, which means that hydroxyl radicals will be needed in the upper atmosphere. Also, when hydroxyl radicals react with methane, they create carbon dioxide and water vapor.

The question lies in whether or not a threshold has been reached or will soon be reached as a result of increasing greenhouse gas levels and persistent ozone layer depletion. Should the Earth become unable to effectively cleanse itself, it would take only a short amount of time for the Earth's equilibrium to be knocked out of balance.

MOMMY AND DADDY, DOES MY BREATH CAUSE GLOBAL WARMING?

The reasons the Earth's greenhouse gas levels rise and fall, the reason the Earth's global temperature changes, and the reason the Earth experiences shifts in its steady-state equilibrium are many and varied. But no matter what the cause of such changes, greenhouse gases and global temperature move in tandem, always.

At the beginning of the 21st century, greenhouse gases and the global temperature are once again changing in tandem. We have a great understanding of the mechanisms that play a role in the regulation of the Earth's temperature. We have learned that greenhouse gas cycles play a role, that ozone dynamics play a role, that the Earth's cleansing through oxidation plays a role, that the Sun plays a role, that chemicals and aerosols play a role, that cloud cover plays a role, that ice cover plays a role, that the many positive feedbacks play a role.

What is natural? What is anthropogenic? What is the difference and does it really matter? The present trend of global warming may have been initiated by the collective force of human activity during the recent past that continues today, but it is well documented and accepted that the greenhouse effect is now being enhanced through a combination of anthropogenic emissions and the Earth feeding emissions back unto itself.

Does my breath cause global warming? Yes.

Does your breath? Yes.

Does the breath of our ancestors? Yes.

But our breath is not the concern here.

The concern is the train that passes by my window, four times a day, 100 traincars long, filled with a million tons of coal. The concern is the global rush-hour traffic fueled by gasoline. The concern is the 800 tons of garbage packed into my small town's landfill each day. The concern is the global economy fueled by oil. The concern is the removal of tracts of forest the size of Texas. The concern is the release of carbon dioxide and methane from thawing permafrost, something that we have no control over but do influence.

Eight billion of us releasing greenhouse gases that have been locked away for a long time are now being set free. As more of the gases are released, the Earth will continue to adjust accordingly, just as it has for billions of years.

For the Earth, the future holds a fate similar to what it experienced in the past when global greenhouse gas levels and the average global surface temperature changed in tandem. For our settlements, the future presents a situation that only our ancient ancestors know.

As in the past, so in the future.

We are approaching the rocky coast at ever-greater speeds.

Hold on tight, we are almost there.

°**STUDY: EARTH TO WARM EVEN IF GREENHOUSE GAS CUT**
REUTERS
19 SEPTEMBER 2002

Earth's climate will warm up over the next 50 years, whether or not greenhouse gases are curbed soon, U.S. researchers reported on Thursday in a NASA study. If nations cut back on emissions, it will not heat up as much, but it will still be hotter than it is now, according to a computer climate model.

"Some continued global warming will occur ... even if the greenhouse gases in the air do not increase further, but the warming could be much less than the worst-case scenarios," lead researcher James Hansen said in a statement. If emissions continue to increase at the current rate, global temperatures may increase by 2-4 degrees Fahrenheit (1-2 Celsius) the study found...

6
EMOCEANAL BREAKDOWN

HEAT AND LET RISE

°GREENHOUSE GASES
'DO WARM OCEANS'
BBC NEWS ONLINE
17 FEBRUARY 2005

Scientists say they have "compelling" evidence that ocean warming over the past 40 years can be linked to the industrial release of carbon dioxide. US researchers compared the rise in ocean temperatures with predictions from climate models and found human activity was the most likely cause.

In coming decades, the warming will have a dramatic impact on regional water supplies, they predict.

..."This is perhaps the most compelling evidence yet that global warming is happening right now and it shows that we can successfully simulate its past and likely future evolution," said lead author Tim Barnett, of the climate research division at the Scripps Institution of Oceanography in San Diego, California.

"If you take this data and combine it with a decade of earlier results, the debate about whether or not there is a global warming signal here and now is over at least for rational people."

The team fed different scenarios into computer simulations to try to reproduce the observed rise in ocean temperatures over the last 40 years.

They used several scenarios to try to explain the oceanic observations, including natural climate variability, solar radiation and volcanic emissions, but all fell short. "What absolutely nailed it was greenhouse warming," said Dr Barnett.

This model reproduced the observed temperature changes in the oceans with a statistical confidence of 95%, conclusive proof - say the researchers - that global warming is being caused by human activities...

As the Earth experiences its present warming trend, all of its surface warms, both land and water. As ocean waters warm, there are several things that happen that affect global greenhouse gas cycles and the settlements of humankind.

As ocean temperatures rise, the photosynthetic properties of phytoplankton are reduced, leaving a considerable amount of carbon dioxide in the atmosphere and removing the ability of oceans to act as a carbon dioxide sink.

As ocean temperatures rise, evaporation increases and more water vapor enters the atmosphere. This increased evaporation could lead to more cloud cover, which would help keep the planet cooler, but could also enhance the greenhouse effect further.

As ocean temperatures rise, the potential for dramatic weather phenomena increases, because warmer surface water is more active and lighter, leading to increased occurrences of hurricanes, typhoons, and El Niño seasons.

Water also expands when it warms, which makes the oceans rise. This is called **THERMAL EXPANSION**. As a result of warming oceans, it is estimated that sea levels will rise between 10-100 centimeters during the 21st century from thermal expansion. On average, each centimeter of sea level rise corresponds with one meter of retreating shoreline, so a modest rise would put anyone living 10-100 meters from a shoreline under water.

As the planet's land and water warm, the ice melts. If the ice is already in the water, called **SEA ICE**, then there is no rise in the average sea level, because the volume is already there. However, if the ice is land-based, such as the case with **MOUNTAIN GLACIERS** and land-based **ICE SHEETS**, then as the ice melts and enters the oceans, it adds to the average rise in sea level. An **ICE SHELF** is a large body of ice which lies at the fringe between land and sea and recent research shows that as an ice shelf disintegrates, the glaciers it was holding back come careening into the sea at a much faster rate.

So now we see there are two ways the sea level rises, one through thermal expansion and one through the melting of land-based ice. Many scientists believe the current projections for sea level rise during the 21st century are underestimated, because both causes have not been adequately taken into account. The average sea level has been rising about 8 inches per century and is projected to rise 1-3 feet during the 21st century as a result of thermal expansion and ice melt. This is quite a bit for anyone living anywhere near a shoreline.

Add another 6 meters when the ice in Greenland melts, add another 60 for Antarctica. Although these scenarios are unlikely during this century, no one is certain.

POSITIVE FEEDBACK

ocean temperature increase
reduced phytoplankton productivity
increase in greenhouse gases
enhanced greenhouse effect

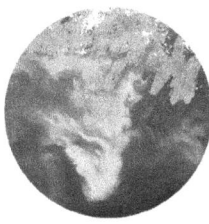

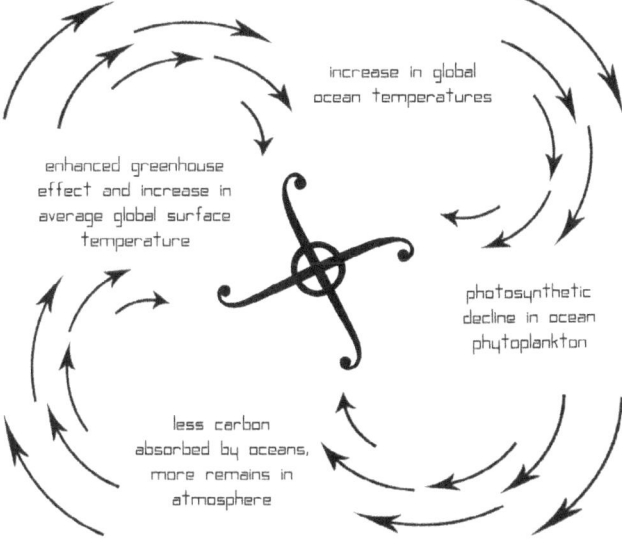

increase in global
ocean temperatures

enhanced greenhouse
effect and increase in
average global surface
temperature

photosynthetic
decline in ocean
phytoplankton

less carbon
absorbed by oceans,
more remains in
atmosphere

°PLUMMETING PLANKTON LINKED TO WARMER OCEANS
CNN NEWS ONLINE
14 AUGUST 2002

Concentrations of microscopic plants that comprise the foundation of the ocean's food supply have fallen during the past 20 years as much as 30 percent in northern oceans, according to a satellite checkup of planetary health.

...The reason for the lower numbers could be that warming water layers on top prevent cooler and more nutrient-rich waters down below from reaching the surface-dwelling plankton, which are crucial to the survival of many ocean species, the scientists said.

...Besides fueling the food chain, phytoplankton plays a crucial role in cycling a primary greenhouse gas through the environment. Phytoplankton accounts for 50 percent of the transfer of carbon dioxide from the atmosphere back into the biosphere through photosynthesis, the process through which plants absorb carbon dioxide gas to grow...

°MOST IMPORTANT GREENHOUSE GAS GOES WAY UP IN PAST 50 YEARS
UNISCI.COM
25 APRIL 2001

Scientist know that atmospheric concentration of greenhouse gases such as carbon dioxide have risen sharply in recent years, but a study released Tuesday in Paris reports a surprising and dramatic increase of the most important greenhouse gas – water vapor – during the last half-century.

The buildup of other greenhouse gases (those usually linked with climate change) is directly attributable to human activity, and the study indicates the water vapor increase also can be traced in part to human influences, such as the buildup of atmospheric methane...

POSITIVE FEEDBACK

ocean and air temperature increase
increased evaporation
increase in atmospheric water vapor
enhanced greenhouse effect

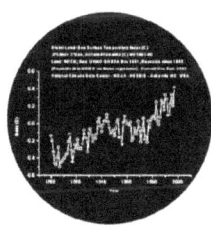

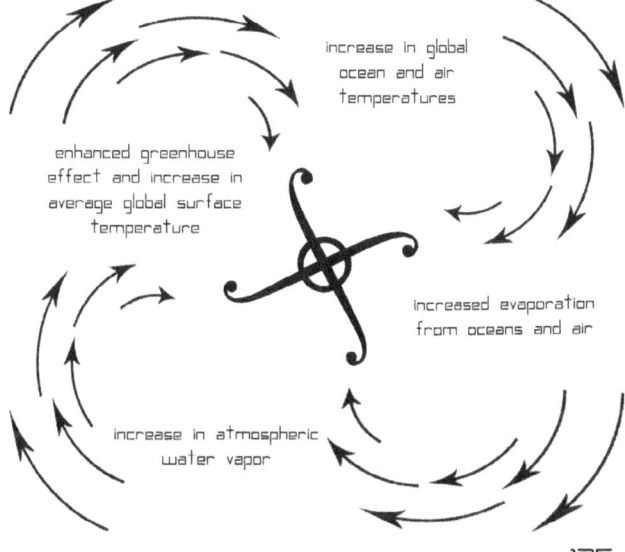

increase in global
ocean and air
temperatures

enhanced greenhouse
effect and increase in
average global surface
temperature

increased evaporation
from oceans and air

increase in atmospheric
water vapor

°TROPICAL WATERS IN NORTHERN HEMISPHERE HEATING UP, DATA SHOWS
CNN NEWS ONLINE
28 JULY 2000

Tropical waters in the Northern Hemisphere have been heating at an enhanced rate since 198, scientists at the National Oceanic and Atmospheric Administration said Friday.

The rate is nearly +0.5 degrees Celsius (+1 degrees Fahrenheit) per decade, 10 times the global rate, NOAA said, and has contributed to unprecedented coral bleaching over the past decade...

°IN POLAR WATERS, A SURGE IN TEMPERATURES TAKES SCIENTIST BY SURPRISE
CHRISTIAN SCIENCE MONITOR
21 FEBRUARY 2002

A new study using seven decades of temperature data shows that mid-depth water around Antarctica has warmed nearly twice as much as the world ocean as a whole. That wasn't supposed to happen.

Geophysicists expect global warming to be strongest in polar regions. However, as Sarah Gille at the University of California, San Diego, explains: "We thought the ocean between 700 and 1,100 meters [2,300 and 3,600 feet] was pretty well insulated from what's happening at the surface. But these results suggest that the mid-depth Southern Ocean is responding and warming more rapidly than global ocean temperatures [generally]." How this unexpected finding fits into global-warming forecasts is unclear, but it could be significant...

The seas are not going to rise several feet this year, they are most likely not going to rise several feet in the next few decades, but by the end of the century, there is a strong likelihood that this will be the case. If changes in sea level do happen more quickly, as they have in the past, there will simply not be enough time to build dams, dykes, or much of anything else that will keep the oceans from inundating coastal cities.

The problem for our settlements is that coastal metropolitan areas are usually the centers of commerce, politics, business, and communication. Should many of these centers be put in harm's way at the same time, there is a greater risk of social disruptions within the whole of society.

So, if you live in New York City, London, Amsterdam, Miami, or any of the Island Nations and you see a noticeable difference over the course of five years, you will know it is time to seek higher ground.

The big problem is that many will have no place to go.

It is not so much that the world is unstable, it has always been unstable. The problem is the rate at which changes may occur, the severity of these changes, and the unpreparedness of our large and fragile settlements.

A small change in sea level rise will have disruptive effects on the stability and security of the daily functioning of our settlements.

Big changes or abrupt changes could have catastrophic effects.

For the Earth, sea level rise is simply one of many changes it will see as a result of an enhanced greenhouse effect during the 21st century.

Let us take a look at the receding hairline of the Earth's icy regions, because as the world continues to warm, the planet continues to thaw.

Some of the ice has been melting for the past 10,000 years.

Some of the ice has been melting for the past 100 years.

Most of the ice will have disappeared during the next 100.

Be it from natural or human-made causes, the ice is melting and there is nothing we can do to halt the retreat.

All we can do is deal with the changes.

°OCEAN IS WARMING, STUDY FINDS
ENVIRONMENTAL NEWS NETWORK
24 MARCH 2000

The world ocean has experienced a net warming of 0.11 degrees Fahrenheit from the sea surface through about 10,000 feet of depth over the past 35-45 years. The upper 1,000 feet has warmed by 0.56 degrees Fahrenheit.

"Although these may seem like small changes, it represents a large change in the heat content of the ocean," said Sydney Levitus, an oceanographer with the National Oceanographic and Atmospheric Administration. "Water is very effective at absorbing heat without undergoing much of a temperature change."

The finding, reported in today's issue of *Science*, answers the question of why the Earth's atmosphere has not warmed as much as climate model simulations have predicted, a point made repeatedly by critics of global warming.

...The ocean warming itself is not conclusive evidence of global warming due to human-induced emissions of greenhouse gases. Combined with lower atmosphere and sea surface warming over the past 100 years and the thinning and retreat of Arctic sea ice during the past 35 to 45 years, however, builds a strong case for anthropogenic global warming...

°THE RISING OCEANS
ENVIRONMENTAL NEWS NETWORK
14 AUGUST 2000

Coastal sea levels have risen a foot in the past century. Scientists expect them to rise still more. The culprit is global warming, say scientists. But just how much of that global warming is fuelled by natural events and how much is man-made? That mystery is lighting another fire: a scientific debate, which could change the face of our shorelines.

The seas are rising nearly one-tenth of an inch each year, fed by rivers of melting glaciers and ice sheets around the globe, according to scientists at the United States Geological Survey. At the current rate of melting, warn many scientists, the seas could rise another foot over the next 50 years. Iceland's glaciers could disappear by 2200...

°SEA LEVEL RISES 'UNDERESTIMATED'
BBC NEWS ONLINE
17 FEBRUARY 2002

Scientists may have seriously underestimated the likely rise in sea levels this century. The claim comes from a research team that has examined the rate at which glaciers and ice caps are melting because of rising temperatures on Earth.

They say new data show these areas to be retreating far faster than previously thought, with the run-off waters set to lift the height of the oceans well above that recently predicted by the UN's Intergovernmental Panel on Climate Change.

"The glacier wastage at the moment is unprecedented," Professor Mark Meier of the University of Colorado at Boulder, told the annual meeting of the American Association for the Advancement of Science in Boston...

°MELTING GLACIERS IN ANTARCTICA ARE RAISING OCEANS, EXPERTS SAY
NEW YORK TIMES
11 DECEMBER 2001

Antarctica appears to be melting and contributing to the slow rise in the oceans, scientists reported to their colleagues here today. Using two sets of radar data from the European Remote Sensing Satellite, two scientists said they had found that about 36 cubic miles of ice had melted from glaciers in West Antarctica over the past decade.

That is enough water to raise sea levels worldwide by about one-sixtieth of an inch, they said. "These glaciers are thinning rapidly," said one of the scientists, Dr. Eric Rignot from NASA's Jet Propulsion Laboratory in Pasadena, Calif.

...Ocean levels have been rising at a rate of about eight inches a century. Half of that is attributable to the fact that water expands as temperatures rise; 20 percent appears to be water running down mountain glaciers. The remaining 30 percent is a mystery, but the new data suggests it is coming from Antarctica...

°OCEANS TO RISE ONE METER BY 2100-ARCTIC EXPERT
REUTERS
9 NOVEMBER 2004

Global warming is melting the Arctic ice faster than expected, and the world's oceans could rise by about a meter (3 feet) by 2100, swamping homes from Bangladesh to Florida, the head of a study said on Tuesday.

Robert Corell, chairman of the eight-nation *Arctic Climate Impact Assessment (ACIA)*, also told a news conference there were some hints of greater willingness by the United States, the world's top polluter, to take firmer action to slow climate change. Speaking at the start of a four-day scientific conference in Reykjavik, Corell said global warming was melting the Greenland ice sheet and Arctic glaciers from Alaska to Norway quicker than previously thought.

...About 17 million people in Bangladesh live less than one meter above sea level. Pacific islands like Tuvalu could be swamped and much of Florida south of Miami would be inundated by a one meter rise.

are you gettin' this?

CHILLING THAW

The present face of the planet is slathered with ice, not as much as during an ice age, and less of it each year, but more than during times of very low ice coverage. There is ice at the tippy tops of mountains around the world, glacial ice that has been around for thousands of years, ice in the seas, and ice on the land at the North and South Poles.

It is true that we are coming out of a cooler geologic period called The Little Ice-Age and have been for the past few hundred years. During this time, glaciers and polar regions have been experiencing a global retreat. However, the Earth's ice is presently melting at a faster rate than any other time in the past 5,000 years.

That is a problem for us.

Glaciers are in global retreat, the mainland of Greenland is thawing and Antarctica is experiencing a rapid disintegration of ice shelves, which has been happening since the end of the last ice age, but not as rapidly as today. As the Earth warms, high altitudes and latitudes warm at a rate three times that of equatorial regions and lower latitudes. As the Earth's average global surface temperature has risen 1° Fahrenheit during the last 150 years, high latitudes and altitudes have risen an average of 3° Fahrenheit.

The changes in planetary cryospheric coverage is being documented by satellites, historical records, and the naked eye. Yet, there are many who still contend the unprecedented rate of glacier retreat and polar ice shelf calving is not a sign of global warming, but is within natural variability.

Sure, there are a few glaciers which are growing and there are areas in Antarctica which are thickening, but more than not, the ice of the Earth is melting. Even in the places where glaciers are growing, the reason for their growth is related to how the present trend of global warming affects the weather patterns in that area.

The chilling thaw is happening and predicted to continue through this century and beyond. It really does not matter if the melting is naturally induced, human induced, global warming induced, or induced by martians with giant laser beams, because the melting will affect us no matter what is to blame.

Maybe the increase in ice melt is being caused by the Pentagon sponsored High Frequency Active Auroral Research Program (HAARP) in Alaska which was

°NASA STUDY FINDS RAPID CHANGES IN POLAR ICE SHEETS
NASA - JPL NEWS RELEASE
31 AUGUST 2002

Recent NASA airborne measurements and a new review of space-based measurements of the thickness of Earth's polar ice sheets concludes they are changing much more rapidly than previously believed, with unknown consequences for global sea levels and Earth's climate.

Large sectors of ice in southeast Greenland, the Amundsen Sea Embayment in West Antarctica and the Antarctic Peninsula are changing rapidly by processes not yet well understood, said researchers Dr. Eric Rignot of NASA's Jet Propulsion Laboratory, Pasadena, Calif., and Dr. Robert Thomas of EG&G Services at NASA's Wallops Flight Facility, Wallops Island, Va.

...Rignot said understanding how polar ice sheets evolve is vital to society. "The Antarctic and Greenland ice sheets together hold enough ice to raise sea level by 70 meters (230 feet)," he said. "Even a small imbalance between snowfall and discharge of ice and melt water from ice sheets into the ocean could be a major contributor to the current sea level rise rate of 1.8 millimeters (0.07 inches) a year and impact ocean circulation and climate. During past periods of rapid deglaciation, ice sheet melting raised sea level orders of magnitude faster than today. This is the real threat of the ice sheets."

...The review reports Greenland's ice sheet is losing 50 cubic kilometers (12 cubic miles) of mass a year due to rapid thinning near its coasts. That's enough to raise sea level 0.13 millimeters (0.005 inches) annually. "Rapid coastal thinning cannot be explained by a few warm summers and is attributed to a dynamic ice sheet response," Rignot said. "A possible contributor to the observed trend is increased lubrication from additional surface melt water reaching glacier beds through crevasses and moulins."

developed to study the physical and electrical nature of the ionosphere with the purpose of 'controlling ionosphere processes in such a way as to greatly improve the performance of military command, control, and communications systems.' Depending on who you ask, one of the applications of HAARP is to better understand the activities that affect communication and navigation devices; another is to exert control over weather phenomena at strategic military operation sites.

Maybe the increase in erratic weather is not related to global warming at all, maybe it is a result of ionospheric tom-foolery?

I doubt it.

Currently, there are over 160,000 glaciers in the world and 40% of them have been documented to be in retreat. The fact that the planet's glaciers are retreating and its polar fringes are disintegrating would not be a cause for concern, except for the settlements affected when ice cover changes. When a population lives down slope from a melting glacier, there is only one place for the water and ice to go.

The back door.

Another problem with melting glaciers is the change in meltwater rates for those downstream who depend on it for their drinking, agriculture, and energy uses. As glacier producing weather decreases, so does the steady water source.

The same is true for those living in mountain regions who depend on snowfall for their water uses. A warmer climate changes snowfall precipitation to rainfall precipitation. An undependable annual snowfall makes the flow needed for drinking, agriculture, and energy undependable too.

What about the communities who depend on snowfall as the base of their ski industry?

No snow = no ski.

There are quite a few problems for such a small change.

A more global concern is this: as the Earth loses the reflective properties of ice cover, more land is exposed, which absorbs more solar radiation, which increases the average global surface temperature, which melts more ice, which exposes more land...

Another positive feedback is unleashed.

As long as modern civilization has been around, say the past 10,000 years, glaciers have been a part of many people's cultural identity and identity of place.

By the end of the 21st century, they will all be gone or significantly diminished... and they will not be coming back for a long time.

°MELTING GLACIERS: UNEXPECTED BOOST TO RISING OCEANS
THE CHRISTIAN SCIENCE MONITOR
25 MARCH 2004

Few effects from global warming raise more red flags than rising sea levels. The topic has led to a growing pile of conflicting research trying to answer the questions: How fast, and why.

Now a pair of US scientists conclude that the oceans rose at a global average rate of 1.5 to 2 millimeters a year (6 to 8 inches a century), confirming a hotly debated, decade-old estimate. But their work also points to the key driver of this change: water from melting glaciers and not, as some have argued, a natural swelling of the oceans caused by higher temperatures. Pinpointing glacial melt as the leading source of the oceans' rise is a new finding that contradicts several past studies.

...More than 100 million people worldwide live within a mile of a coastline and would be first affected by any rise...

SO BEFORE THEY ARE GONE, LET US PAY OUR RESPECTS AND WISH THEM FAREWELL...

DOKRIANI BAMAK GLACIER: INDIA
retreating 20 meters per year.

SPEKA GLACIER: UGANDA
retreated 150 meters 1977–
1990 compared to 40 meters
1958–1977.

CAUCUS MOUNTAINS: RUSSIA
50 percent of glacial
volume gone since 1900.

MEREN CARSTENZ AND
NORTHWALL FIRN GLACIERS:
INDONESIA
retreating at 45 meters/year.

...DRIP...
...THUMP...
...DRIP...DROP...
...THUMP...THUMP...
...DRIP...DROP...DRIP...
...THUMP...THUMP...THUMP...
...DRIP...DROP...DRIP...DROP...DRIP...
...THUMP...THUMP...THUMP...THUMP...
...DRIP...DROP...DRIP...DROP...DRIP...DROP...
...DRIP...DROP...DRIP...DROP...DRIP...DROP...
...THUMP...THUMP...THUMP...THUMP...THUMP...THUMP...THUMP...THUMP...THUMP...
...DRIP...DROP...DRIP...DROP...DRIP...DROP...DRIP...

DUOSUOGANG PEAK: CHINA
has shrunk 60 percent
since 1970s.

TIEN SHAN MOUNTAINS:
CENTRAL ASIA
22 percent of glacial
volume gone since 1960.

MOUNT KENYA:
KENYA
largest
glacier
has lost 92
percent of
mass since
late 1800s.

south cascade glacier: oregon, usa
40% retreat even with glacier producing weather

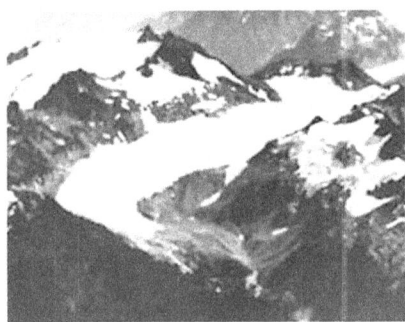

1928

2000

ALPS:
EUROPE
35–40
percent of
glacial
area and
50 percent
of glacial
volume
gone since
1850.

...THUMP...THUMP...THUMP...THUMP...THUMP...THUMP...THUMP...THUMP...THUMP...THUMP...THUMP...
...DRIP...DROP...DRIP...DROP...DRIP...DROP...DRIP...DROP...
...THUMP...THUMP...THUMP...THUMP...THUMP...THUMP...THUMP...THUMP...THUMP...THUMP...THUMP...THUMP...
...DRIP...DROP...DRIP...DROP...DRIP...DROP...DRIP...DROP...DRIP...
...THUMP...THUMP...THUMP...THUMP...THUMP...THUMP...THUMP...THUMP...THUMP...THUMP...THUMP...THUMP...
...DRIP...DROP...DRIP...DROP...DRIP...DROP...DRIP...DROP...DRIP...DROP...
...THUMP...THUMP...THUMP...THUMP...THUMP...THUMP...THUMP...THUMP...THUMP...THUMP...THUMP...THUMP...
...DRIP...DROP...DRIP...DROP...DRIP...DROP...DRIP...DROP...DRIP...DROP...
...THUMP...THUMP...THUMP...THUMP...THUMP...THUMP...THUMP...THUMP...THUMP...THUMP...THUMP...THUMP...THUMP...

QUELCCYA GLACIER: PERU
30 meter/year retreat
since 1990 vs. 3 meters/
year 1970–1990.

COLUMBIA
GLACIER: USA
retreated 13
km since 1982,
presently
retreating 35
meters/day.

UPSALA GLACIER:
ARGENTINA
retreated 60 meters/
year during past 60
years and retreat is
accelerating.

GLACIER NATIONAL PARK: USA
number of glaciers reduced
from 150 to fewer than 50
since 1850.

G'BYE MELTING MATES

°GLOBAL WARMING MELTS AUSTRALIA'S GLACIERS
REUTERS
31 MAY 2001

Australia's glaciers are melting. In the land of outback deserts this is not as strange as it sounds. Scientists say the shrinking of Australia's little-known glaciers on remote, sub-Antarctic Heard Island in the Indian Ocean reveals global warming now stretches from the tropics to the edge of Antarctica.

"The recession of many glaciers during the past 50 years has been unprecedented in modern times for Heard Island," glaciologist Andrew Ruddell, with the Australian Antarctic Division, told Reuters on Friday.

...Since 1947 the temperature has risen 1.3F causing glaciers to melt rapidly. The island's 34 glaciers have decreased by 11 percent in area and 12 percent in volume -- half the loss occurred in the 1980s.

..."It's a very significant retreat. Glaciers are very sensitive to climate change," said Ruddell. "The rate of retreat is similar to New Zealand alps and European alps and Central Asia."

Recent studies of glaciers and ice caps in Tibet, Africa and Peru have shown dramatic reductions over the past few years due to global warming. Photographs of Kenya's Mount Kilimanjaro in February showed its had lost 82 percent of its ice since 1912 and scientists calculate Kilimanjaro will lose its snow between 2010 and 2020...

ADIOS ANDES

°SMALL GLACIERS OF THE ANDES MAY VANISH IN 10-15 YEARS
UNISCI.COM
17 JANUARY 2001

In ten to fifteen years' time, the small glaciers of the Andes, which constitute 80% of all the glaciers in the tropical regions of that mountain chain, are likely to have vanished. That is the main conclusion reached after studies conducted on the Chacaltaya glacier in Bolivia and the Antizana glacier in Ecuador.

...The researchers stress that the recent quickening of glacial melt in the Andes corresponds with the greater intensity and frequency of El Niño events observed over the same period. Andean glaciers are particularly sensitive to climatic fluctuations owing to the position of most of them in the tropics and to the specific mechanisms by which they function. (In contrast with the glaciers in the Alps, which undergo a long period of accumulation in winter, glaciers in the tropics are subject to a permanent regime of ablation in the lower half of their mass, which makes them sensitive to the slightest sudden variation in climate.)

...How are tropical glaciers affected by climatic variations? At a time when an acceleration in global warming is being predicted, the question is especially vital in the Andes, where some regions are largely dependent on the Cordillera's glaciated mountain tops for their water supply...

AMUCHI MAY PATAGONIAN PEAKS

patagonia:
argentina and
southern chile
as seen by
NASA MODIS

°SCIENTISTS: PATAGONIAN GLACIERS MELTING
ASSOCIATED PRESS
17 NOVEMBER 2003

When Borge Ousland and Thomas Ulrich trekked across the vast and wild Patagonian glaciers, they braved heavy snows and bitterly cold temperatures - nothing to make them think the ice was melting beneath their feet.

"We had ... a lot of snow. We got a meter (3 feet) of snow in one night," said Ousland after the pair completed the 54-day crossing of the icefield. While the two European adventurers became the first to cross the icefields unaided on Oct. 16, a day later the glaciers made news of their own: A study published Oct. 17 in the journal *Science* found that the Patagonian glaciers melted at twice the rate in 1995-2000 - when compared with measurements from 1975 to 2000.

Three top scientists used maps, satellite imagery and digital elevation models to measure the change in velocity of 63 Patagonian glaciers over 25 years. They found the glaciers are losing the equivalent of 10 cubic miles of ice every year - enough to raise the world's sea level an estimated four-one thousandths of an inch.

The shrinking is reflective of rising global temperatures, scientists say, and is happening to glaciers around the world. As the melting causes sea levels to rise and freshwater supplies to disappear, scientists warn of potential worldwide economic and environmental disaster if the process isn't reversed...

°SOUTH AMERICAN GLACIERS' BIG MELT
BBC NEWS ONLINE
17 OCTOBER 2003

The Patagonia glaciers of Chile and Argentina are melting so fast they are making a significant contribution to sea-level rise, say scientists. They report ice was lost at a rate sufficient to push up ocean waters by 0.04 millimetres per year during the period from 1975 through to 2000. This is equal, the researchers say, to 9% of the total annual global sea-level rise from all mountain glaciers.

..."The Patagonia ice fields are dominated by so-called 'calving' glaciers," Rignot said. "Such glaciers spawn icebergs into the ocean or lakes and have different dynamics from glaciers that end on land and melt at their front ends.

"Calving glaciers are more sensitive to climate change once pushed out of equilibrium, and make this region the fastest area of glacial retreat on Earth," he said.

°RECORD RETREAT IN SWISS GLACIERS IN 2003 DUE TO CLIMATE CHANGE: SCIENTISTS

AGENCE FRANCE-PRESSE
13 JANUARY 2004

Switzerland's glaciers melted by a record amount during 2003 under the onslaught of long-term climate change, a top Swiss science academy said Tuesday.

The retreat of the glaciers in the Swiss Alps reached up to 150 metres, with an overall melting exceeding that observed in any year since measurements began in the 19th century, according to the Swiss Academy of Natural Sciences.

...Overall, glaciers in the heart of Europe's biggest mountain range stopped advancing about 50 years ago, [Andreas] Bauder pointed out.

The Swiss length measurements were based on regular data recorded on 96 Alpine glaciers.

Climate change has been blamed on global warming caused by the rise in air pollution from greenhouse gases such as carbon dioxide. Bauder said scientists were not able to predict longer term trends for the ice floes but felt confident enough to forecast that the Swiss glaciers would again shrink in 2004.

"The glaciers will retreat, just on the signals we had in the last couple of years," he observed.

ADIEU
ARRIVEDERCI
AND
AUF
WIEDERSEHEN
EUROPEAN ALPS

°'MELTING SWISS GLACIERS THREATENING THE ALPS'

REUTERS
16 NOVEMBER 2004

Switzerland's glaciers are melting faster than expected, shrinking by as much as one-fifth of their size over the 1985-2000 period alone, scientists at Zurich University said on Monday.

Hot summers in the 1990s in particular, prompted a glacier melt-down which has outpaced previous forecasts. It could impact tourism and cause more environmental hazards such as flash floods, scientist Frank Paul said.

"It is amazing what huge masses of ice have been lost," he told Reuters in a telephone interview. "Every hiker in the Alps knows about it: the changes in recent years have been dramatic."

While Swiss glaciers shrank a meagre one percent in the 12 years to 1985, they lost 18 percent of their area in the 1985-2000 period, the research showed...

This suggests they are melting faster than earlier estimates which put the loss at 30 percent between 1980 and 2025.

..."When glaciers are in the retreating phase they normally lose about 30 centimetres of snow and ice a year. In the 1990s they lost about 70 centimetres a year; in 2003, they lost three metres," Paul said.

°MELTING SNOWS OF KILIMANJARO
NASA NEWS RELEASE
FEBRUARY 2000

Africa's "Shining Mountain" may soon shine no more. The snow and ice on the summit of Mount Kilimanjaro is melting so fast that some scientists believe its ice cap could be gone by the year 2015. The ice cap formed more than 11,000 years ago.

Researchers say the ice fields on Africa's highest mountain shrank by 80 percent over the past century.

About a foot and a half of the summit's glacial ice is lost each year due to rising surface temperatures. There is concern that the loss of Kilimanjaro's ice cap could impact both the local climate as well as the availability of fresh water for local populations who depend upon the glacial melt runoff, particularly during the dry seasons...

KWAHERI
KILIMANJARO

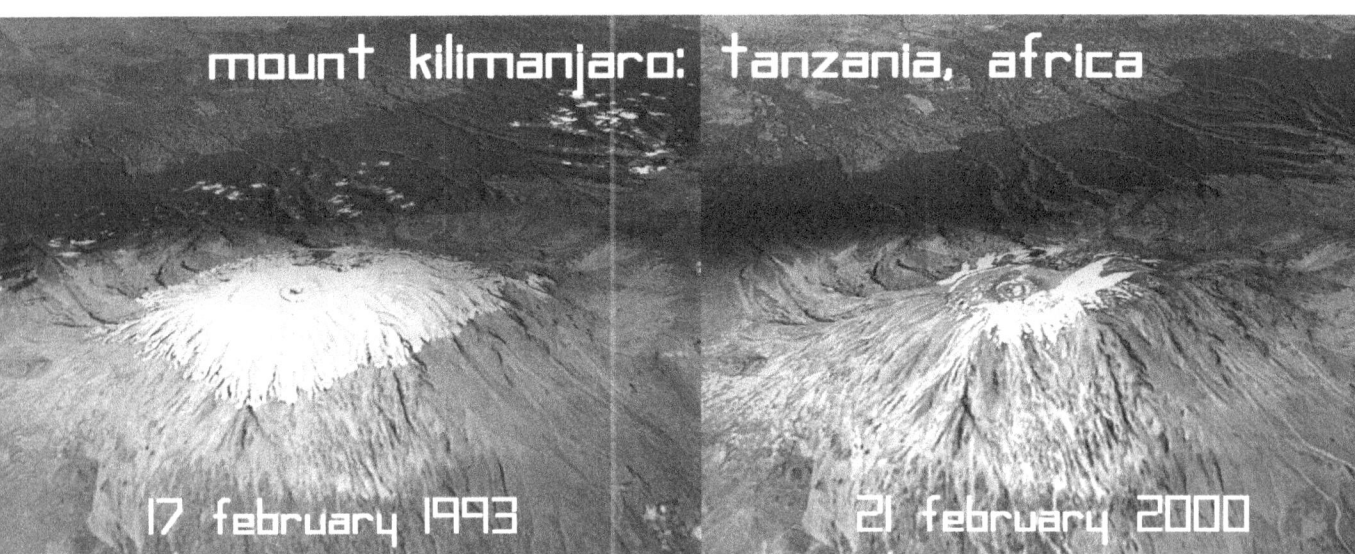

mount kilimanjaro: tanzania, africa

17 february 1993   21 february 2000

NAMASTE
NEPALESE
NATIVES

°GLOBAL WARMING BLAMED FOR MELTING EVEREST GLACIER
REUTERS
6 JUNE 2002

A glacier from which Sir Edmund Hillary and Tenzing Norgay set out to conquer Mount Everest nearly 50 years ago has retreated three miles up the mountain due to global warming, a U.N. body says. A team of climbers, backed by the United Nations Environment Program (UNEP), reported after their two-week visit last month that the impact of rising temperatures was everywhere to be seen.

The landscape bears the scars of sudden glacial retreat, while glacial lakes are swollen by melted ice, UNEP spokesman Michael Williams told Reuters on Thursday.

...UNEP recently warned that more than 40 Himalayan glacial lakes were dangerously close to bursting, threatening the lives of thousands of people, because of ice melt caused by global warming...

TLAA ALASKAN ICE

muir glacier 1899-2003

bering glacier 1986-2002

carroll glacier 1906-2003

°MANY ALASKAN GLACIERS ARE THINNING, USGS STUDY SAYS
UNISCI.COM
11 DECEMBER 2001

Of Alaska's several thousand valley glaciers, including nearly 700 that are named, fewer than 20 are advancing, according to a major study. Many of the rest are retreating...

..."The Earth recently emerged from a global climate event, called the 'Little Ice Age' during which Alaskan glaciers expanded significantly. The Little Ice Age began to wane in the late 19th century. In some areas of Alaska, glacier retreat started during the early 18th century, prior to the beginning of the industrial revolution," explains Molnia. "At the peak of the Little Ice Age, glaciers covered about 10 percent more area in Alaska than they do today."

"During the 20th century, most Alaskan glaciers receded and, in some areas, disappeared. But it is important to note that our data do not address whether or not any of these changes are human induced," says Molnia, who warns against drawing any quick conclusions.

°GLACIERS ON THIN ICE: EXPERT SAYS MELTING TO BE FASTER THAN EXPECTED
SAN FRANCISCO CHRONICLE
17 FEBRUARY 2002

New data from melting glaciers in Alaska, and other signs of polar warming, point to a catastrophic rise in the level of the world's oceans by the end of this century, a leading authority in the field told scientists here yesterday.

University of Colorado professor Mark Meier, a renowned expert on glaciers, predicts the level of the world's oceans will rise between 7 and 11 inches by the end of this century. His prediction is more than twice that adopted in 2000 by the International Panel on Climatic Change.

...While climatologists have focused their attention on the great ice sheets of Greenland and Antarctica for signs of melting that could raise sea levels, Meier said the impact of "smaller" glacial melting has been underestimated. "In many environments, the 'little' glaciers are disappearing," he said. "In 20 years, you'll be able to visit 'Non-Glacier National Park' (in Montana). If you want to see the snows on Kilimanjaro, you'd better go soon."

...Outside of Greenland and Antarctica, all the smaller glaciers of the world contain enough water to raise ocean levels nearly 20 inches, according to Meier.

Sea level increases are driven by both water temperature -- warmer water is less dense and so it takes up more space -- and by the melting of glaciers' ice...

°85ˣ OF ALASKAN GLACIERS MELTING AT 'INCREDIBLE RATE'
THE GUARDIAN
19 JULY 2002

Glaciers in Alaska are melting at "an incredible rate" according to US researchers. They report in *Science* today that 85% of the glaciers they examined had lost vast portions of their mass in the last 40 years. Some were now thinning at double the rates of the 1950s.

"Most glaciers have thinned several hundred feet at low elevation in the last 40 years and about 60 feet at higher elevations," said Keith Echelmayer of the University of Alaska at Fairbanks, who with four colleagues has systematically flown over 67 of Alaska's ice streams, and checked glacier thickness against measurements made by the US geological survey in the 1950s.

The ice cover in the Arctic ocean itself is shrinking by an area the size of the Netherlands each year. The ice cap has also thinned from three metres on average, to two metres, in 30 years of nuclear submarine measurements. Ecologists have warned that Arctic bird populations are threatened, and that polar bear, seal and caribou poulations are losing their natural habitat...

IT HAS BEEN GREAT TO HAVE YOU AROUND GLOBAL GLACIERS

°WORLD'S GLACIERS CONTINUE TO SHRINK
ENVIRONMENTAL NEWS NETWORK
27 MAY 1998

The world's glaciers outside of Antarctica and the Greenland Ice Sheet are shrinking at an increasingly accelerated rate, according a University of Colorado at Boulder study.

"In the last century, there has been a significant decrease in the area and volume of glaciers, especially at middle and low latitudes," said Professor Emeritus Mark Meier of the geological sciences department. "The disappearance of glacier ice is more pronounced than we previously had thought."

The smaller, low-latitude glaciers seem to be taking the biggest hit, said Meier, noting the largest glacier on Africa's Mount Kenya lost 92 percent of its mass in the last century and Mount Kilimanjaro glaciers have shrunk by 73 percent in that time period. While there were 27 glaciers in Spain in 1980, that number has since dropped to 13. "I think we might find statistics similar to Spain in places like Africa, New Guinea and parts of South America," said Meier.

Middle latitude glaciers also are showing significant shrinking, he said. In the European Alps, the ice loss has been about 50 percent in the past century, and New Zealand glaciers have shrunk about 26 percent since 1890. In the Caucasus Mountains of Russia, the volume of glacier ice has decreased by about 50 percent in a century, according to calculations by Meier and CU-Boulder researcher Mark Dyurgerov. In the Tien Shan Mountain Range bordering China and Russia, 22 percent of the ice volume from the thousands of glaciers there has disappeared in the past 40 years...

°MOUNTAIN GLACIERS AROUND THE WORLD RECEDING, AGU TOLD
UNISCI.COM
31 MAY 2001

Mountain glaciers around the world are receding, geophysicists said Wednesday at the annual spring meeting of the American Geophysical Union (AGU) in Boston, MA. In a finding he calls "dramatic," Dr. Rick Wessels from the United States Geologic Survey (USGS) presented research that compared new satellite data with historical records and photographs of glaciers on mountains worldwide, showing that the majority of glaciers have decreased in size.

Wessels is part of the Global Land Ice Measurement from Space (GLIMS) project at USGS, which is using NASA's Advanced Spaceborne Thermal Emission and Reflection Radiometer (ASTER) to monitor mountain glaciers around the world. ASTER is one of the instruments on the TERRA satellite, launched in December 1999.

Using ASTER data from the last year and comparing it to historical glacier data, Wessels says his team was able to get preliminary results that show significant reduction in glacier size over the past decade.

For example, Wessels showed images of glaciers in the Andes Mountains in South America, which have decreased by almost a kilometer in the past 13 years, and a glacier in Columbia that the team is watching closely because it is losing meters of ice every week...

°GLACIERS MELTING WORLDWIDE, STUDY FINDS
CONTRA COSTA TIMES
21 AUGUST 2002

New surveys from satellites and aircraft document an alarming acceleration in the melting of glaciers around the world. The swift retreat of these great ice streams is helping to raise ocean levels and is threatening significant changes in human, animal, and plant life - some good, but mostly bad.

Like a canary in a coal mine, the dwindling of the glaciers is visible evidence that the earth really is getting hotter.

"Receding and wasting glaciers are a chief telltale sign that global climate change is real and accelerating," said Jeffrey Kargel, a glacier expert with the U.S. Geological Survey in Flagstaff, Arizona.

Most of Earth's 160,000 glaciers have been slowly shrinking and thinning for more than a century as the climate warms up from both natural causes and human activity.

But scientists say the melt rate has accelerated dramatically since the mid-1990s, which was the hottest decade in a thousand years, according to data from ancient ice cores and tree rings.

By the middle of this century, the Rockies, the Cascades, and Glacier National Park will have lost almost all their ice, Kargel predicted.

"I expect to see absolutely no glaciers in the Swiss or European Alps by the end of the century," he added. "The huge valleys of the Himalayas will be completely deglaciated."...

Up and down the icy spine of South America, the glaciers are melting, the white mantle of the Andes Mountains washing away at an ever faster rate.

"Look. You can see. Chacaltaya has split in two," scientist Edson Ramirez said as he led a visitor up toward a once-grand ice flow high in the thin air of the Bolivian cordillera. In the distance below, beneath drifting clouds, sprawled 2-mile-high La Paz, a growing city that survives on the water running off the shoulders of these treeless peaks. Chacaltaya, a frozen storehouse of such water, will be gone in seven to eight years, said Ramirez, a Bolivian glaciologist, or ice specialist.

"Some small glaciers like this have already disappeared," he said as melting icicles dripped on nearby rock, exposed for the first time in millennia. "In the next 10 years, many more will." They'll disappear far beyond Bolivia. From Alaska in the North, to Montana's Glacier National Park, to the great ice fields of wild Patagonia at this continent's southern tip, the "rivers of ice" that have marked landscapes from prehistory are liquefying, shrinking, retreating.

In east Africa, the storied snows of Mount Kilimanjaro are vanishing. In the icebound Alps and Himalayas of Europe and Asia, the change has been stunning. From South America to south Asia, new glacial lakes threaten to overflow and drown villages below. In the past few years, space satellites have helped measure the global trend, but scientists such as Rajendra K. Pachauri, a native of north India, have long seen what was happening on the ground.

"I know from observation," Pachauri told a reporter at an international climate conference in Argentina. "If you go to the Himalayan peaks, the rate at which the glaciers are retreating is alarming. And this is not an isolated example. I've seen photographs of Mount Kilimanjaro 50 years ago and now. The evidence is visible."

"Ample" evidence indicates that global warming is causing glaciers to retreat worldwide, reports the Intergovernmental Panel on Climate Change, a U.N.-sponsored network of climate scientists led by Pachauri. Global temperatures rose about 1 degree Fahrenheit in the 20th century. French glaciologists working with Ramirez and other scientists at La Paz's San Andres University estimate that the Bolivian Andes are warming even faster, currently at a half-degree Fahrenheit per decade.

...An international study concluded in November that winter temperatures have risen as much as 7 degrees Fahrenheit over 50 years in the Arctic, where permafrost is thawing and sea ice is shrinking. Pacific islands are losing land to encroaching seas, oceans expanding as they warm and as they receive runoff from the Greenland ice cap and other sources.

Those sources include at least one gushing new river of meltwater in western China, where thousands of Himalayan and other glaciers are shrinking.

In the Italian Alps, 10 percent of the ice melted away in the European heat wave of 2003 and experts fear all will be gone in 20 to 30 years. Such rapid runoff would do more than feed rising seas. It would end centuries of reliable flows through populated lands, jeopardizing water supplies for human consumption, agriculture and electricity.

In Peru, endowed with vast Andean ice caps and glaciers, 70 percent of the power comes from hydroelectric dams catching runoff, but officials fear much of it could be gone within a decade. Meanwhile, new mountainside lakes are bulging from the melt, threatening to break their banks and devastate nearby towns...

...Whatever the regional wrinkles, "it's a global view," said Lonnie Thompson, one of the world's foremost glaciologists. "What we see in the Andes is happening in Kilimanjaro and in the Himalayas. We've just been in southeast Alaska, and 1,987 out of 2,000 glaciers are retreating there," the Ohio State University scientist said in a telephone interview from Columbus. "It's a very compelling story," he said. The glaciers - "water towers of the world" - are the most visible indicators that we are now in the first phase of global warming, Thompson said.

BYE

BYE

BYE

BYE

BYE

IT WAS

NICE TO

HAVE

YOU

AROUND

FOR A

WHILE

°DECLINE OF WORLD'S GLACIERS EXPECTED TO HAVE GLOBAL IMPACTS OVER THIS CENTURY
NASA-GSFC PRESS RELEASE
29 MAY 2002

The great majority of the world's glaciers appear to be declining at rates equal to or greater than long-established trends, according to early results from a joint NASA and United States Geological Survey (USGS) project designed to provide a global assessment of glaciers. At the same time, a small minority of glaciers are advancing.

Jeff Kargel, a USGS scientist who will discuss glacier changes and their potential political and economic impacts at the American Geophysical Union (AGU) Spring Meeting in Washington, suggests that accelerating climate change over the next century will directly impact the rate that glaciers retreat.

..."Glaciers in most areas of the world are known to be receding," said Kargel, who is also the international coordinator for Global Land Ice Measurements from Space (GLIMS). "But glaciers in the Himalaya are wasting at alarming and accelerating rates, as indicated by comparisons of satellite and historic data, and as shown by the widespread, rapid growth of lakes on the glacier surfaces."

While ice reflects the Sun's rays, lake water absorbs and transmits heat more efficiently to the underlying ice, kicking off a feedback that creates further melting.

...Glacier changes in the next 100 years could significantly affect agriculture, water supplies, hydroelectric power, transportation, mining, coastlines, and ecological habitats. Melting ice may cause both serious problems and, for the short term in some regions, helpful increases in water availability, but all these impacts will change with time, Kargel said...

SEE YOU AGAIN IN

A FEW HUNDRED

YEARS

OR A FEW THOUSAND

SOUTHERN OVEREXPOSURE

Antarctica contains 80% of the planet's ice pack on an area three times the size of North America. It is huge and frozen. Being more or less isolated from other climate factors in the rest of the world, it creates its own regional climate with freezing ocean waters that swirl around its borders and freezing winds that circulate throughout its interior all helping to build ice pack here and disintegrate it there.

Antarctica is in a fairly stable steady-state, perhaps the most stable of anywhere else on the planet and because of this, there is great difficulty in deciphering exactly how it is or will be affected by global temperature change.

The majority of ice in Antarctica has been in retreat for the past 10,000 years, long before an actively industrializing human species came on the scene. This brings up the dispute of whether the present trend of accelerated ice melt seen in parts of Antarctica is natural or caused by a human influenced greenhouse effect. Again, placing the blame is fairly irrelevant, because any melt from this continent for any reason could have severe consequences for our settlements.

Antarctica is bombarded by intense ultraviolet radiation for half of the year, which affects surface temperature and ice cover. As is the case with high altitudes and latitudes, Antarctica's average temperature has been rising at a rate three times that of equatorial regions. However, recent data also shows that some areas have cooled on average. These conflicting temperature readings again add an ounce of confusion and debate to present measurements and future projections.

Some glaciers in Antarctica are flowing into the oceans faster than before, some glaciers are thickening. Some of its ice shelves are growing, some are disintegrating. In 2004, the western peninsula of Antarctica recorded a release of 250 billion tons of ice into the oceans, enough to raise global sea levels two millimeters per year and an eight-fold increase of the discharge rate from 2002-2003.

That is a lot of ice and a big increase in a short amount of time.

The WEST ANTARCTIC ICE SHEET or WAIS is the big daddy of Antarctica's ice, a land-based chunk the size of Texas. Some reports say this ice is completely stable and has no possibility of collapse in the near future. Other reports claim that the WAIS is becoming unstable and is at risk of collapsing.

If the WAIS did collapse, sea levels would rise six meters.

That is more than a lot of ice.

Although a complete meltdown of the WAIS is possible, most scientists agree that nothing will occur for another few thousand years.

But something BIG could happen this century.

Nobody is absolutely certain.

°GRASS FLOURISHES IN WARMER ANTARCTIC
THE TIMES ONLINE
26 DECEMBER 2004

Grass has become established in Antarctica for the first time, showing the continent is warming to temperatures unseen for 10,000 years. Scientists have reported that broad areas of grass are now forming turf where there were once ice-sheets and glaciers.

Tufts have previously grown on patches of Antarctica in summer, but the scientists have now observed bigger areas surviving winter and spreading in the summer months.

...Pete Convey, an ecologist conducting research with the British Antarctic Survey (BAS), said: "Grass has taken a grip. There are very rapid changes going on in the Antarctic's climate, allowing grass to colonise areas that would once have been far too cold."...

°ICE FLOW DIRECTION CHANGE IN INTERIOR WEST ANTARCTICA
SCIENCE
24 SEPTEMBER 2004

Upstream of Byrd Station (West Antarctica), ice-penetrating radar data reveal a distinctive fold structure within the ice, in which isochronous layers are unusually deep.

The fold has an axis more than 50 kilometers long, which is aligned up to 45 degrees to the ice flow direction. Although explanations for the fold's formation under the present flow are problematic, it can be explained if flow was parallel to the fold axis approximately 1500 years ago.

This flow change may be associated with ice stream alterations nearer the margin. If this is true, central West Antarctica may respond to future alterations more than previously thought.

°ANTARCTIC ICE SHELF CRUMBLING
ENVIRONMENTAL NEWS NETWORK
10 JUNE 1998

The Antarctic Peninsula's Larsen B ice shelf retreated past its historical minimum in March 1998, when a 200-square-kilometer block of ice collapsed into the sea. Scientists believe the ongoing loss of such massive quantities of ice has permanently destabilized the ice shelf.

According to U.S. National Snow and Ice Data Center research associate Ted Scambos: "This could be the beginning of the end" for Larsen B.

...The retreat of the ice shelves coincides with a 2.5-degree Celsius increase in the average temperature on the peninsula since the mid-1940s. During the dog days of the Southern Hemisphere summer, temperatures on the peninsula now rise above 0 C for a longer time. Scientists expect that about two-thirds of Larsen B could crumble away during the next warm season...

THE ANTARCTIC FRINGES ARE MELTING

°ANTARCTIC ICE CRUMBLING RAPIDLY
BBC NEWS ONLINE
8 APRIL 1999

...Two ice shelves on the Antarctic peninsula are crumbling far faster than anyone had predicted. The shelves, Larsen B on the eastern side of the peninsula and Wilkins on the southwest, have together lost nearly 3,000 sq km in the last year. The British and US scientists who made the discovery say the two shelves are in "full retreat".

They are based at the British Antarctic Survey in Cambridge, UK and the National Snow and Ice Data Center at the University of Colorado at Boulder in the US. Despite plans to publish their findings in a scientific journal, the researchers have decided to release them now because of their alarm at what they found.

Satellite photographs show that Larsen B has continued to crumble after an initial small retreat in spring 1998. Since last November the shelf, which is at least four centuries old, has lost a further 1,714 sq km. The Wilkins shelf has lost almost 1,100 sq km in the last year. The scientists had expected the break-up to happen, but more gradually. Ted Scambos, of NSIDC, said: "It happened much faster than we thought." "It was nearly as much activity in a single year as we've seen in 10 or 15 years up to now on average." David Vaughan, of BAS, said that over the last 50 years the shelves had lost about 7,000 sq km. "To have retreats totalling 3,000 sq km in a single year is clearly an escalation."...

THE ANTARCTIC PENINSULA IS CRUMBLING

°CLIMATE CHANGE: HUNDREDS OF ANTARCTIC GLACIERS IN RETREAT, SAYS STUDY
AGENCE FRANCE-PRESSE
21 APRIL 2005

Scientists have issued a fresh warning about the effect of climate change on Antarctica, saying that more than 200 coastal glaciers are in retreat because of higher temperatures. Of the 244 marine glaciers that drain inland ice on the Antarctic peninsula, a region previously identified as vulnerable to global warming, 87 percent have fallen back over the last half century, according to research by British experts.

...Over the last half century, during which time regional temperatures have risen by around 2 C (3.6 F), these glacier fronts have reversed direction, the authors note in a study published on Friday in the US weekly journal *Science*.

Until the mid-1950s, most of the glaciers advanced. For the next decade after that, they were roughly stable. Since then, though, most have been shrinking. In the past five years, the retreat has accelerated, and the pattern of retreat is widening. It started in the warmer northern tip of the peninsula and is heading progressively to the colder south as atmospheric temperatures rise.

THE ANTARCTIC GLACIERS ARE IN RETREAT

IS HUMAN INDUCED GLOBAL WARMING TO BLAME FOR THIS? YES

°STUDY: GLOBAL WARMING LINKED TO MELTING ICE CAP
REUTERS
21 FEBRUARY 2001

A study of melting ice caps has shown that global warming is the cause of rising sea levels and flooding in coastal communities, a team of international scientists said on Wednesday.

Scientists in Canada, the United States and England used computer models to devise patterns, or fingerprints, of how melting ice caps affect sea levels around the globe.

Their results, published in the science journal *Nature*, show global warming will continue to increase sea levels and cause more flooding late into this century.

But sea level rises will not be uniform around the globe.

"We've really strengthened the link between today's sea level changes and ice melting and we've found a way of unraveling the details of this link," said Jerry Mitrovica, a professor of geophysics at the University of Toronto and lead author of the study.

"We've also strengthened extrapolations being made about the future effect of climate warming. And these extrapolations show continued acceleration of sea level rise late into the present century, leading to more flooding of coastal communities," he added in a statement...

IS THE MELTING A LONG-TERM NATURAL PROCESS? YES

°ANTARCTICA'S ICE SHEET MELTING NATURALLY
BBC NEWS ONLINE
3 JANUARY 2003

The West Antarctic Ice Sheet (WAIS) has been melting naturally and releasing water to the ocean for the last 10,000 years. Research published in the journal *Science* suggests that the last Ice Age never really ended in that part of the world.

If the melting continues at its current rate then the WAIS could disappear in 7,000 years, possibly raising worldwide sea levels by five metres.

However, scientists warn that a sudden rapid melting of the WAIS could cause serious problems for some coastal regions.

...Professor John Stone from the University of Washington, US, led the teams behind the work. He said: "In all cases we got very tight, consistent correlations of age with altitude, so we are able to track the margins of the ice sheet coming down the mountain sides with this approach."

The most surprising finding though is how recently the ice has thinned in West Antarctica. Ice sheets in North America and Europe had nearly all melted 10,000 years ago, but this process had only just started in West Antarctica at that time.

"The Ice Age never really came to an end in that part of the world," Professor Stone said...

IS IT A LITTLE BIT OF BOTH? YES

°RETREAT OF ANTARCTIC ICE SHELVES NOT NEW
ASSOCIATED PRESS
24 FEBRUARY 2005

The current retreat of ice shelves in the Antarctic due to global warming is nothing new — but this time the problem is manmade and therefore potentially more serious, according to research released Wednesday.

Writing in the latest issue of the journal "Geology," British scientists said a survey had shown that ice shelves had retreated thousands of years ago as a result of rising air and ocean temperatures.

"What this tells us is that ice shelves don't just break up because they get too big — as the global warning skeptics argue," said Dominic Hodgson, a scientist with the British Antarctic Survey and one of the leading investigators.

He said previous periods of warming — about 9,500 years ago and some 2,000 years to 4,000 years ago — were caused by natural causes, including the ending of ice ages, rather than man's emissions and the ice shelves had been able to reform.

"This time, the problem is man-made and if we don't take steps, the damage will be worse," he said. "There is no room for complacency."

...The study is particularly relevant for other surveys on the West Antarctic Ice Sheet where scientists have found that a relatively warm current, Circumpolar Deep Water, is causing high melt rates on the underside of an ice shelf in Pine Island Bay..The gradual removal of this ice shelf may be causing the glaciers inland to flow faster, which could lead to enhanced drainage of part of the West Antarctic Ice Sheet, and a rise in sea level...

DOES THE CAUSE MATTER? NOT REALLY

°WEST ANTARCTIC ICE SHEET NOT IN JEOPARDY
ENVIRONMENTAL NEWS NETWORK
1 DECEMBER 1998

The West Antarctic ice sheet is not melting rapidly, is reasonably stable and has been so for more than a century, according to an international team of scientists. The ice sheet is the largest grounded repository of ice on the planet and some scientists caught up in the debate over global warming have argued that the melting of this ice sheet would lead to a dramatic rise in sea levels.

The international team of scientists, who reported their findings in the journal *Science*, analyzed five years of satellite radar measurements covering a large part of the Antarctic ice sheet in an effort to determine if there is any direct evidence of the ice sheet melting. While the scientists generally agree that global warming is occurring, their study suggests that it is not having an effect on the Antarctic ice sheet.

...“We assume that global warming is under way now,” said Shum, “and it may be enhanced by human activities but, until now, its effect on ice loss in Greenland and the Antarctic has been mostly speculation. We wanted to look at ice sheets to see how they contribute to sea level rise.”...

SO THERE IS NOTHING TO WORRY ABOUT?

°POLAR ICE SHEET SHOWS SHRINKAGE
ASSOCIATED PRESS
1 FEBRUARY 2001

Scientists have worried for decades that the Antarctic ice sheet was shrinking, threatening a global rise in sea level. Now, satellite studies show that about 7.5 cubic miles of ice have eroded from a key area in just eight years.

Melting of that much ice doesn't mean that it is time to get into boats, said one researcher, but the finding may be a “yellow warning flag” that confirms long-term changes are under way in the ice fields covering the South Polar region.

The study, which appears Friday in the journal *Science*, involved altitude measurements of the West Antarctica Ice Sheet, the smaller of two major ice sheets. It covers 740,000 square miles of the frozen continent.

Based on satellite measurements, said Andrew Shepherd, a University College London geologist and first author of the study, it appears that since 1992 the ice sheet has lost ice principally through the speeded-up movement of the Pine Island Glacier, an ice stream that drains about a third of the ice sheet.

“The Pine Island Glacier is key,” said Shepherd. “It is totally exposed to the sea, and people have identified it as the weak underbelly of the West Antarctica Ice Sheet.” Melting of the entire sheet theoretically could cause a global sea level rise of 25 to 45 feet, but Shepherd said that at the present rate of change it would take centuries for the Pine Island Glacier, which is only about 10 percent of the ice sheet, to affect sea level seriously...

JUST A LITTLE SHRINKAGE? ONLY 10% OF THE ICE SHEET? CENTURIES TO MELT?

°RAPID ANTARCTIC WARMING PUZZLE
BBC NEWS ONLINE
6 SEPTEMBER 2001

UK scientists say parts of Antarctica have recently been warming much faster than most of the rest of the Earth. They believe the warming is probably without parallel for nearly two thousand years. They suggest three possible mechanisms that may account for what is happening. But they say they cannot identify a cause with certainty, nor can they predict whether the warming will continue.

...Trends in mean annual air temperature for 1950-98 show three areas of especially rapid regional warming: northwestern North America and the Beaufort Sea; an area around the Siberian plateau; and the Antarctic peninsula and the adjoining Bellingshausen Sea. For all Antarctic stations, the mean temperature trend for 1959-96 is +1.2 degrees C per century, but there are marked regional variations.

...They suggest three possible mechanisms: changing ocean currents may have brought warmer deep water on to the continental shelf, reducing sea-ice; warmer air may have come into the region; or a unique sea-ice-atmosphere feedback may be at work. Not knowing the cause of the changes so far, the authors say they cannot predict the future.

°ANTARCTIC EXPERTS WARN OF GLOBAL WARMING MELTDOWN
REUTERS
28 DECEMBER 2001

There is a one in 20 chance of a dramatic rise in world sea levels over the next century due to global warning, according to a new risk assessment published on Friday. The survey -- by the British Antarctic Survey (BAS) and Norwegian environmental safety organization, Det Norske Veritas -- said there was a five percent chance of the giant West Antarctic Ice Sheet disintegrating due to climate change and raising sea levels by one meter (yard) in the next 100 years.

"You have to balance the likelihood against the severity of the impacts, and in this case even a five percent chance of this happening is really damn serious," said scientist David Vaughan of BAS, responsible for British scientific research in Antarctica.

Scientists have already predicted a rise in sea levels of 50 cm (20 ins) over the next century due to a combination of climate change and increased extraction of ground water, even with no contribution from melting Antarctic ice.

"So we might be looking at something like one and a half meters in the next century," Vaughan told Reuters...

°WEST ANTARCTIC ICE GETTING THICKER
ASSOCIATED PRESS
17 JANUARY 2002

New measurements show the ice in West Antarctica is thickening, reversing earlier estimates that the sheet was melting. Scientists concerned about global warming had worried that higher temperatures could melt the massive ice sheet, causing a rise in sea levels worldwide. But new flow measurements for the Ross ice streams, using special satellite-based radars, indicate that movement of some of the ice streams has slowed or halted, allowing the ice to thicken, according to a paper in Friday's issue of the journal *Science*.

...Their finding comes less than a week after a separate paper in *Nature* reported that Antarctica's harsh desert valleys - long considered a bellwether for global climate change - have grown noticeably cooler since the mid-1980s. Air temperatures recorded continuously over a 14-year period ending in 1999 declined by about 1 degree Fahrenheit in the polar deserts and across the White Continent, that paper said.

The cooling defies a trend spanning more than 100 years in which average land surface temperatures have increased worldwide by about 1 degree Fahrenheit

The scientists said Antarctica is the only continent that is cooling.

They could not say why.

I THOUGHT ANTARCTICA WAS COOLING?

A 1 IN 20 CHANCE OF THE WAIS MELTING DURING THE NEXT 100 YEARS?

THE WAIS IS GETTING THICKER?

I THOUGHT IT WAS MELTING?

IS ANTARCTICA COOLING OR WARMING? WHAT ARE THE TRENDS?

WELL THIS DOESN'T SEEM LIKE A GOOD TREND

°ANTARCTIC ICE SHELF COLLAPSES IN LARGEST EVENT OF LAST 30 YEARS
UNIVERSITY OF COLORADO
18 MARCH 2002

Recent satellite imagery analyzed at the National Snow and Ice Data Center at the University of Colorado at Boulder has revealed that the northern section of the Larsen B ice shelf, a large floating ice mass on the eastern side of the Antarctic Peninsula, has shattered and separated from the continent in the largest single event in a 30-year series of ice shelf retreats in the peninsula.

"This breakup gave us the information we need to reassess the stability of ice shelves around the rest of the Antarctic continent," said glaciologist Ted Scambos. "They are closer to the limit than we thought."

...Over the last five years, the Larsen B shelf has lost a total of 5,700 square kilometers -- 2,200 square miles -- and is now about 40 percent the size of its previous minimum stable extent.

...International cooperation between Argentinian, American, British, Austrian and German scientists has resulted in detailed information on the breakup from field observations, shipboard studies and a variety of satellite sensors. Scientists attribute the retreats to strong regional climate warming.

Antarctic temperatures have increased about 2.5 degrees Celsius since the late 1940s. Since 1974 ice shelf extent in the Antarctic Peninsula has declined by about 13,500 square kilometers, or 5,200 square miles...

°ANTARCTIC GLACIERS ACCELERATING IN RESPONSE TO 2002 ICE SHEET COLLAPSE
NATIONAL SNOW AND ICE DATA CENTER
21 SEPTEMBER 2004

Glaciers in Antarctica's most rapidly warming region have quickened their pace following the collapse of a Delaware-sized ice shelf in March 2002, according to a new study led by the University of Colorado at Boulder and a related study by NASA's Jet Propulsion Laboratory.

Landsat 7 satellite images taken before, during and after the break-up of the Larsen B ice shelf in March 2002 show that several of the glaciers are now moving at up to five times their previous speed, said University of Colorado at Boulder researcher Ted Scambos. Other satellite data show that the glaciers also have thinned significantly since the disintegration of the Larsen B, he said.

The recent events underscore the potential for sea-level rise as a result of climate warming over the Earth's polar caps. "The Larsen area can be looked at as a miniature experiment, showing how warming can dramatically change the ice sheets, and how fast it can happen," he said. "At every step in the process, things have occurred more rapidly than we expected."

...The area, located at the far northern tip of the Antarctic just south of Chile and Argentina, has seen a rise in mean annual temperatures of up to 4.5 degrees Fahrenheit in the past 60 years -- faster than almost any region in the world. In the past 30 years, ice shelves in the region have decreased by more than 5,200 square miles.

...The study highlights the sensitivity of the poles to climate change, Scambos said. "As temperatures crossed the threshold of melting in the summer months, ice shelves in the area rapidly disintegrated. Not only do the ice shelves collapse rapidly, but the subsequent effects on the glaciers are immediate," he said...

31 January 2002

17 February 2002

23 February 2002

5 February 2002

disintegration of the larsen b ice shelf in antarctica

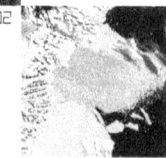
7 March 2002

°ANTARCTIC ICE MELT POSES WORLDWIDE THREAT
REUTERS
14 MAY 2002

The Antarctic Peninsula ice shelves are cracking up and, on the face of things, it is the most serious thaw since the end of the last ice age 12,000 years ago. The break-up of the ice shelves in itself is a natural process of renewal, but the size and rate of production of icebergs -- some the size of major cities -- is alarming scientists, who blame global warming.

The break-off last month of a 500 billion ton chunk of the Larsen Ice Shelf -- 650 feet thick and with a surface area of 1,250 sq. miles -- is the second big break since a giant iceberg broke away in 1995 and is well beyond normal activity, scientists say. The production of vast amounts of icebergs is a threat to the world's climate and the way the ocean's function, they say. And the process, once started, cannot be reversed.

The fear is that a snowball effect will lead to disintegration of the vast West Antarctic ice shelf, kilometers thick in parts…A longer-term effect would be if the disintegration led to a meltdown of the grounded West Antarctic ice sheet, which would cause the world's oceans to rise by up to five meters (17 feet).

…"These patterns are beyond natural variability," he said.

POSITIVE FEEDBACK

high latitude warming
accelerated ice melt
reduced albedo
increased heat absorption

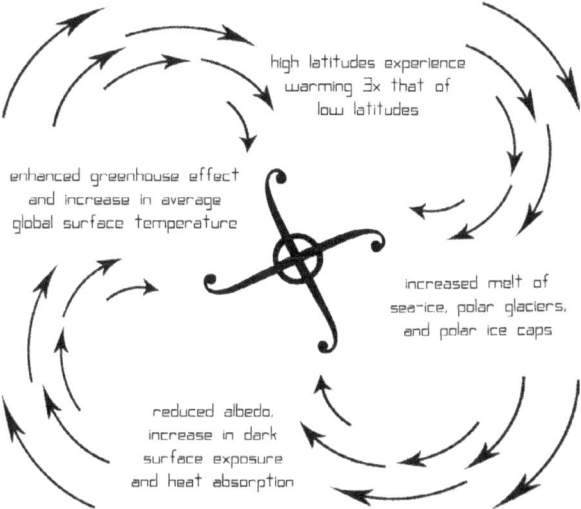

high latitudes experience warming 3x that of low latitudes

enhanced greenhouse effect and increase in average global surface temperature

increased melt of sea-ice, polar glaciers, and polar ice caps

reduced albedo, increase in dark surface exposure and heat absorption

WELL THESE DON'T SEEM LIKE GOOD TRENDS

°ANTARCTIC ICE SHELF IS MELTING RAPIDLY, SCIENTISTS WARN
THE GUARDIAN
31 OCTOBER 2003

A giant ice shelf the size of Scotland is melting rapidly in the Antarctic, scientists have warned today. Two sections of the Larsen ice shelf collapsed in 1995 and 2002.

Now satellite measurements have confirmed that it has thinned by as much as 18 metres more than usual in the past decade, because of a warmer ocean.

The report comes a day after a University College London report in the journal *Nature* confirmed a 40% thinning of the ice in the Arctic Ocean in the past 30 years. Andrew Shepherd of Cambridge University and colleagues from University College London, Bristol, and the Antarctic Institute of Argentina, report in the journal *Science* that they have been mapping changes in the Larsen shelf, on the east of the Antarctica peninsula, since 1992.

The shelf covers approximately 27,000 sq miles and is on average 300 metres deep. Its faster rate of melting is releasing an extra 21bn tonnes of icy water into the oceans each year. This is equivalent to eight times the annual flow of the river Thames. Warmer oceans, as well as warmer air, are now thought to play an important part in the process,

POSITIVE FEEDBACK

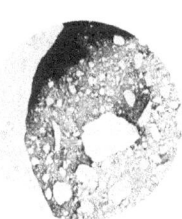

high latitude warming
warmer water and polar fringe melt
acceleration of ice slide
reduced albedo

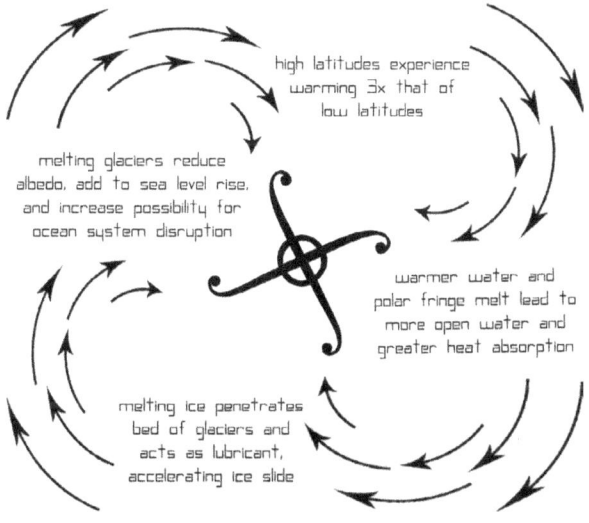

high latitudes experience warming 3x that of low latitudes

melting glaciers reduce albedo, add to sea level rise, and increase possibility for ocean system disruption

warmer water and polar fringe melt lead to more open water and greater heat absorption

melting ice penetrates bed of glaciers and acts as lubricant, accelerating ice slide

°ANTARCTIC ICE FRINGE 'MELTING FASTER'
BBC NEWS ONLINE
13 JUNE 2002

US scientists say the floating fringes of the Antarctic ice sheet are melting faster than previous studies had suggested. They say the rate of melting is linked to the temperature of the surrounding seawater. They estimate that each 0.1 Celsius rise in sea temperature can increase the rate of melting by one metre annually. The scientists say their findings could have implications for the stability of the West Antarctic Ice Sheet (WAIS).

...The US team examined 23 glaciers. They say: "Most of the melt rates we calculate near glacier grounding lines exceed the area-average rates for the largest ice shelves by one to two orders of magnitude."

They say the rates "are strongly correlated with ocean thermal forcing... Our results demonstrate that bottom melting near an ice shelf grounding line is strongly leveraged by the temperature of seawater that comes into contact with the ice in that region."

An increase in seawater temperature, they say, which increases the bottom melting rate, will steepen the ice thickness gradient near the grounding line. "That will increase the driving stress and flow velocity, and it will reduce the ice shelf resistance to ice discharge, potentially causing the glacier to accelerate and its grounding line to retreat...

°GLACIERS ARE FLOWING FASTER
NATURE NEWS SERVICE
23 SEPTEMBER 2004

Ice is sliding off the Antarctic continent much faster than it did during the 1990s, several teams of researchers have found.

On the Antarctic Peninsula that sticks out from the western side of the continent, the speeds at which several glaciers are surging into the sea have increased eight-fold between 2000 and 2003. And below the peninsula, in Western Antarctica, glaciers are now releasing 250 billion tonnes of ice into the Amundsen Sea each year - enough to raise global sea levels by 2 millimetres per decade.

The results, which draw on satellite and aircraft measurements of glacier thickness and flow, highlight the disturbing potential consequences of recent climate warming in the Antarctic region.

The thick sheet of ice covering Western Antarctica is a focal point for fears about the effects of climate change on the frozen poles. It is particularly vulnerable because it rests on land that lies below sea level, and there is a danger that if the ice shelves surrounding it were to disintegrate, the entire ice sheet could slide into the sea.

If that happened, global sea level would rise by an awesome 5 metres: five times greater than the highest current prediction for the increase in sea level over the next century...

°STUDY SHOWS POTENTIAL FOR ANTARCTIC CLIMATE CHANGE
NASA-GSFC NEWS RELEASE
6 OCTOBER 2004

While Antarctica has mostly cooled over the last 30 years, the trend is likely to rapidly reverse, according to a computer model study by NASA researchers. The study indicates the South Polar Region is expected to warm during the next 50 years.

Findings from the study, conducted by researchers Drew Shindell and Gavin Schmidt of NASA's Goddard Institute of Space Studies (GISS), New York, appeared in the Geophysical Research Letters. Shindell and Schmidt found depleted ozone levels and greenhouse gases are contributing to cooler South Pole temperatures.

Low ozone levels in the stratosphere and increasing greenhouse gases promote a positive phase of a shifting atmospheric climate pattern in the Southern Hemisphere, called the Southern Annular Mode (SAM). A positive SAM isolates colder air in the Antarctic interior.

In the coming decades, ozone levels are expected to recover due to international treaties that banned ozone-depleting chemicals. Higher ozone in the stratosphere protects Earth's surface from harmful ultraviolet radiation. The study found higher ozone levels might have a reverse impact on the SAM, promoting a warming, negative phase. In this way, the effects of ozone and greenhouse gases on the SAM may cancel each other out in the future. This could nullify the SAM's affects and cause Antarctica to warm.

...Shindell said the biggest long-term danger of global warming in this region would be ice sheets melting and sliding into the ocean. "If Antarctica really does warm up like this, then we have to think seriously about what level of warming might cause the ice sheets to break free and greatly increase global sea levels," he said...

as open water warms and increases

fringe melt and glacial slide accelerate

MAYBE TRENDS ARE BETTER AT THE NORTH POLE

MAYBE NOT

as more water and land are exposed

more solar energy is absorbed and the temperature of land and water increases

NORTHERN LITE

Once upon a time, the North Pole was green and lush.

Once upon a time, the Northern Hemisphere was coated with ice.

Today, the North Pole is also warming at a rate three times that of equatorial regions and many changes are happening as result of this warming trend.

For starters there are about 50 cubic kilometers of Greenland's icy fringe slipping into the ocean each year; a steady retreat of about 3 feet per year in some places. At a place called the Jacobshavn Isbrae, the speed of ice entering the ocean has doubled during the past 10 years. This important glacier helps balance the continental ice sheet and disruptions here could strongly effect the stability of Greenland's land-based ice.

The North Pole has a much smaller land mass and a much larger amount of annual sea ice change than Antarctica. Throughout the year, the sea ice shifts around and circulates throughout the Arctic Ocean and recently, the annual average of sea ice has declined. More open water means a greater area available to absorb heat energy, increasing the average temperature of a larger area for a longer period of time. As the average temperature increases throughout the year, Greenland and other areas up North are becoming greener and as they green, several things happen.

First of all, as the average temperature of the North Pole increases, summers last longer which perpetuates warmer waters, both accelerating and increasing the amount of ice melt on land and in the sea.

Second, as temperatures increase, the precipitation patterns change and the hydrologic cycle becomes more active. Until now, the North Pole has been generally too cold for much precipitation to occur, cold winds scour the land and keep the climate relatively dry. As it warms, there is more evaporation, rain, and snow.

Third, as the permafrost thaws, anything built on top of it: roads, buildings, oil production plants, pipelines; all of their foundations begin to buckle, crack, and sag with the softening surface of the Earth.

Fourth, the greenhouse gases carbon dioxide and methane locked in the ice crystals of the icy tundra for thousands of years are being set free. The stores of greenhouse gases in this permafrost are massive, trillions and trillions of cubic tons, far more than the annual emissions from our activities. As huge amounts of stored methane and carbon dioxide are released into the atmosphere, there is a greater likelihood that the greenhouse effect could be strongly enhanced in a short amount of time.

Fifth, ice melting at the North Pole adds to rising seas everywhere else. If all of Greenland's land based ice were to melt, it would add six meters to the present sea level.

°STUDY BLAMES ARCTIC ICE DECLINE ON HUMANS
ENVIRONMENTAL NEWS NETWORK
6 DECEMBER 1999

Humans are contributing more to the retreat of sea ice in the Arctic than is natural climate change, according to a recent study.

For the first time, scientists have placed satellite-based observations of Arctic sea ice declines into a larger context, by using a computer model on climate change.

"For climate studies, you would ideally like to see hundreds of years of records, but satellite data only goes back a couple of decades," explains Claire Parkinson, a climatologist at NASA's Goddard Space Flight Center.

Satellite data between November 1978 and March 1998 reveals that that the Arctic ice overall has shown a downward trend of 37,000 square kilometers a year. That means a loss of an ice chunk that exceeds the combined areas of the states of Maryland and Delaware, Parkinson said.

While some scientists might argue that such a loss could occur naturally over time, Parkinson said results of the recent study strongly suggest that humans have played a large role in this decline...

Finally, as more freshwater from increased precipitation, melting glaciers, and land-based ice comes flowing into the North Atlantic, there is a very strong likelihood that this disrupts the planet's ocean circulation system which provides us all with the climates and weather patterns we are accustomed to and our settlements are dependent upon.

Dramatic changes would probably not happen overnight like they do in the movies, but they could happen in 10 years, or 20 years, or 100 years.

All of which are overnight to the Earth.

SO...THE NORTH POLE IS MELTING

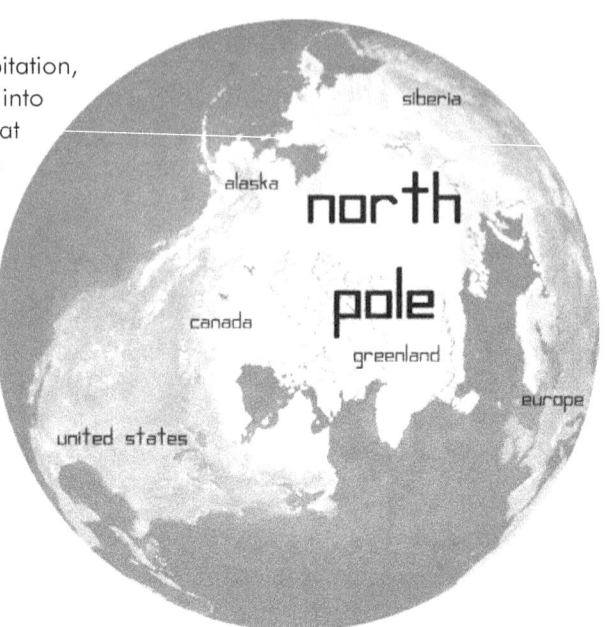

°NORTH POLE FACES A MAJOR MELTDOWN
NEW YORK TIMES
11 JULY 2000

The mythic icescape that stretches south in all directions from the North Pole is melting so fast that Norwegian scientists say it could disappear entirely each summer beginning in just 50 years, radically altering the Earth's environment, the global economy and the human imagination.

Climatologists have warned for a decade that the northern ice cap is retreating. But researchers at the University of Bergen's Nansen Environmental and Remote Sensing Center are apparently the first to predict the disorienting specter of a watery North Pole open to cruise ships and the Polar Bear Swim Club within the lives of today's young people.

"The changes we've seen have been much faster and more dramatic than most people imagine," said Tore Furevik, 31, a polar researcher and co-author of the article "Toward an Ice-Free Arctic?" in the latest issue of the Norwegian science journal *Cicerone*...

°SUBSTANTIAL CONTRIBUTION TO SEA-LEVEL RISE DURING THE LAST INTERGLACIAL FROM GREENLAND ICE SHEET
NATURE
6 APRIL 2000

During the last interglacial period (the Eamian), global sea level was at least three metres, an probaly more than five metres, higher than at present.

...We conclude that the high sea level during the last interglacial period most probably included a large contribution from Greenland meltwater and therefore should not be interpreted as evidence for a significant reduction of the West Antarctic ice sheet...

°U.N. WARNS GLOBAL WARMING IS MELTING ARCTIC SOIL
REUTERS
7 FEBRUARY 2001

U.N. scientists said Wednesday that global warming was melting the Arctic's permafrost, causing it to release greenhouse gases that could in turn raise temperatures even higher.

"This is very alarming," said Svein Tveidtal, a prominent scientist with the United Nations Environment Program. "The Arctic is an area where temperature changes are going to cause tremendous problems."

The vicious cycle could accelerate the so-called greenhouse effect and lead to the disintegration of the permafrost, causing serious damage to buildings, roads, pipelines and other infrastructure in Arctic areas like Alaska and Siberia.

...For thousands of years, the permafrost has mopped up carbon dioxide from the atmosphere and stored it in its soil, mainly because the decomposition of dead vegetation is extremely slow in such low temperatures.

However, with rising temperatures in the Arctic, microbes decompose dead plant matter at a higher rate, releasing carbon dioxide that then adds to the problem of global warming.

U.N. scientists say the vicious cycle appears to have already begun.

"There are indications that at least some parts of the Arctic have switched from being sinks of carbon dioxide to being sources," scientists monitoring the melting of the permafrost said in a recent report...

WHICH INCLUDES ICE IN THE ARCTIC OCEAN

°ARCTIC SEA ICE GETS THINNER
BBC NEWS ONLINE
16 NOVEMBER 1999

There has been a "striking" decline in the thickness of Arctic sea ice according to scientists who have studied data gathered by US Navy submarines.

The researchers say the average draught of the sea ice in the region has declined by 1.3 metres (4.3 ft) compared with the 1960s and 1970s. By draught they mean the difference between the surface of the ocean and the bottom of the ice pack - just like the draught of a boat.

This amounts to a 40% reduction, says Dr Andrew Rothrock of the University of Washington, Seattle, and colleagues, who report their findings in the journal *Geophysical Research Letters*...

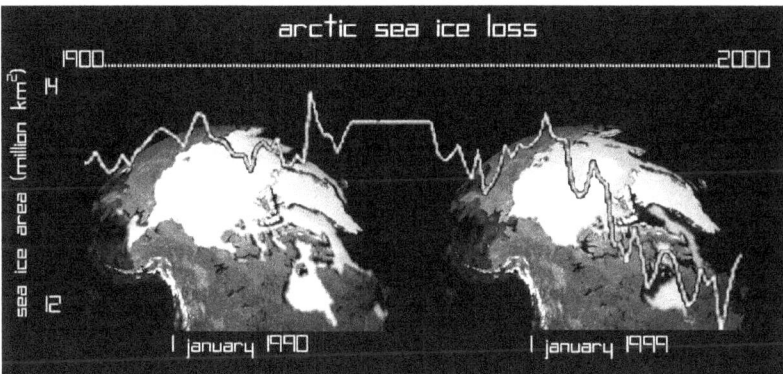

°ARCTIC ICE 'MELTING FROM BELOW'
BBC NEWS ONLINE
27 MARCH 2002

Scientists believe they have identified a mechanism which can explain the thinning of the Arctic sea ice. They say the thinning, which in summer reaches more than 40% in some areas, has two causes.

Rising air temperatures, possibly the consequence of global warming, are melting the ice from above. And warmer water is also rising from the depths to attack the ice from below.

Professor Peter Wadhams, of the Scott Polar Research Institute in Cambridge, UK, said in 2000 that he had established the degree of thinning using measurements from submarines in 1976 and 1996.

...He told BBC News Online: "People say global warming can't be raising air temperatures enough to melt the ice, because the Arctic winter temperature is around -30C anyway, and a one-degree warming would be irrelevant.

"But it's the summer temperatures that matter. Arctic summers are getting longer, so there is longer for the warmer air to melt the snow and affect the ice beneath.

...Professor Wadhams thinks the Arctic could be virtually ice-free during the summer by about 2080...

°ARCTIC SEA ICE DECLINES AGAIN IN 2004, ACCORDING TO CU-BOULDER RESEARCHERS
NATIONAL SNOW AND ICE DATA CENTER PRESS RELEASE
4 OCTOBER 2004

Researchers at the University of Colorado at Boulder have found that the extent of Arctic sea ice, the floating mass of ice that covers the Arctic Ocean, is continuing its rapid decline.

The latest satellite information indicates the September 2004 sea ice extent was 13.4 percent below average, a reduction in area nearly twice the size of Texas, said Mark Serreze of CU-Boulder's National Snow and Ice Data Center, or NSIDC. In 2002, the decline in arctic sea ice during September - which traditionally marks the end of the summer melt season - was about 15 percent, a record low, said CU-Boulder researcher Walt Meier of NSIDC.

The decline in sea ice extent during September has averaged about 8 percent over the past decade, said Serreze, who is part of a CU-Boulder team monitoring Arctic sea-ice conditions. "This is the third year in a row with extreme ice losses, pointing to an acceleration of the downward trend," he said.

...One complicating factor is the atmospheric circulation pattern known as the Arctic Oscillation, which may be contributing to the loss of the much thicker "multi-year" ice that has accumulated over many years. "As winds and currents force this ice southward, more of it melts," said Stroeve. "And while new ice is still forming in the winters, it is thinner, and therefore melts faster in the summer than older ice."...

∧ND GREENLAND

°NASA SCIENTISTS DETECT RAPID THINNING OF GREENLAND'S COASTAL ICE
NASA PRESS RELEASE
20 JULY 2000

Scientists who want to monitor the state of our global climate may have to look no farther than the coastal ice that surrounds the Earth's largest island.

A NASA study of Greenland's ice sheet reveals that it is rapidly thinning. In an article published in the July 21 issue of *Science*, Bill Krabill, project scientist at the NASA Goddard Space Flight Center's Wallops Flight Facility, Wallops Island, VA, reports that the frozen area around Greenland is thinning, in some places, at a rate of more than three feet per year. Any change is important since a smaller ice sheet could result in higher sea levels.

"A conservative estimate, based on our data, indicates a net loss of approximately 51 cubic kilometers of ice per year from the entire ice sheet, sufficient to raise global sea level by 0.005 inches per year, or approximately seven percent of the observed rise," Krabill said.

"This amount of sea level rise does not threaten coastal regions, but these results provide evidence that the margins of the ice sheet are in a process of change," Krabill said. "The thinning cannot be accounted for by increased melting alone. It appears that ice must be flowing more quickly into the sea through glaciers."...

°GREENLAND IS A CANARY
SCRIPPS HOWARD NEWS SERVICE
23 OCTOBER 2000

Just how much global warming is the Earth undergoing?

There are many reasons to care about this question, but the reason that most interests Bill Krabill, a scientist at the Goddard Space Center at Wallops Island, Va., is that global warming could mean the melting of water stored in the polar icecaps.

According to Krabill, three-quarters of the world's population lives within a couple of miles of a coastline. Even a small rise in the sea level would mean a devastating impact on people in these areas.

Knowing the immense importance of the issue, Krabill and his colleagues have been trying to assess the likelihood of its actually happening. One way of getting a handle on the issue is by studying the ice sheet that covers Greenland.

Why Greenland?

It turns out that Greenland may be better suited for tracking climate change than any other region in the world. Greenland's ice cover is only one seventh the size of the ice sheet in Antarctica, and it's far more sensitive to climate change.

The Antarctic ice sheet is so large that it has its own climate system and interacts with the rest of the world's climate system at a relatively slow rate. Greenland, on the other hand, protrudes into the temperate latitudes and responds much more rapidly to climate change.

..."In a nutshell," he summarizes, "we've determined that the high elevation portions of the Greenland Ice Sheet are remarkably stable. But all around the edges, more than 70 percent is thinning fairly rapidly, more than 3 feet per year, over thousands of miles of coastland."...

°SCIENTISTS ALARMED AT INCREASE IN MELT RATE OF ICE
THE SCOTSMAN ONLINE
4 AUGUST 2004

Greenland's cover of ice is melting ten times quicker than previously thought, an increase that could lead to floods across the world, scientists have found.

Newly published research shows an alarming rise in the rate of collapse of the massive Greenland ice-sheet as a result of global warming. Scientists now believe the ice-sheet is shrinking at a rate of ten metres a year, not the one metre previously thought.

If the entire ice-sheet melts, the resulting flood waters would raise the level of global seas by seven metres, submerging large areas of land, including sea-level cities such as London.

...One medium-term side-effect of the destruction of the Greenland ice-sheet could be the loss of the Gulf Stream, which keeps Europe warm and temperate. The fresh water from the ice mixes with the salt water in the sea, altering the salinity and changing the direction and behaviour of major currents...

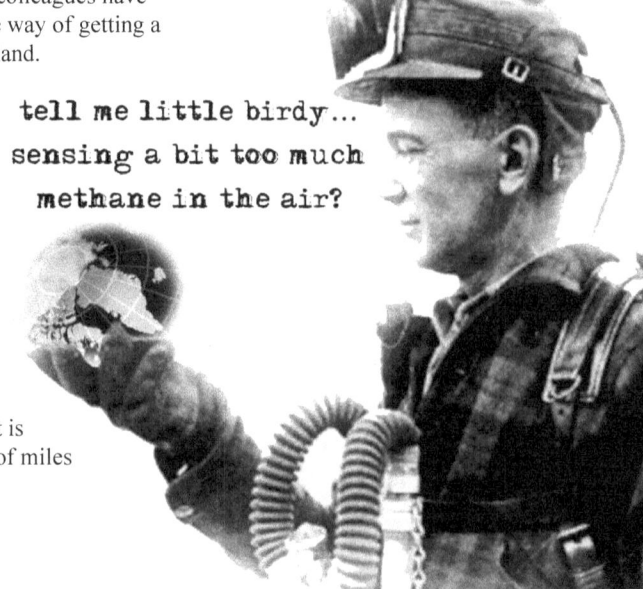

tell me little birdy...
sensing a bit too much
methane in the air?

AND

ICELAND

°ICELANDIC GLACIER IN RAPID BREAKUP
UNITED PRESS
INTERNATIONAL
22 OCTOBER 2000

A British newspaper reported Sunday new research shows Europe's biggest glacier is about to disintegrate. The Observer report - based on an interview with David Evans, of Glasgow University, who has spent decades studying the glacier - says the mighty Breidamerkurjökull in southern Iceland is breaking apart. Evans said that the glacier "is beginning to disintegrate and in the next few years will collapse" into the North Atlantic...

°EUROPE'S LARGEST GLACIER MELTING AWAY - EXPERT
REUTERS
17 JANUARY 2002

The largest glacier in Europe, Iceland's Vatnajokull, is melting away and getting thinner by an average of about three feet per year because of a warmer climate, an expert said on Thursday.

"If the shrinking continues at this pace, there won't be much left of Vatnajokull by the end of the century," glaciologist Helgi Bjornsson told Reuters...

°ARCTIC ICE CAP MELTING AT WORRYING RATE: NASA
AGENCE FRANCE-PRESSE
24 OCTOBER 2003

The polar ice cap is melting at an alarming rate due to global warming, according to NASA scientists, with satellite images showing the ice cap has been shrinking by 10 percent per decade over the past quarter century.

"It is happening now. We cannot afford to wait a long period of time for technological solutions," said David Rind of NASA's Goddard Institute for Space Studies in New York...

AND NORTHERN CANADA

°CANADA'S SHRINKING ICE CAPS
NASA-GSFC NEWS RELEASE
4 MARCH 2005

Earth's ice-covered polar regions help to keep our climate cool and hold tremendous amounts of fresh water locked up in their glaciers, ice caps, and ice sheets. The ice contained in these vast freshwater reservoirs is the equivalent of nearly 220 feet of sea level. However, when most people think of polar ice, they usually do not think of Canada, the location of only a small percentage of the Arctic's polar land ice.

Recent research conducted by NASA scientists has revealed that Canada's ice caps and glaciers have important connections to Earth's changing climate, and they have a strong potential for contributing to sea level rise.

Canada's Arctic region is covered by approximately 150,000 square kilometers (57,660 square miles) of land ice. This is much smaller than Antarctica's 13.5 million square kilometers (5.2 million square miles), and Greenland's 1.7 million square kilometers (650,000 square miles) of ice coverage, but still quite significant. In the next 100 years, melting glaciers and ice caps outside of Greenland and Antarctica, a significant portion of which includes those in Canada, are expected to raise global sea levels by 20 to 40 centimeters (7.9 to 15.8 inches).

Waleed Abdalati, Head of the Cryospheric Sciences Branch at NASA's Goddard Space Flight Center (GSFC), Greenbelt, Md. published research recently in the Journal of Geophysical Research showing that Canada's Arctic ice is one of the more significant and immediate sources of world-wide changes in sea levels.

Abdalati and his colleagues say that Canada's Arctic ice is important because the wide area covered by these ice caps and the dramatic changes that have taken place in the Arctic climate in recent years. Studying this region will help researchers understand how much and in what ways Arctic glaciers and ice caps are contributing to sea level rise.

"The ice-covered parts of the world, and, in particular, of the Arctic are considered to be very sensitive to change," said Abdalati...

SO...THE NORTH POLE IS MELTING

°RECORD MELT IN ARCTIC AND GREENLAND: ICE SHEETS SHRANK BY 1 MILLION SQUARE KILOMETRES THIS SUMMER
NATURE NEWS SERVICE
9 DECEMBER 2002

Ice covering the Arctic Ocean and Greenland shrank by record amounts this summer, new research shows. The rise in seasonal melting has led some experts to estimate that 20% of Arctic sea ice could be lost by 2050.

Using satellite-based microwave surveys, Mark Serreze and colleagues at the National Snow and Ice Data Center in Boulder, Colorado, spotted that Arctic sea ice, which usually covers around 6.5 million square kilometres in the summer, fell to around 5.5 million square kilometres this year. Similarly, Konrad Steffen, of the University of Colorado at Boulder, found that 16% more Greenland ice melted this summer than in 1979, when satellite monitoring began.

These are the largest decreases ever seen...

PERMAFROST IS THAWING TOO WHICH IS NOT GOOD

POSITIVE FEEDBACK

high latitude warming
permafrost thaw
greenhouse gas release
enhanced greenhouse effect

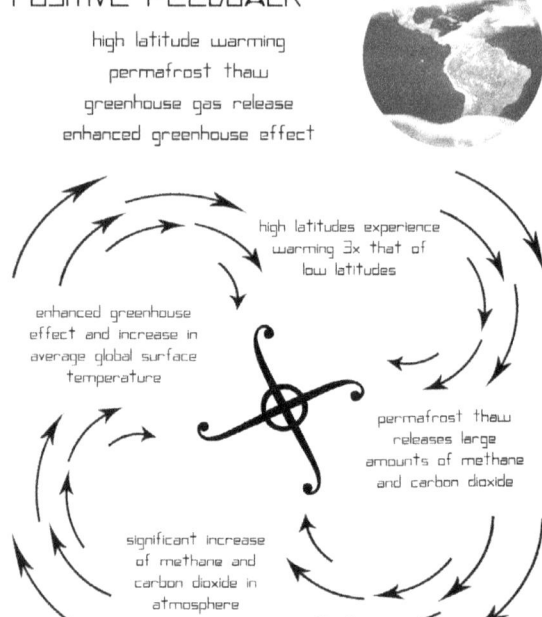

high latitudes experience warming 3x that of low latitudes

enhanced greenhouse effect and increase in average global surface temperature

permafrost thaw releases large amounts of methane and carbon dioxide

significant increase of methane and carbon dioxide in atmosphere

°THAWING SUBARCTIC PERMAFROST INCREASES GREENHOUSE GAS EMISSIONS
AMERICAN GEOPHYSICAL UNION- LUND UNIVERSITY JOINT PRESS RELEASE
24 FEBRUARY 2004

The permafrost in the bogs of subarctic Sweden is undergoing dramatic changes. The part of the soil that thaws in the summer, the so-called active layer, has become thicker since 1970, and the permafrost has disappeared altogether in some locations. This has lead to significant changes in the vegetation and to a subsequent increase in emission of the greenhouse gas methane. Methane is 25 times more potent than carbon dioxide as a greenhouse gas.

...Methane is released from the breakdown of plant material under wet soil conditions. The disappearance of permafrost and subsequent wetter soil conditions have lead to the observed increases in methane emissions. "At a particular mire, Stordalen, we have been able to estimate an increase in methane emissions of at least 20 percent, but maybe as much as 60 percent, from 1970 to 2000," says the lead researcher, Torben R. Christensen of Lund University's GeoBiosphere Science Centre...

°RESEARCHERS FIND FROZEN NORTH MAY ACCELERATE CLIMATE CHANGE
NASA-GSFC NEWS RELEASE
12 OCTOBER 2004

NASA-funded researchers have found that despite their sub-zero temperatures, a warming north may add more carbon to the atmosphere from soil, accelerating climate warming further.

"The 3 to 7 degree Fahrenheit rise in temperature predicted by global climate computer models could cause the breakdown of the arctic tundra's vast store of soil carbon," said Michelle Mack, an ecologist at the University of Florida, Gainsville, Fla., and one of the lead researchers on a study published in last week's issue of *Nature*. It would release more of the greenhouse gas carbon dioxide into the air than plants are capable of taking in.

The study results suggest that climate warming in the arctic tundra may cause the release of much more carbon dioxide than previously expected. This type of positive feedback will make the Earth's climate change even more rapidly.

The findings were collected in a 20-year experiment of the effects of fertilization on the arctic tundra at the Arctic Long-Term Ecological Research site near Toolik Lake, Alaska. The National Science Foundation and NASA provided funding for the research.

One-third of the Earth's soil carbon is locked in northern latitudes because low temperatures and water-saturated soil slow the decomposition of organic matter by bacteria, fungi and other organisms.

...It has long been thought that global warming would have two opposing effects on arctic soils. First, would increase the breakdown of soil organic matter, releasing carbon dioxide, the major cause of warming, into the atmosphere. Second, the breakdown of soil organic matter would liberate nutrients that would enhance rates of plant growth, thereby removing carbon dioxide from the atmosphere.

Peter Vitousek, a professor of biological sciences at Stanford University, said "This work demonstrated beautifully that there is another, even stronger effect, that an increase in nutrients also enhances the breakdown of soil organic matter." The overall effect of warming especially in the Arctic will be to release carbon dioxide to the atmosphere, enhancing the likelihood of further warming.

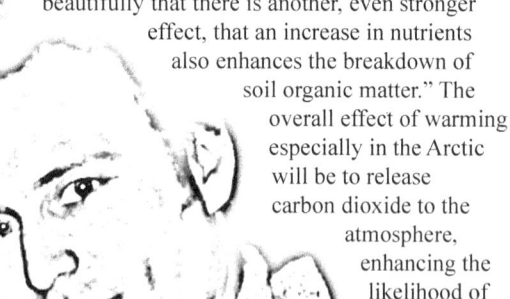

hey...

...are you gettin this

VACILLATE WILDLY

The circulation of the planet's oceans is largely responsible for driving the climate systems and weather patterns we have today. But before looking at this global circulation system, there are two hemisphere-wide phenomena that also affect climate and weather patterns on a more regional scale. There is a lot of energy involved with these two systems, which help to drive changes in atmospheric pressure, terrestrial winds, stratospheric winds, ocean temperatures, and land temperatures.

First, there is the media darling named **EL NIÑO/SOUTHERN OSCILLATION CYCLE**, also known as **ENSO**. The ENSO cycle strongly affects climates and weather patterns for extended periods of time from Indonesia to the west coast of North and South America. The ENSO cycle produces a change in water temperature, wind movement, and weather patterns in the equatorial Pacific Ocean. El Niño events are characterized by warmer than normal ocean temperatures in this region. El Niño's chilly little sister, **LA NIÑA**, is characterized by cooler than normal ocean temperatures along this same stretch of the equatorial Pacific.

The surface waters moving West to East in the Pacific Ocean are strongly driven by **TRADE WINDS**. The trade winds are produced by a hemisphere wide pattern of high pressure in the Eastern Pacific and a low pressure in the Western Pacific. The change in pressure measured over a given distance is called the **PRESSURE GRADIENT** and the result of this change from high to low pressure produces a force called the **PRESSURE GRADIENT FORCE**. It is this force which drives the trade winds and strongly drives ocean currents in the Pacific. As the pressure gradient is reduced by a shorter distance between high and low pressures, the pressure gradient force is reduced, the trade winds are reduced, and the movement of water in this part of the world is reduced.

As these ocean currents lose their boost, the sea level flattens out, allowing more warm water to move eastward along the surface of the ocean. Upwelling of cooler, deep water is also reduced, due to a lower **THERMOCLINE** layer, the transitional layer between the surface water and the deep water. As the surface waters drift across the Pacific Ocean, they stay exposed to solar radiation for a longer time, which increases the average surface temperature of the water and creates a warmer than normal ocean temperature.

A larger amount of warmer surface water flowing from West to East has several consequences for marine life and human settlements. For marine life, it means living in more nutrient-poor water, because nutrient-rich deep water upwelling is reduced. Coral reef bleaching and death also happen, as the warmer waters of El Niño verge on the maximum tolerance of temperature in which coral reefs can survive.

°GLOBAL WARMING MAY WORSEN EL NIÑO EFFECTS
ENVIRONMENTAL NEWS NETWORK
23 FEBRUARY 1999

El Niño has exhibited some peculiar behavior over the last 20 years, leading one meteorologist to hypothesize that global warming is exacerbating the effects of the weather phenomenon.

Kevin Trenberth, a meteorologist at the National Center for Atmospheric Research, notes that El Niños have been unusually frequent since the mid-1970s. That same time period has seen a dramatic rise in global temperatures, suggesting a correlation between global warming and the El Niño Southern Oscillation.

The global mean temperature peaks three to four months after the peak in El Niño, said Trenberth. "It is no coincidence that the exceptional warmth in the first seven months of 1998 occurred as the Pacific Ocean lost heat following the peak of the 1997-98 El Niño in December 1997," he said.

...Trenberth theorizes that much of the additional heat trapped by increasing amounts of greenhouse gases may be going into the oceans. It is later released through El Niños that are larger, more frequent, or less efficient in releasing the ocean-stored heat...

°CYCLE LOCK CAUSES QUICK CHANGE
NATURE NEWS SERVICE
8 JUNE 2001

The processes that produce El Niño events may also cause abrupt shifts in global climate, according to new research. This could explain why the thaw after the last ice age was interrupted by a frigid spell lasting hundreds of years.

Amy Clement and colleagues of the Lamont-Doherty Earth Observatory in Palisades, New York, are challenging conventional wisdom about the reasons for the Younger Dryas event, a return to near-ice-age conditions that took place about 11,500 years ago. Clement's group provide a new explanation for the sudden shifts in temperature that can occur across the globe.

...The changeover from normal to locked ENSO behaviour happens in a matter of decades. Very few climate phenomena are known to be able to produce such a big change so quickly. The current ENSO cycle seems to be in a phase close to that in which these sudden switches can occur...

El Niño events strongly alter the climate systems and disrupt the regular and predictable weather patterns across the Pacific Ocean. Rains are brought to the usually barren South American deserts. Hurricanes and typhoon events occur more frequently in Polynesia and other Pacific Island states. Drought conditions occur in Australia and Indonesia which can lead to an increase in wildfires. The risk of torrential flooding increases in Central America, which can bring the disease carriers that reside in stagnant waters.

Such weather usually leads to economic and social disruptions by reducing farm yields, disrupting fish migrations and harvesting, destroying property, and producing disease outbreaks. The El Niño event of 1997-98 caused an estimated 24,000 deaths and $3.5 billion in damages in Peru alone; $34 billion in damages worldwide.

Since modern measurements began, El Niño events have been recorded on a regular cycle, appearing every 3-7 years. The most recent El Niño events have occurred in 1986-1987, 1991-1992, 1993, 1994, 1997-1998, 2002-2003, and 2004-2005. But, as the oceans warm on average as a result of an enhanced greenhouse effect, El Niño events are predicted to become more common and bring with them more intense, frequent, and prolonged severe weather events.

In the past, the ENSO cycle has shut down for centuries and helped plunge the planet into a deep freeze. Should the ocean-atmospheric system that produces the ENSO cycle breach a threshold of temperature and stay locked in a permanent El Niño phase, extreme weather events from Indonesia to America will become the norm, rather than the exception. If this were to happen, there would not only be the concern of extreme weather events, but the concern that Melbourne burns during a 100 year drought. Should the ENSO cycle shut down, which it certainly could if ocean currents are significantly disrupted, there would be a whole new set of climate change concerns.

That is one temperamental boy-child.

°NIGHT TEMPERATURES SOAR
AUSTRALIAN ASSOCIATED PRESS
22 FEBRUARY 2005

Global climate change has pushed night temperatures almost four degrees above average in eastern Australia and triggered a wave of heat records, a scientist says.

Griffith University environmental scientist Associate Professor Clyde Wild said recorded temperatures as high as 3.6 degrees above night averages, were part of an overall temperature rise of 0.7 degrees in eastern Australia over the last 35 years.

Prof Wild said temperature records were now breaking at a rate of almost one a month with most of eastern Australia experiencing similar conditions.

..."It is a remarkable increase in temperature, quite dramatic."

Prof Wild said changes in climate were also leading to a "persistent" el Niño period characterised by droughts and bushfires in Australia.

He said deep el Niño could cost Australia $5 billion in lost agricultural production in 12 months. The alternative, la Niña brought good rain and cyclones, but they were few and far between, Prof Wild said.

"We used to get la Niña conditions reliably but since 1974 they have been quite rare and when they do occur are shortlived and relatively mild," he said.

Both the North Pole and the South Pole have an oscillating atmospheric-land-ocean phenomenon which help drive climates and weather patterns in these areas. The system in the North is called the **NORTH ATLANTIC OSCILLATION** or **NAO**, which is thought to be a regional manifestation of what is known as the **ARCTIC OSCILLATION** or **AO**, and both are sometimes referred to in the context of what is called the **NORTHERN ANNULAR MODE** or **NAM**. To avoid confusion, I will refer to this northern hemispheric driver of climate as the **NAO-AO**.

Besides ocean circulation, the NAO-AO is perhaps the largest determiner of weather patterns across the northern hemisphere, making its presence known from North America, to Europe, to Asia. It is also one of the oldest known phenomena to be documented by our ancestors. In the late 1770's, the Vikings noted that when severely cold winters struck Greenland, milder winters persisted in Denmark and vice versa.

The NAO-AO is sensitive to activities happening from the bottom of the ocean, through changes in the land and sea ice, all the way up to changes in the amount of ice particles in the stratosphere, and all of the gas, water, land, ice, wind, and water vapor in between.

The NAO-AO is called an oscillation because it alternates between two operating modes. These modes can change on an annual cycle, as well as frequent weekly and monthly flip-flops. Depending on the power of jet streams, pressure gradients, stratospheric temperature, and surface temperatures, the NAO-AO operates in either a **POSITIVE MODE** or **NEGATIVE MODE**. The Positive Mode is characterized by a strong Icelandic low pressure and strong subtropical high pressure. The Negative Mode is characterized by a weaker Icelandic low pressure and a weaker subtropical high pressure.

A **POLAR VORTEX** resides in a ring of air extending from the lower stratosphere to the surface of the Earth at both poles as well. In the North, this helps drive westerly winds in the North Atlantic, influences weather patterns in Europe and Asia, and influences ozone layer depletion.

Under the Positive Mode, the pressure gradient force between the Icelandic low pressure and the subtropical high pressure is strong, which produces a strong polar vortex, strong westerly winds, and keeps cold polar air to the North. This allows the warm Pacific Jet Stream air to flow across the globe offering both the eastern coast of North America and Europe a warmer and wetter winter, but also a more stormy one. Northern Canada and Greenland experience dryer and colder conditions as the cold polar air stays locked up north.

Under the Negative Mode, a weaker pressure gradient force produces weaker trade

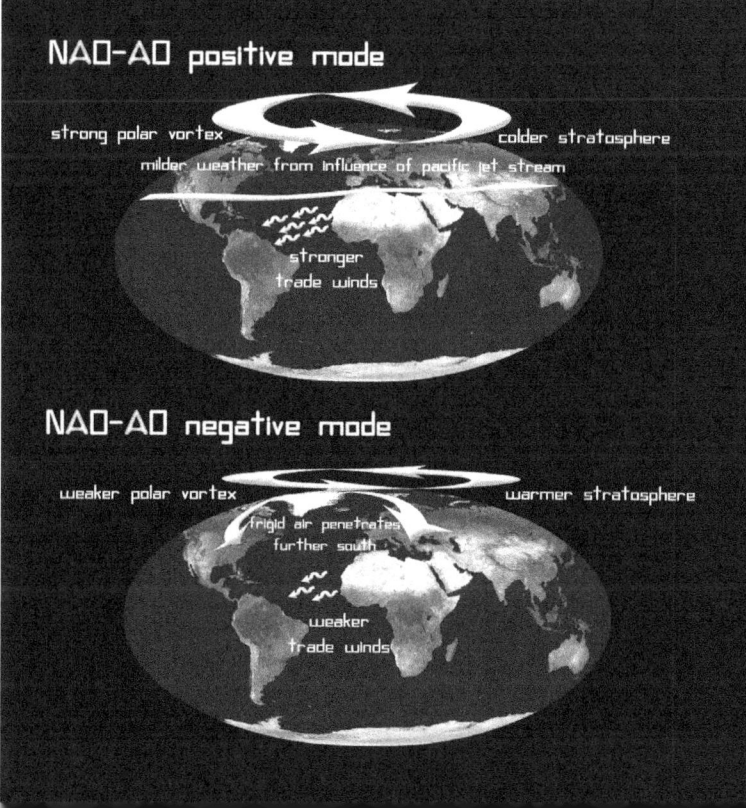

NAO-AO positive mode

strong polar vortex   colder stratosphere

milder weather from influence of pacific jet stream

stronger trade winds

NAO-AO negative mode

weaker polar vortex   warmer stratosphere

frigid air penetrates further south

weaker trade winds

°THOUSANDS STILL WITHOUT POWER IN EASTERN CANADA
REUTERS
15 NOVEMBER 2004

Tens of thousands of people in Eastern Canada could be without electricity until the end of the week, officials said on Monday, after an early winter blizzard downed power lines and crumpled transmission towers.

Up to 50 cm (20 inches) of heavy, wet snow blanketed Nova Scotia over the weekend, leaving more than 100,000 homes and businesses without power and temporarily shutting down Halifax International Airport.

Helicopters were sent out on Monday to inspect thousands of kilometers of transmission line in remote areas. But current weather conditions are hampering those efforts, said Christine Cragg, a spokeswoman for Nova Scotia Power, which supplies most of the electricity in the province...

°VIOLENT STORM CAUSES CHAOS ACROSS EUROPE
AGENCE FRANCE-PRESSE
20 NOVEMBER 2004

A storm accompanied by violent gusts of wind, heavy snowfalls and chilling temperatures has knifed into the centre of Europe after earlier causing major disruption in Scandinavia and Poland.

Winds gusting at up to 180 kilometres an hour have been recorded at Wendelstein in Bavaria. Fallen trees disrupted traffic in several regions, including Stuttgart in the southeast. Heavy snow fell on Lower Saxony, obstructing traffic and cutting of domestic electricity supplies.

...Road and rail services were seriously affected and hundreds of trees were uprooted. The Czech-German border at Cinovec/Altenburg was closed for several hours due to heavy snowfalls.

...Earlier, at least seven people, including a six-month-old child, were killed in gales in Poland that widely disrupted road traffic and left tens of thousands of homes without electricity.

The massive storm also swept across Scandinavia disrupting land, sea and air traffic.

°THE NORTH ATLANTIC OSCILLATION AND GREENHOUSE-GAS FORCING
GEOPHYSICAL RESEARCH LETTERS
11 FEBRUARY 2005

...The observed temporal trend in the NAO in recent decades lies beyond the natural variability found in the model control runs. For the majority of the models, there is a significant increase in the NAO trend in the forced runs relative to the control runs, suggesting that the NAO may intensify with further increases in greenhouse-gas concentrations.

winds and a weaker polar vortex, which allows colder air to penetrate into lower latitudes. This brings colder and less stormy weather to the eastern coast of North America and Europe. In Greenland and northern Canada, temperatures are milder and the climate becomes wetter.

Except for the period between 1900-1930, the NAO-AO has see-sawed back and forth between Positive and Negative Modes on a fairly regular basis, never crossing what would be considered 'normal' variation. Then after 1970, the NAO-AO began to experience prolonged periods in the Positive Mode, which meant a colder stratosphere, stronger westerly winds, warmer European winters and a warmer Arctic Ocean.

This suggests that as the average global surface temperature rises, the NAO-AO operates more frequently in the Positive Mode as a result of a positive feedback perpetuating a warming of the surface and cooling of the stratosphere. As the NAO-AO stays in a Positive Mode there are other phenomena to take notice of:

COLDER STRATOSPHERE AND STRONGER POLAR VORTEX
=
INCREASED POSSIBILITY FOR ACCELERATED RATE OF OZONE LAYER DEPLETION AT NORTH POLE

STRONGER WESTERLY WINDS AND WARMER WINTERS
=
INCREASED OCCURRENCE OF SEVERE WEATHER EVENTS ACROSS THE NORTHERN HEMISPHERE

WARMER ARCTIC OCEAN
=
INCREASED RATE OF SEA ICE THINNING AND MELT

INCREASED RATE OF SEA ICE THINNING AND MELT
=
LARGER AMOUNT OF FRESHWATER ENTERING ARCTIC OCEAN AND INCREASED POSSIBILITY OF OCEAN SYSTEM DISRUPTION

Whether or not the NAO-AO will begin a more regular oscillation between Positive and Negative Modes is yet to be seen. Whether or not the NAO-AO is more frequently in the Positive Mode as a result of the measured trend of global warming is also still under examination. But as more data comes in, it becomes more clear that the more frequent NAO-AO Positive Mode of the past few decades goes beyond natural variability.

As global warming continues, the NAO-AO is projected to persist in a Positive Mode for years to come. As the NAO-AO stays in a Positive Mode, the average temperature of the northern hemisphere will continue to rise and extreme weather events will become more frequent in the parts of the world affected by the NAO-AO.

It is possible that the average global surface temperature will reach a level where a return to a Negative Mode is not possible for many years. It is also possible that a persistent Positive Mode would aid in melting enough Arctic ice to disrupt ocean systems, at which point the NAO-AO could shift into a persistent Negative Mode, allowing a lot of frigid air to penetrate the Northern Hemisphere and helping to plunge it into a deep freeze.

In Antarctica, the oscillating positive and negative mode is called the **ANTARCTIC OSCILLATION** or **SOUTHERN ANNULAR MODE (SAM)** and it functions like the NAO-AO. The SAM does not garner as much attention as its northern counterpart, because there are no people or settlements directly affected by it. However, it has been receiving attention lately for its connection with ozone layer dynamics.

Just like over the North Pole, a polar vortex swirls in the stratosphere above the South Pole. Scientists believe that as the ozone layer over the Southern Hemisphere recovers, it will promote a Negative Mode and warmer stratosphere. This regional warming along with an enhanced greenhouse effect would make the average temperature over Antarctica increase rapidly. A rapidly warming Antarctica would have big consequences for people and settlements far, far away.

Choices, choices, choices.

An ozone hole or a warming continent with enough water to bury five miles of coastline throughout the entire world? We just cannot win, can we.

Time to set sail on the rivers of the ocean.

Don't forget to pack the life preserver.

°**THE ARCTIC OSCILLATION: A KEY TO THIS WINTER'S COLD -- AND A WARMER PLANET**
NATIONAL CENTER FOR ATMOSPHERIC RESEARCH PRESS RELEASE
25 FEBRUARY 2004

Why has the Arctic warmed so dramatically in recent years? How does the Arctic's circulation keep frigid air over the poles and sometimes allow it to spill across the United States? And how might global change affect the behavior of this circulation?

...Although the AO has been recognized by various names for many years, it has become a topic of keen research interest only in the last decade. The AO index describes the relative intensity of a semipermanent low-pressure center over the North Pole.

A band of upper-level winds circulates around this center, forming a vortex. When the AO index is positive and the vortex intense, the winds tighten like a noose around the North Pole, locking cold air in place.

A negative AO and weak vortex (as has been the case most of this winter) allow intrusions of cold air to plunge southward into North America, Europe, and Asia. Apart from variations like this winter's, the index has been mostly positive in wintertime since the late 1980s.

"The Arctic Oscillation has strengthened in recent decades, contributing to the unusual warmth over the Northern Hemisphere land masses," says (Clara) Deser. "This has led to the speculation that the AO may be affected by increasing greenhouse gas concentrations and that it may, in turn, enhance global warming."...

SUPERFLOWES

Ocean circulation.
The great conveyor belt of water.
The Mothership of climate systems and weather patterns.
The smoking gun of abrupt global temperature shifts.
The heartbeat of the planet.

Although atmospheric currents, land use, volcanism, and regional climate dynamics play a roll in weather production, the flow of the oceans strongly drive the planet's climate regimes that produce the consistent and predictable weather patterns we all depend on for a life tomorrow like we had today.

Regularities and predictabilities in weather patterns allow us to plan our lives, they always have. So, when nature suddenly offers us something else, it can ruin our day. A lot of them.

The oceans are moved by many things including solar influences, wind pressures and systems, the rotation of the planet, the tug of the moon, and primarily by the sinking of cold and heavy saltwater. Of the planet's ocean water, 90% of it moves deep under the surface of the ocean, the other 10% floats at the surface.

The movement of cold, dense, salty, deep water under warmer, lighter, less salty surface water is called THERMOHALINE CIRCULATION or THC.

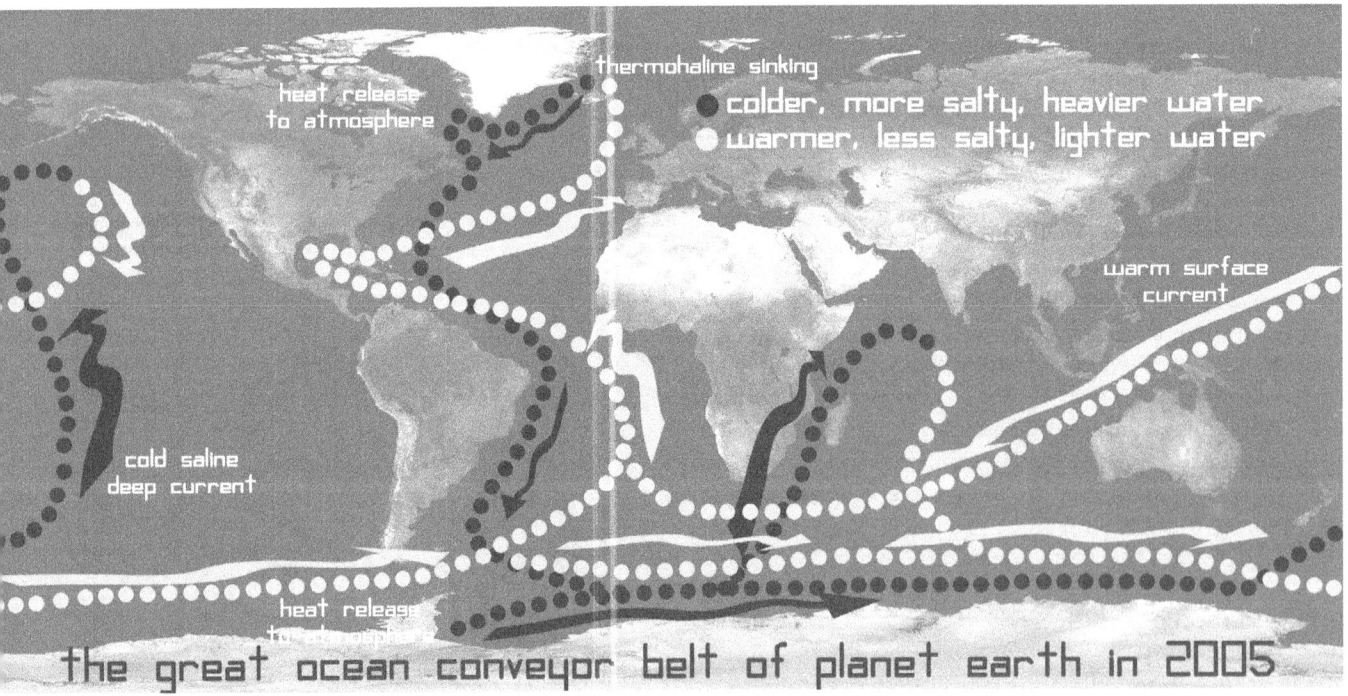

the great ocean conveyor belt of planet earth in 2005

Thermohaline refers to the movement of water with different temperatures and different salinity. In certain places, cold, salty and dense water sinks and gives the deep current a powerful thrust that makes the entire ocean conveyor run. In other places, warmer, less salty and less dense water rises and releases tremendous amounts of heat energy.

One of the critical currents for our settlements in the present ocean system is called the GULF STREAM/NORTH ATLANTIC DRIFT. This current brings warm water from the equatorial region of the Atlantic Ocean northeast across Europe. The people in Britain, western Europe and Scandinavia enjoy a temperate climate 9° Celsius warmer than their Canadian counterparts at the same latitudes. The climates in this part of the world are partially produced by the Gulf Stream, which releases heat energy as it reaches the North Atlantic Ocean near Great Britain, and partially produced by the Pacific jet stream and other atmospheric currents which travel the planet as part of the NAO-AO.

The Little Ice Age period that occurred between the 17th and 19th century was a minor cooling event believed to be largely caused by an enhanced Negative Mode of the NAO-AO. To put it into perspective, the Little Ice Age dropped average temperatures across Eurasia .6°-.9° Celsius compared with present-day conditions. Back then, winters were cold enough to go ice skating on the Thames River in London.

Were the Gulf Stream disrupted today, temperatures on the British Isles, Scandinavia, and western Europe would plummet an average of 5° Celsius. Not only would ice skating once again become common on the Thames, one might be able to do it during the summer months.

Although the Gulf Stream seems very stable and constant, it really is not. It shifts in size, temperature, and location on a decadal basis, on an annual basis, and sometimes on an hourly basis. It is an amoebic monster which is very sensitive to the perturbations of surrounding waters, especially those at the Arctic Circle. And the Arctic Circle has recently seen some pretty significant changes, would you not agree?

The crux of the biscuit for thermohaline circulation is located in the Labrador and Greenland seas: one thermohaline pump is located between the northeast corner of Canada and southwest Greenland, the other is located between Iceland, the tip of Great Britain, and western Scandinavia.

This is the home of the SUBPOLAR GYRE, where waters flowing North lose the rest of their heat, mix with cold Arctic waters, become saltier and heavier, and plunge beneath surface waters to begin their journey southward.

Many scientists will tell you this area in the Arctic Circle is the Achilles' Heel of the ocean's circulation system and the main power source behind the global ocean conveyor belt. North Atlantic thermohaline sinking is the pump that pushes the conveyor belt along, eventually pushing the Gulf Stream northward again.

For this reason, the amount of freshwater flowing in from glacier melt around Greenland is of great concern. The concern being that an accelerated influx of freshwater will dilute salinity levels and significantly weaken the thermohaline pumps, disrupting global climate systems and

°LAMONT'S BROECKER WARNS GASES COULD ALTER CLIMATE: OCEANS' CIRCULATION COULD COLLAPSE COLUMBIA UNIVERSITY RECORD 5 DECEMBER 1997

On the eve of the international meeting on global warming that opened Dec. 1 in Kyoto, Japan, one of the world's leading climate experts warned of an underestimated threat posed by the buildup of greenhouse gases—an abrupt collapse of the oceans' prevailing circulation system that could send temperatures across Europe plummeting in a span of 10 years.

If that system shut down today, winter temperatures in the North Atlantic region would fall by 20 or more degrees Fahrenheit within 10 years. Dublin would acquire the climate of Spitsbergen, 600 miles North of the Arctic Circle.

"The consequences could be devastating," said Wallace S. Broecker, Newberry Professor of Earth and Environmental Sciences at Columbia's Lamont-Doherty Earth Observatory, and author of the new research, which appeared in the Nov. 28 issue of the magazine *Science*.

The Conveyor "is the Achilles heel of the climate system," Broecker wrote in *Science*. "The record ... indicates that this current has not run steadily, but jumped from one mode of operation to another. The changes in climate associated with these jumps have now been shown to be large, abrupt and global."...

subpolar gyre and thermohaline pumps

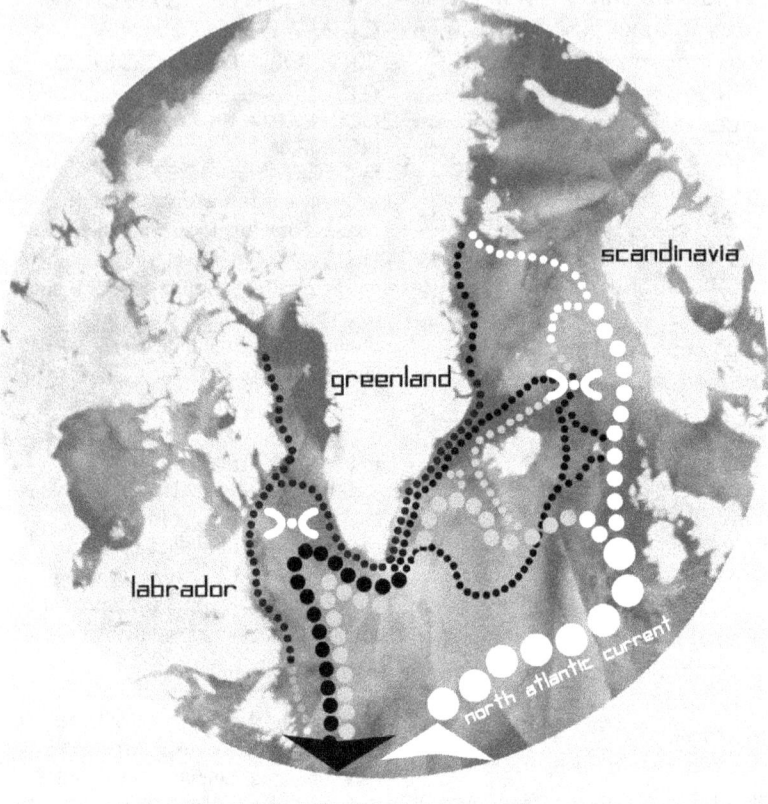

scandinavia

greenland

labrador

north atlantic current

⬤ warm gulf stream waters flowing north

⬤ fresh cold water flow at surface

⬤ cold, deep, dense water flowing south

thermohaline circulation pumps and underturning

size of ⬤ = volume of water

weather patterns for those living in this part of the in the Northern Hemisphere.

The history of thermohaline circulation weakening as a result of freshwater influx is well documented. Dramatic and abrupt temperature changes in the Northern Hemisphere have been a common occurrence throughout the history of the Earth, because the North Pole is especially sensitive to changes in global temperature, changes in freshwater flow entering the North Atlantic, and is a cornerstone for stable ocean circulation. At least during the past 100,000 years.

Many studies have shown that, in the past, the smallest increases in freshwater discharge into this sensitive area have weakened the thermohaline pumps enough to stop the movement of the Gulf Stream. It is believed that this could happen again in the near future, as measurements are now recording a rapid increase in the amount of freshwater now entering this area.

Many studies also show a historic relationship between salinity levels in the North Atlantic and salinity levels in the Tropical Atlantic, and record that salinity levels in the Tropical Atlantic are rising and North Atlantic salinity levels are plummeting, suggesting reduced circulation and a weakening of thermohaline pumping. However, other studies have shown that increased evaporation from an enhanced greenhouse effect might increase salinity levels in the Northern Hemisphere, which may offset the dilution caused by freshwater.

Along with a lot of water of different salinity levels and temperature coursing through the ocean, there is a tremendous amount of heat energy circulating and being transferred as well. Heat energy is transferred between deep water and surface water, as well as between surface water and the lower atmosphere that can be seen in phenomena like the El Niño/Southern Oscillation and the North Atlantic Oscillation.

Tropical waters transport tropical heat, Antarctic waters create an isolated, frozen island climate, and North Atlantic waters keep the conveyor belt pumping along.

There is enough heat energy in the ocean's pinky finger, the Gulf Stream, to keep western Europe from freezing.

Since the changing level of greenhouse gases in the atmosphere so closely corresponds with a changing average global surface temperature, most scientists agree that a rise in greenhouse gas levels will enhance the greenhouse effect and cause a change in the average global surface temperature.

But the planet's thermotaxis is not that straightforward.

Abrupt hemispheric temperature shifts have not always been in sync with one another, nor have they always been global in reach. Even today we see that some parts of Antarctica are experiencing a cooling trend, while many parts of the North are experiencing a dramatic warming trend. For this reason, more and more scientists are beginning to agree that a possible reason for swift and yet differing hemisphere-wide changes in temperature can only be explained by a disruption in ocean circulation and **OCEAN-HEAT TRANSFER**. These scientists do not rule out atmospheric greenhouse gas levels as a cause for temperature shifts, but rather integrate ocean-heat transfer into the twisted fabric of atmospheric greenhouse gas levels, the greenhouse effect, thermohaline circulation, and the average global surface temperature.

Although past changes in the Northern Hemisphere may not have been global in reach, the fact remains that the North Atlantic's powerful climate-driving thermohaline pumps are very sensitive to the slightest of changes at the North Pole. When a threshold is breached in the North Atlantic and the thermohaline pumps shut down, the results are as swift as falling dominos. When the waters in the North Atlantic again reach the density and salinity needed to sink, the pumps return just as quickly and circulation resumes. However, the time in between the pumps shutting down and starting back up again can be a very cold thousand years.

The temperature in Greenland changed sharply 84,000 years ago; about 83,000 years ago; about 75,000 year ago; 74,000 years ago; 70,000 years ago; 68,000 years ago; 62,000 years ago; 59,000 years ago; 58,000 years ago; 57,000 years ago; 55,000 years ago; 54,000 years ago; 53,000 years ago; 52,000 years ago; 51,000 years ago; 50,000 years ago; 42,000 years ago; 41,000 years ago; 40,000 years ago; 39,000 years ago; 38,000 years ago; 37,000 years ago; 36,000 years ago; 35,000 years ago; 34,000 years ago; 33,000 years ago; 32,000 years ago; 31,000 years ago; 30,000 years ago; 17,000 years ago; 15,000 years ago...up, down, up, down, up, down, up, down, up, down, up, down, up, down, up, down, up, down, up, down, up, down, up, down, up, down, up, down, up, down, up, down, up, down, up, down, up............

it gets a little freaky when the sea level rises a few millimeters a year

even freakier should the sea level rise a few feet during the 21st century

crazy freaky should the sea level rise a few meters during the 21st century

and it gets superfreaky should the fresh water flowing into the Arctic tip the delicate balance of ocean circulation

get ready for a shakedown

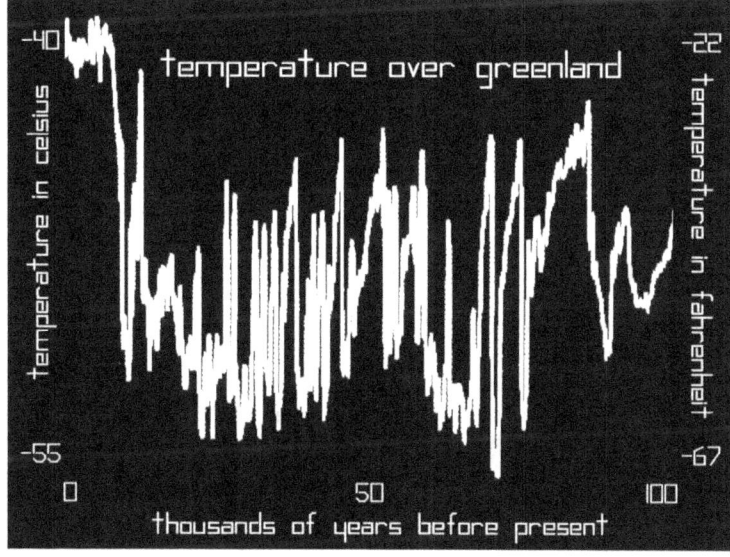

temperature over greenland

temperature in celsius

temperature in fahrenheit

thousands of years before present

°UC STUDY SHEDS NEW LIGHT ON CLIMATE-CHANGE PROCESSES
UNIVERSITY OF CALIFORNIA-DAVIS PRESS RELEASE
10 MARCH 2004

A new study from the University of California shows, for the first time, that the deep-ocean circulation system of the north Atlantic, which controls ice-age cycles of cold and warm periods in the Northern Hemisphere, is integrally coupled to salinity levels in the Caribbean Sea. This research reinforces concerns that global warming, by melting the glacial ice of Greenland, could quickly and profoundly change salinity and temperatures in the North Atlantic Ocean.

The study is published by the journal *Nature* in its online edition today (March 10) and its print edition tomorrow. The authors are graduate student Matthew Schmidt and geology professor Howard Spero of UC Davis, and geology professor David Lea of UC Santa Barbara.

...The authors hypothesize that elevated Caribbean salinity, which is transported via the Gulf Stream to the north Atlantic, amplifies the heat transport system by increasing the deep-ocean circulation rate. When the North Atlantic cools, Caribbean salinity builds up because the deep ocean circulation drops to a fraction of its previous rate and the Gulf Stream no longer transports salty water away.

...A paper that refers to similar events occurring today was recently published in the journal *Nature* by an international group of scientists led by Ruth Curry of Woods Hole Oceanographic Institute in Massachusetts. Curry's study showed that in the past 50 years, salinity in the tropical Atlantic Ocean has risen and salinity in the north Atlantic has fallen. That is the pattern that Schmidt, Spero and Lea say has led to colder climates in the past. "We present a historical analog to the salinity buildup that is observed in the tropics today by Dr. Curry and others," said Spero and Lea.

°ABRUPT CLIMATE CHANGE: NEW RESEARCH SUPPORTS HYPOTHESIS THAT OCEAN CURRENTS REDISTRIBUTED HEAT DURING RAPID WARMING AND COOLING
GEORGIA INSTITUTE OF TECHNOLOGY PRESS RELEASE
24 JUNE 2004

A paper published this week in the journal *Science* supports the hypothesis that heat transfer by ocean currents – rather than global heating or cooling – may have been responsible for the global temperature patterns associated with the abrupt climate changes seen in the North Atlantic during the past 80,000 years.

Authored by the University of Bremen's Frank Lamy and colleagues, the paper provides new evidence that Southern Hemisphere climate may not have changed in step with Northern Hemisphere climate. Though these new measurements of ocean surface temperature off Chile are consistent with information from Antarctic ice core samples, they still contradict measurements made on land in the Southern Hemisphere – suggesting additional research will be needed to resolve the issue.

Scientists have found evidence of rapid and dramatic climate change that took place in a matter of decades during cool periods of the last 80,000 years in the North Atlantic. Knowing whether climate changes took place simultaneously in the Northern and Southern Hemispheres is vital to understanding the mechanism involved – and assessing whether similar abrupt climate change could be a threat today.

"People are very interested in these dramatic climate changes because they occur on very human time scales," said Jean Lynch-Stieglitz, associate professor in the School of Earth and Atmospheric Sciences at the Georgia Institute of Technology and author of a "Perspectives" article accompanying the Lamy paper in *Science*. "It's really important to understand what is causing them and what conditions are necessary for the climate to rapidly transition from cold to warm and back again."

..."The real significance of this paper is that it gets us closer to understanding the mechanism causing these rapid climate changes," she said. "Earlier sediment core work at lower resolution has suggested that the Southern Hemisphere has been doing its own thing. The record from Antarctica is nicely resolved and shows that the Southern Hemisphere is not participating either in magnitude or timing with the climate changes that have occurred in the North Atlantic."

...Both hemispheres warming and cooling at the same time would imply global changes caused by rising levels of greenhouse gases. But one hemisphere cooling while the other warmed would suggest simple heat transfer, accomplished by changes in ocean or atmospheric currents.

"You can make the climate cool in certain places just by redistributing the heat through changes in ocean currents, atmospheric circulation or both," said Lynch-Stieglitz. "The most fully developed theory to account for these rapid climate changes is that they do represent changes in the transport of heat into the North Atlantic by what we call overturning circulation of the ocean."

In that scenario, warm water flows northward from the Southern Hemisphere into the North Atlantic, where it gives up its heat. Being denser, the cooled water then sinks and flows back south. The scenario accounts for both heating in the north and cooling in the south.

It's possible, Lynch-Stieglitz notes, that both global warming and changes in ocean heat transport occurred simultaneously, though records of carbon dioxide concentrations do not show concentration increases that would be enough by themselves to account for the climate change...

The period between 12,900 years ago and 11,500 years ago is called the YOUNGER DRYAS and was a time when the temperature of Greenland dropped between 7°-10° Celsius within a few decades. During this time, it is believed an influx of freshwater from a rapidly deglaciating North America weakened the thermohaline pumps and brought the Gulf Stream to a standstill, causing a rapid drop in the average surface temperature of the Northern Hemisphere. This cool period lasted about 1,300 years and was followed by just as abrupt of a warming.

An event in which the temperature changes 7°-10° Celsius within 30 years is difficult to comprehend, almost unfathomable. If we did not know it was a common occurrence, we probably would not believe it was possible. To put the temperature change in perspective, the last time the average global surface temperature was 7°-10° Celsius cooler than today, it was covered in ice. To put the rate and speed of change in perspective, the Northern Hemisphere went from being relatively temperate to being locked in ice in the amount of time it took for me to be born, grow, and write this book.

That is a really big and really fast change.

The settlements during the time of the Little Ice Age were sensitive to a small change in the average surface temperature that affected climate patterns and weather patterns significantly. Our settlements are no less sensitive to small changes, perhaps even more so. And all settlements are sensitive to big changes.

The predicted consequences of gradual global temperature change are severe. The predicted consequences of a swift hemispheric or global temperature change are difficult to imagine. The shutdown of the Gulf Stream and global ocean circulation disruption is what analysts call a HIGH-RISK, LOW-PROBABILITY event, although it is definitely possible, it is not likely to happen this century or for centuries to come.

But nobody is certain.

The critical threshold may be a decade away.

These are the forecasts for our planet at the beginning of the 21st century presented to us by scientific investigation. They are troubling to the world's business leaders, political leaders, scientists, public interest groups, and even military strategists.

All of our settlements are built and depend upon the climate systems and weather patterns produced by a strong circulation of ocean waters, strong thermohaline pumping in the North Atlantic, and the presence of a strong Gulf Stream.

As a result of freshwater influx from melting glaciers at the North Pole, thermohaline pumping has weakened and circulation has shut down in the past. No one can say when the Earth's ocean systems will breach a threshold in which the conveyor belt breaks down again.

It could take a lot less time and a lot less disturbing than we think.

As the North Pole continues to warm.

As more ice becomes liquid.

As more freshwater enters the ocean systems.

As salinity levels drop in the North Atlantic.

Why would the Earth not react as it has in the past?

dryas octopetala
A creeping evergreen shrub with large white flowers that is widely distributed in northern portions of Eurasia and North America.

abrupt climate changes at the north pole

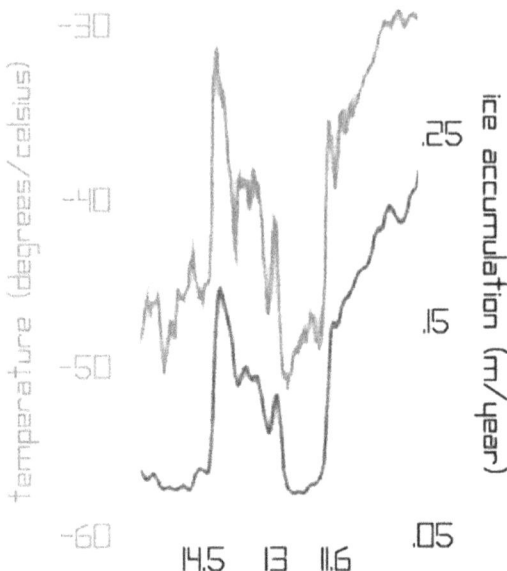

temperature (degrees/calsius)

-30
-40
-50
-60

ice accumulation (m/year)

.25
.15
.05

14.5   13   11.6

thousands of years before present

AS THE PLANET'S
HEARTBEAT GROWS WEAKER.

IT SEEMS HISTORY
IS BEGINNING TO
REPEAT ITSELF.

SEVEN TO
TEN DEGREES
CELSIUS IN
30 YEARS.

NO WONDER THE
WORLD IS WORRIED.

strong thc - good

weak thc - not so good

PICK A CARD...ANY CARD

°NATURE 401, 458-461
30 SEPTEMBER 1999

RAPID CHANGES IN THE MECHANISM OF OCEAN CONVECTION DURING THE LAST GLACIAL PERIOD

TROND M. DOKKEN AND EYSTEIN JANSEN

°NATURE 402, 511-514
2 DECEMBER 1999

WARMING OF THE TROPICAL ATLANTIC OCEAN AND SLOWDOWN OF THE THERMOHALINE CIRCULATION DURING THE LAST DEGLACIATION

CARSTEN RUEHLEMANN, ET AL

°NATURE 408, 567-570
30 NOVEMBER 2000

CHANGES IN DEEP-WATER FORMATION DURING THE YOUNGER DRYAS EVENT INFERRED FROM ^{10}BE AND ^{14}C RECORDS

RAIMUND MUSCHEIER, ET AL

°NATURE 409, 153-158
11 JANUARY 2001

RAPID CHANGES OF GLACIAL CLIMATE SIMULATED IN A COUPLED CLIMATE MODEL

ANDREY GANOPOLSKI AND STEFAN RAHMSTORF

°NATURE 400, 740-743
19 AUGUST 1999

RAPID ENVIRONMENTAL CHANGE IN SOUTHERN EUROPE DURING THE LAST GLACIAL PERIOD

JUDY R. M. ALLEN, ET AL

°NATURE 402, 644-648
9 DECEMBER 1999

WEAKER GULF STREAM IN THE FLORIDA STRAITS DURING THE LAST GLACIAL MAXIMUM

JEAN LYNCH-STIEGLITZ, ET AL

°NATURE 409, 171-175
11 JANUARY 2001

AN ABRUPT CLIMATE EVENT IN A COUPLED OCEAN-ATMOSPHERE SIMULATION WITHOUT EXTERNAL FORCING

ALEX HALL AND RONALD J. STOUFFER

°NATURE 428, 834-837
22 APRIL 2004

COLLAPSE AND RAPID RESUMPTION OF ATLANTIC MERIDIONAL CIRCULATION LINKED TO DEGLACIAL CLIMATE CHANGES

J.F. MCMANUS, ET AL

PICK ONE

°RESEARCHERS FEAR
DECLINE IN SEA ICE IS
CHANGING CLIMATES
SCRIPPS HOWARD NEWS
SERVICE
24 OCTOBER 2003

NASA scientists released new evidence yesterday that the Arctic region is warming up and its sea ice cover is diminishing, with implications for further climate change worldwide.

Satellite data show that compared with the 1980s, surface temperatures across most of the Arctic warmed significantly in the last decade, with the biggest temperature increases occurring over North America.

When compared with ground-based surface temperatures, the rate of warming in the Arctic between 1981 and 2001 was eight times the rate of warming over the last 100 years, said Josefino Comiso, a senior research scientist at NASA's Goddard Space Flight Center who compiled the data. "The Arctic is in the process of being transformed," he said.

"Previously, similar studies used data from very few points scattered in various parts of the Arctic region," said Comiso, whose work will be published Nov. 1 in the American Meteorological Society's *Journal of Climate*. "These results show the large spatial variability that only satellite data can provide."

...Another NASA-funded researcher, Mark Serreze of the University of Colorado-Boulder, reported yesterday that the extent of Arctic summer sea ice in 2002 reached the lowest level ever recorded by satellites.

"It appears that the summer of 2003, if it does not set a new record, will be very close to the levels of last year," Serreze said.

"How much of this warming is due to natural fluctuations and how much is caused by human activity, we don't really know. But the fact is, the climate is changing, and in the Arctic it is changing rapidly."...

°SATELLITES RECORD WEAKENING NORTH
ATLANTIC CURRENT
NASA-GSFC NEWS RELEASE
15 APRIL 2004

A North Atlantic Ocean circulation system weakened considerably in the late 1990s, compared to the 1970s and 1980s, according to a NASA study. Sirpa Hakkinen, lead author and researcher at NASA's Goddard Space Flight Center, Greenbelt, Md. and co-author Peter Rhines, an oceanographer at the University of Washington, Seattle, believe slowing of this ocean current is an indication of dramatic changes in the North Atlantic Ocean climate.

...The current, known as the sub polar gyre, has weakened in the past in connection with certain phases of a large-scale atmospheric pressure system known as the North Atlantic Oscillation (NAO). But the NAO has switched phases twice in the 1990s, while the subpolar gyre current has continued to weaken. Whether the trend is part of a natural cycle or the result of other factors related to global warming is unknown.

"It is a signal of large climate variability in the high latitudes," Hakkinen said. "If this trend continues, it could indicate reorganization of the ocean climate system, perhaps with changes in the whole climate system, but we need another good five to 10 years to say something like that is happening." Rhines said, "The subpolar zone of the Earth is a key site for studying the climate. It's like Grand Central Station there, as many of the major ocean water masses pass through from the Arctic and from warmer latitudes...

°ARCTIC RIVERS DISCHARGE MORE FRESHWATER
INTO OCEAN, REFLECTING CHANGES TO
HYDROLOGIC CYCLE
AMERICAN GEOPHYSICAL UNION PRESS RELEASE
19 JANUARY 2005

Far northern rivers are discharging increasing amounts of freshwater into the Arctic Ocean, due to intensified precipitation caused by global warming, say researchers at the Hadley Centre for Climate Prediction and Research in the United Kingdom.

Water exchange between the ocean, atmosphere, and land is called the global hydrological cycle. As Earth's climate warms, the rate of this exchange is expected to increase. As part of this process, high-latitude precipitation and, consequently, river runoffs are also expected to increase. This could change the distribution of water on Earth's surface, with important social and economic consequences.

It could also alter the balance of the climate system itself, such as the Atlantic thermohaline circulation, a kind of conveyor belt. Cold water flows southward in the Atlantic at great depths to the tropics, where it warms, rises, and returns northward near the surface.

...Researchers Peili Wu, Richard Wood, and Peter Stott of the Hadley Centre compared observational data reported in *Science* in 2002 by Peterson and others with model simulations, produced by Hadley, part of the United Kingdom's Met Office. Writing in the journal *Geophysical Research Letters* (21 January), they note that increased human-caused greenhouse gas emissions are expected to intensify the Arctic hydrologic cycle, that is, the cycle of water as it rains onto land and sea, runs off into rivers, and evaporates to continue the cycle.

...Seeking to determine the source of the upward trend of recent decades, the researchers asked first whether it could be the early part of the predicted increase in the global hydrological cycle, caused by global warming.

...They concluded that had there been no human inputs, the hydrological cycle would have shown no trend at all in the 20th century...

°RATE OF OCEAN CIRCULATION DIRECTLY LINKED TO ABRUPT CLIMATE CHANGE IN NORTH ATLANTIC REGION
WOODS HOLE OCEANOGRAPHIC INSTITUTION PRESS RELEASE
21 APRIL 2004

A new study strengthens evidence that the oceans and climate are linked in an intricate dance, and that rapid climate change may be related to how vigorously ocean currents transport heat from low to high latitudes. A new study, reported April 22 in the journal *Nature*, suggests that when the rate of the Atlantic Ocean's north-south overturning circulation slowed dramatically following an iceberg outburst during the last deglaciation, the climate in the North Atlantic region became colder. When the rate of the ocean's overturning circulation subsequently accelerated, the climate warmed abruptly.

Study author Jerry McManus and colleagues Roger Francois, Jeanne Gherardi, Lloyd Keigwin and Susan Brown-Leger at the Woods Hole Oceanographic Institution and in France report that the coldest interval of the last 20,000 years occurred when the overturning circulation collapsed following the discharge of icebergs into the North Atlantic 17,500 years ago. This regional climatic extreme began suddenly and lasted for two thousand years. Another cold snap 12,700 years ago lasting more than a thousand years and accompanied another slowdown of overturning circulation. Each of these two cold intervals was followed by a rapid acceleration of the overturning circulation and dramatically warmer climates over Northern Europe and the North Atlantic region.

...The research team found that the rate of ocean circulation varied remarkably following the last ice age, with strong reductions and abrupt reinvigorations closely tied to regional climate changes. McManus says this is the best demonstration to date of what many paleoclimatologists and ocean scientists have long suspected. "Strong overturning circulation leads to warm conditions in the North Atlantic region, and weak overturning circulation leads to cold conditions," he said. "We've known for some time from changes in the chemistry of the seawater itself that something was different about the ocean's circulation at times of rapid climate changes, and it now appears that the difference was related to changes in the rate of ocean circulation. One big question is why the circulation would collapse in the first place and possibly trigger abrupt climate change. We think it is the input of fresh water to the surface ocean at a particularly sensitive location."

°SHUTDOWN OF CIRCULATION PATTERN COULD BE DISASTROUS, RESEARCHERS SAY
UNIVERSITY OF ILLINOIS AT URBANA-CHAMPAIGN NEWS RELEASE
14 DECEMBER 2004

If global warming shuts down the thermohaline circulation in the North Atlantic Ocean, the result could be catastrophic climate change. The environmental effects, models indicate, depend upon whether the shutdown is reversible or irreversible.

"If the thermohaline shutdown is irreversible, we would have to work much harder to get it to restart," said Michael Schlesinger, a professor of atmospheric sciences at the University of Illinois at Urbana-Champaign and a co-author of a report presented at the American Geophysical Union meeting in San Francisco. "Not only would we have the very difficult task of removing carbon dioxide from the atmosphere, we also would have the virtually impossible task of removing fresh water from the North Atlantic Ocean."

...Paleoclimate records constructed from Greenland ice cores have revealed that the thermohaline circulation has, indeed, shut down in the past and caused regional climate change. As the vast ice sheet that covered much of North America during the last ice age finally receded, the meltwater flowed out the St. Lawrence and into the North Atlantic. "The additional fresh water made the ocean surface less dense and it stopped sinking, effectively shutting down the thermohaline circulation," Schlesinger said.

"As a result, Greenland cooled by about 7 degrees Celsius within several decades. When the meltwater stopped, the circulation pattern restarted, and Greenland warmed."

Since the system has previously shut down by itself, "it is not unlikely that it will do so again, especially with our help in pouring greenhouse gases into the atmosphere," Schlesinger said. "Higher temperatures due to global warming could add fresh water to the northern North Atlantic by increasing the precipitation and by melting nearby sea ice, mountain glaciers and the Greenland ice sheet. This influx of fresh water could reduce the surface salinity and density, leading to a shutdown of the thermohaline circulation

Schlesinger and his team simulated the potential effects with an uncoupled ocean general circulation model and with it coupled to an atmosphere general circulation model. They found that the thermohaline circulation shut down irreversibly in the uncoupled model simulation, but reversibly in the coupled model simulation.

"The different results occurred because of a crucial feedback mechanism that appeared only in the coupled model simulation," Schlesinger said. "Enhanced evaporation increased the salinity and density of the ocean surface, offsetting the effects of additional fresh water."

"The irreversible shutdown of the thermohaline circulation thus appears to be an artifact of the model, rather than a likely outcome of global warming," Schlesinger said. "But, because the possibility of an irreversible shutdown cannot be excluded, suitable policy options should continue to be explored. Doing nothing to abate global warming would be foolhardy if the thermohaline circulation shutdown is irreversible."...

WHAT DO YOU SEE?

°OCEAN DRIFT DISRUPTION MAY CHILL EUROPE
BBC NEWS ONLINE
25 NOVEMBER 1999

Scientists have found evidence that the Atlantic Ocean current which gives Europe its mild climate is being disrupted. If it stopped, then the temperatures in western Europe would plunge by five degrees Celsius, creating bitter winters. The culprit for the changes could, ironically, be global warming.

The current, called the North Atlantic Drift, brings warm water northwards from the Gulf Stream. It is being disrupted by a growing amount of freshwater entering the Arctic Ocean, reports New Scientist magazine.

This increase is a result of changes attributed to climate change and possibly global warming: melting ice, increased rainfall and changing wind patterns.

The North Atlantic Drift is part of a global conveyor belt that brings warm, surface water north from the Gulf of Mexico and sends cold, deep water back.

The belt is driven by two "pumps", one in the Greenland Sea and one in the Labrador Sea, where the surface water cools, becomes denser, sinks and then returns south.

...Svein Osterhus of the University of Bergen in Norway has also discovered that a deep-sea current closer to the Arctic has gone into reverse...

°GLOBAL WARMING COULD TRIGGER OCEAN CURRENT COLLAPSE
ENVIRONMENTAL NEWS NETWORK
9 DECEMBER 1997

The buildup of greenhouse gases could lead to an abrupt collapse of the ocean's prevailing circulation system sending temperatures across Europe plummeting in a span of 10 years.

If that system shut down today, winter temperatures in the North Atlantic region would fall by 20 or more degrees Fahrenheit within 10 years. Dublin would acquire the climate of Spitsbergen, 600 miles north of the Arctic Circle...

°MELTING GLACIERS DIMINISHED GULF STREAM, COOLED WESTERN EUROPE, DURING LAST ICE AGE
NATIONAL SCIENCE FOUNDATION PRESS RELEASE
19 NOVEMBER 2001

At the end of the last Ice Age --11.5 to 13 thousand years ago -- the north Atlantic deep water circulation system that drives the Gulf Stream may have shut down because of melting glaciers that added freshwater into the north Atlantic Ocean over several hundred years, confirm researchers funded by the National Science Foundation (NSF)'s paleoclimate program.

"For the first time, we have shown that realistic additions of glacial meltwater into the north Atlantic would have shut down north Atlantic deep water production over a period of a few hundred years, if the initial ocean circulation was somewhat weaker than that of today," said David Rind, lead author of the study and a senior climate researcher at the NASA Goddard Institute for Space Studies. The study appears in the current issue of *Journal of Geophysical Research - Atmospheres...*

GLACIERS ARE MELTING
POLES ARE THAWING
WATER IS FLOWING
OCEANS ARE WARMING
CURRENTS ARE SHIFTING
CLIMATES ARE CHANGING

THESE ARE NOT ONLY THE PREDICTIONS FOR TOMORROW

THESE ARE THE REALITIES OF TODAY

IT DOES NOT MATTER HOW

HUMANS OR NATURE OR BOTH

THE CHANGES ARE HAPPENING NOW

THEY WILL CONTINUE

AND STRONGLY AFFECT THE THE
LIVING PLANET AND THE STABILITY OF
OUR GROWING SETTLEMENTS DURING
THE 21ST CENTURY AND BEYOND

THIS IS WHAT MATTERS

are you gettin' this
yeah, i'm gettin' this

CLIMATIC CONTORTIONS, SOCIAL DISTORTIONS AND BIOSPHERIC ABORTIONS

Present climate models estimate the average global surface temperature of the Earth will rise between 1-7 degrees Celsius or 3-12 degrees Fahrenheit, during the 21st century as a result of an enhanced greenhouse effect from human and natural disturbances to global greenhouse gas cycles.

A rise of .6 degrees Celsius, 1 degree Fahrenheit, has been recorded since measurements began in 1850 and this meager rise is having a noticeable effect on global climate systems, weather patterns, and the living planet. The projected rise in global surface temperature during the 21st century would have tremendously noticeable effects.

It is substantially agreed upon, that as a result of a climbing average global surface temperature, severe weather related events will increase in frequency, intensity, and duration on average. These events include forest fires, hurricanes, typhoons, stormy weather, blizzards, abnormally high winds, heat waves, cold waves, droughts, and floods.

Recently, certain extreme weather events have been singled out and blamed on global warming. Although one cannot point to a particular weather event here or there and say it was caused by global warming, one can say that such events are an example of a larger overall change. A change in the frequency and intensity of extreme weather events, and this is a result of an enhanced greenhouse effect and a rising average global surface temperature.

Besides extreme weather events, a rising planetary temperature will alter things like growing seasons, agricultural zones and production, invasive species migration, water flow, and water security. Furthermore, as the onset of seasons change, as seasonal extremes become more common, as biozones, migration patterns, and habitats are altered, the rest of the living planet will be affected more dramatically than at any other time in the history of humankind.

Should the change in average global surface temperature, corresponding climate systems, and weather patterns continue to be gradual, the living planet and our settlements might be able to adapt as the problems arise. But even this is questionable.

Should a sudden shift occur, the time for adapting is significantly and suddenly reduced and the changes encountered could render even the most advanced settlements utterly helpless.

For the rest of the living planet, the present and predicted rise in average global surface temperature, coupled with increasing pressure from expanding human settlements, will continue to drive the planet into its sixth mass extinction in the Earth's 4.6 billion year history.

That is a pretty big deal.

There are many estimates, many projections, many consequences, and not all of them should be thought of as bad, just different.

Sure, Canadians might enjoy a warmer winter.

But will they enjoy inviting 250 million refugees from Bangladesh, Africa, and Polynesia?

Because that is what they will be asked to do.

Has the climate always changed?

Yes.

Have human settlements always changed?

Yes.

However, the severity of predicted changes, the possibility of rapid changes, and the inevitability of these changes is what has everyone concerned.

Time to take a walk through our common future.

Ready or not.

Here it comes.

EXTREME WEATHER EVENTS ON AVERAGE WILL BECOME MORE FREQUENT MORE INTENSE LONGER LASTING AND MORE DISRUPTIVE

°REPORT: CLIMATE CHANGE CAUSING JUMP IN NATURAL DISASTERS
REUTERS
29 SEPTEMBER 2000

Climate change is already increasing the frequency and intensity of natural disasters, and the trend is likely to continue according to a report released on Friday by the World Wide Fund for Nature.

The report, *"Climate Change and Extreme Weather Events,"* said global temperatures would increase, sea levels would rise, and few places in the world would be spared an increase in violent rainstorms, droughts, tropical cyclones and other climatic disruptions...

°NATURAL DISASTERS AT RECORD LEVEL IN 2000 - INSURER
REUTERS
29 DECEMBER 2000

The world was hit by a record number of natural disasters in 2000 and global warming and a rising population are likely to make future years even worse, the world's largest reinsurer said on Thursday.

Munich Re said the number of what it categorises as natural disasters rose by more than 100 to 850 in 2000, although the number of deaths was much lower than in 1999 because less populated areas were affected.

It said 10,000 people died as a result of natural disasters in 2000 compared to 75,000 in 1999. Material damage was put at more than $30 billion in 2000...

°WORLD DISASTERS SEEN AS GLOBAL WARMING OUTCOME
REUTERS
19 FEBRUARY 2001

Massive flooding, disease and drought could hit rich and poor countries around the world over coming decades if global warming is not halted, an authoritative U.N. scientific team warned Monday.

The scientists said they foresaw glaciers and polar icecaps melting, countless species of animals, birds and plant life dying out, farmland turning to desert, fish-supporting coral reefs destroyed, and small island states sunk beneath the sea...

°EXTREME WEATHER EVENTS MIGHT INCREASE
WORLD METEOROLOGICAL ORGANIZATION PRESS RELEASE
2 JULY 2003

Record extremes in weather and climate events continue to occur around the world. Recent scientific assessments indicate that, as the global temperatures continue to warm due to climate change, the number and intensity of extreme events might increase, the World Meteorological Organization (WMO) states in a press release issued today.

...New record extreme events occur every year somewhere in the globe, but in recent years the number of such extremes have been increasing. According to recent climate change scientific assessment reports of the joint WMO/UNEP Intergovernmental Panel on Climate Change (IPCC), the global average surface temperature has increased since 1861.

Over the 20th century the increase has been around 0.6° C. This value is about 0.15° C larger than that estimated by the previous reports. New analyses of proxy data for the Northern Hemisphere indicated that the increase in temperature in the 20th century is likely to have been the largest in any century during the past 1000 years...

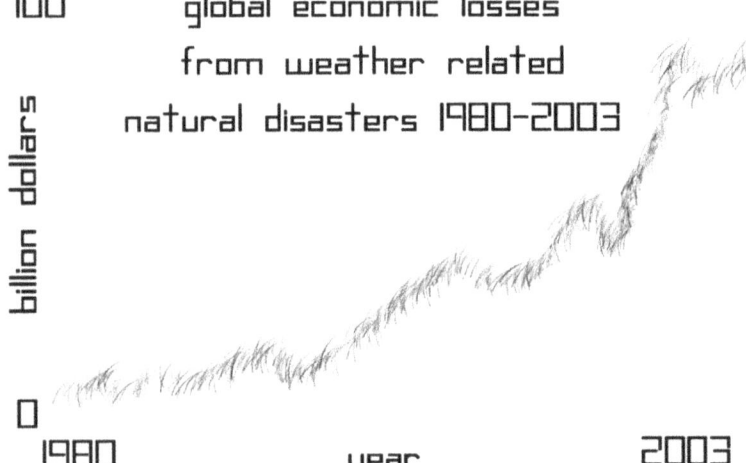

global economic losses from weather related natural disasters 1980-2003

100 · billion dollars · 0 · 1980 · year · 2003

EXTREME EVENTS IN THE FORM OF

FOREST FIRES

°GLOBAL WARMING MAY SPUR MORE WILDFIRES
ASSOCIATED PRESS
12 NOVEMBER 2003

Drought- and beetle-ravaged trees in this mountain community stick up like matchsticks in the San Bernardino National Forest, bypassed by the fires still smoldering, but left like kindling for the next big blaze.

Welcome to the future.

Fires that charred nearly three-quarters of a million acres could presage increasingly severe fire danger as global warming weakens more forests through disease and drought, experts warn.

"You're really going to increase the chances of and prevalence of fire," said Susan Ustin, a professor of environmental and resource science at the University of California, Davis…

°STORMY WEATHER BUFFETED U.S. IN 1999
ENVIRONMENTAL NEWS NETWORK
4 JANUARY 2000

Record-breaking weather stormed across the United States in 1999, killing hundreds of people and wreaking havoc to the tune of billions of dollars in property damage. From drought to hurricanes, the evidence is clear as day in the National Oceanic and Atmospheric Administration's weather wrap-up for the year.

Even before all the numbers are tallied, NOAA is predicting that 1999 will be the second warmest year on record since 1900 in the U.S., with an average of 55.7 degrees Fahrenheit. This figure comes hot on the heels of 1998's record average high of 56.4 degrees Fahrenheit.

can you feel the winds of change?

POSITIVE FEEDBACK

increased forest fire frequency
increased greenhouse gas release
enhanced greenhouse effect
global temperature increase

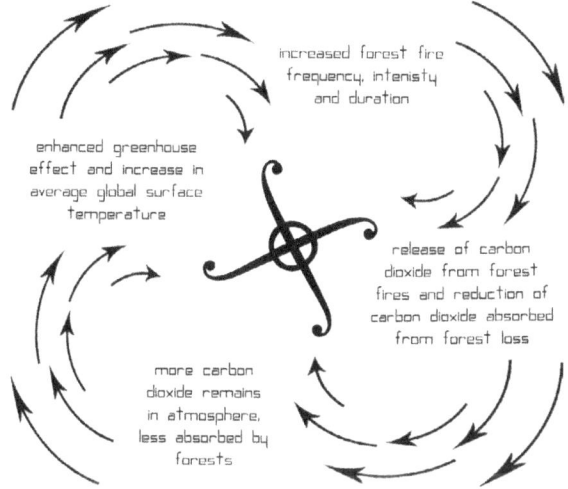

increased forest fire frequency, intensity and duration

enhanced greenhouse effect and increase in average global surface temperature

release of carbon dioxide from forest fires and reduction of carbon dioxide absorbed from forest loss

more carbon dioxide remains in atmosphere, less absorbed by forests

STORMS

TORNADOES

°RECORD NUMBER OF TORNADOES REPORTED IN U.S. DURING AUGUST
NATIONAL OCEANIC AND ATMOSPHERIC ADMINISTRATION PRESS RELEASE
2 SEPTEMBER 2004

A land-falling hurricane and a tropical storm helped push the total number of tornado reports to a record high for the month of August across the United States, according to the NOAA Storm Prediction Center in Norman, Okla. Preliminary numbers indicate a total of 173 tornadoes reported during the month, said Dan McCarthy, SPC's warning coordination meteorologist. Based on tornado records going back to 1950, this significantly tops the previous August record of 126 tornadoes set in 1979. Other high numbers for August include: 120 tornadoes in 1994; 115 in 1992; 112 in 1993; and 108 in 1985.

HEAT WAVES

DROUGHTS

FLOODS

°FUTURE HEAT WAVES WILL BE HOTTER, LONGER, MORE FREQUENT
ENVIRONMENT NEWS SERVICE
13 AUGUST 2004

Heat waves in North America and Europe will become more intense, more frequent and longer lasting during this century, scientists said on Thursday.

A new modeling study shows that an increase in heat absorbing greenhouse gases intensifies an unusual atmospheric circulation pattern already observed during heat waves in Europe and North America. As the pattern becomes more pronounced, severe heat waves occur in the Mediterranean region and the southern and western United States. Other parts of France, Germany and the Balkans also become more susceptible to severe heat waves.

"It's the extreme weather and climate events that will have some of the most severe impacts on human society as the climate changes," said Gerald Meehl, a senior scientist with the Climate and Global Dynamics Division of the National Center for Atmospheric Research (NCAR), a federally funded research center in Boulder, Colorado.

..."This study provides significant insight into the complex response of global climate to possible future worldwide economic and regulatory policies," said Cliff Jacobs, program director in the National Science Foundations division of atmospheric sciences, which funded the research.

°DROUGHT'S GROWING REACH: NCAR STUDY POINTS TO GLOBAL WARMING AS KEY FACTOR
NCAR-UCAR NEWS RELEASE
10 JANUARY 2005

The percentage of Earth's land area stricken by serious drought more than doubled from the 1970s to the early 2000s, according to a new analysis by scientists at the National Center for Atmospheric Research (NCAR).

Widespread drying occurred over much of Europe and Asia, Canada, western and southern Africa, and eastern Australia. Rising global temperatures appear to be a major factor, says NCAR's Aiguo Dai, lead author of the study.

...Dai and colleagues found that the fraction of global land experiencing very dry conditions (defined as -3 or less on the Palmer Drought Severity Index) rose from about 10-15% in the early 1970s to about 30% by 2002. Almost half of that change is due to rising temperatures rather than decreases in rainfall or snowfall, according to Dai.

"Global climate models predict increased drying over most land areas during their warm season, as carbon dioxide and other greenhouse gases increase," says Dai. "Our analyses suggest that this drying may have already begun."...

°CLIMATE STUDIES POINT TO MORE FLOODS IN THIS CENTURY
NATIONAL GEOGRAPHIC NEWS
30 JANUARY 2002

Two separate teams of scientists are predicting more extreme rainfall and greater flooding in this century. According to their projections, it will be particularly striking at northern latitudes - across Canada, Alaska, northern Europe, and northern Asia, regions that already receive the most precipitation. But the equatorial tropics and Southeast Asia are also likely to have increased rainfall and flooding.

Both teams, one from the United States and the other from Europe, attribute the expected pattern to global warming accelerated by human activities. Although people may adapt to gradual climate change, the effects of extreme rain and flooding are often broad, devastating, and costly to society. Landslides, avalanches, and flooding damage infrastructure such as roads, bridges, and buildings, and hurt agricultural productivity because of lost crops and soil erosion. Disaster relief often requires enormous funding, and the loss of human life may also be high...

°RECORD COLD GRIPS MUCH OF THE NATION IN NOVEMBER AND DECEMBER: TWO-MONTH PERIOD IS THE COLDEST ON RECORD IN THE UNITED STATES
NOAA PRESS RELEASE
5 JANUARY 2001

...Following the second coldest November on record in the U.S., below normal temperatures continued to grip much of the nation in December. With an average temperature of 28.9 F, December 2000 was the seventh coldest December since national records began in 1895...Near record cold temperatures for the same period occurred most recently in 1985 and 1983, when the nation's average temperature was 34.6 F and 34.8 F respectively, the 3rd and 5th coldest such two-month periods on record.

...Even though average long-term U.S. and global temperatures are warmer than they were a century ago, dramatic short-term swings in temperature are to be expected due to variability in circulation patterns. This variability can lead to periods of record cold temperatures while long-term trends remain positive...

COLD

WAVES

eye can't wait to sea you again

HURRICANES

°HURRICANES AND CLIMATE CHANGE: IS THERE A CONNECTION?
NCAR STAFF NOTES MONTHLY
OCTOBER 2004

...And so it comes as no surprise that the relentless line-up of storms has ignited debate about the connection between hurricanes and global warming. The issue attracted a fair amount of media attention in September, when newspapers ran headlines saying "Global Warming May Spawn More Super-Storms" and "Ivan May Just Be a Messenger."

From 1970 to 1994, hurricane activity in the Atlantic was fairly mild, generating half as many destructive storms as both the previous period, dating back to the 1920s, and the period since 1995. While the current period is the most active nine consecutive years on record and also contains some of the hottest years on record, climate scientists are divided on whether or not global warming affects hurricane activity.

Kevin Trenberth (CGD) says that although it's controversial, he thinks that global warming is in fact creating conditions that are favorable for hurricanes to be more severe. "Global climate change, and global warming in particular, create a different background environment in which the hurricanes are working," he says. "The sea surface temperatures are a little warmer, the whole environment is a bit wetter, there's more humidity, and that's the main fuel for hurricanes."

...The hurricane season might also start earlier and last longer, Kevin says. Storms could strengthen in some parts of the world where they wouldn't normally be vigorous enough to be considered hurricanes. Brazil was recently struck by the first hurricane ever recorded in the South Atlantic, and Hurricane Juan thrashed the coast of Canada last year.

Bob Gall, former MMM director and current lead scientist of the U.S. Weather Research Program, is less certain that global warming is fueling more severe hurricanes, for now at least. He says that at this point there is no evidence linking global warming to hurricanes. "Changes due to global warming, so far as I see it, are pretty small," he says. "That doesn't mean we aren't on an upward trend, and I think we are, but I don't know if the trend has been significant enough to provide a clear signal that anthropogenic warming is causing it."...

AND AS THE CLIMATE CHANGES

MORE PEOPLE

°HYDRO-QUEBEC CUTS POWER TO 500,000 AFTER FOREST FIRE NEARS TRANSMISSION LINE
CANADIAN PRESS
5 JULY 2002

Nearly 500,000 Quebecers lost power at different times Friday after Hydro-Quebec cut service on one of its James Bay transmission lines when a forest fire got too close. Fires that have been burning in the northern regions of the province forced the public utility to use only one 600-megawatt line in James Bay...

°WARMING TREND WILL DECIMATE ARCTIC PEOPLES, REPORT WARNS
INTER PRESS SERVICE
9 SEPTEMBER 2004

Climate change will soon make the Arctic regions of the world nearly unrecognisable, dramatically disrupting traditional Inuit and other northern native peoples' way of life, according to a new report that has yet to be publicly released.

The dire predictions are just some of the findings by the Arctic Climate Impact Assessment (ACIA), an unprecedented four-year scientific investigation into the current and future impact of climate change in the region.

"This assessment projects the end of the Inuit as a hunting culture," said Sheila Watt-Cloutier, chairwoman of the group that represents about 155,000 Inuit in the Arctic regions of Canada, Russia, Greenland, and the United States.

The report predicts the depletion of summer sea ice, which will push marine mammals like polar bears, walrus and some seal species into extinction by the middle of this century, Watt-Cloutier told IPS.

...The impacts of climate change are already widely felt in the Arctic. Thawing permafrost -- the normally perpetually frozen layer of earth -- has collapsed roads and buildings.

..."Or traditional wisdom on how to survive and thrive on the land is becoming useless because everything is changing and changing fast."...

°CALIFORNIA AGAIN SETS POWER ALERT AMID RECORD HEAT
REUTERS
10 JULY 2002

Urgent pleas to save energy filled California's airwaves on Wednesday as the state's 35 million residents face a second day of scorching heat and heavy air conditioning demand on their electricity grid.

Energy industry officials also expressed concern that southern Nevada, especially the booming gambling town of Las Vegas, could see a serious energy crunch through Thursday as power supplies dwindled across the U.S. West...

°A QUARTER OF CHINA'S POPULATION AT RISK AS GLACIERS START MELTING
AGENCE FRANCE-PRESSE
13 MAY 2004

Global warming may cost China two thirds of its glaciers by mid-century, putting 300 million people at risk, state media reports.

The country's glaciers are melting at an unprecedented rate, threatening the livelihoods of Chinese dependent on the water they provide, the China Daily said, citing local experts.

"Glaciers are much more than scenic gifts from nature," the paper said. "They allow room for bio-diversity and are a crucial source of water by storing snow in the winters and releasing water in hot dry summers."

As many as 64 percent of China's glaciers may be gone by 2050, said Yao Tandong, director of the Institute of Tibetan Plateau Research at the Chinese Academy of Sciences.

This sounds the alarm bells for the 23 percent of the country's 1.3 billion people living in oases in western China, Yao warned.

Desertification, already among China's top environmental worries, could become even worse, he said...

°ICE, SNOW LEAVE 335,000 IN SOUTH POWERLESS FOR 4TH DAY
REUTERS
28 DECEMBER 2001

More than 335,000 residents of the Southern U.S. remained without electricity early Thursday after a Christmas Day ice storm and record snowfall battered the region this week, according to local utilities.

More than 60,000 residents of east Texas, Arkansas and northwest Louisiana were still without power on Thursday after nearly two inches of ice snapped tree limbs and downed transmission lines, a spokesman for AEP-Southwestern Electric Power Co. told Reuters...

°MILLIONS MENACED BY CHINA FLOODS
BBC NEWS ONLINE
20 AUGUST 2002

Officials fear the worst floods in years. More than 10 million people are under threat from potentially devastating floods, as water levels in central China's massive Dongting Lake continue to rise.

With more torrential rain forecast in the coming days, and swollen rivers already emptying into the lake, officials fear it could burst its banks, flooding millions of homes and 667,000 hectares (1.6 million acres) of fertile farmland. Dongting, in the southern province of Hunan, acts as a giant overflow for the flood-prone Yangtze River.

When the Yangtze and the Dongting last broke their banks in 1998, more than 4,000 people were killed. There are fears even more people could perish this year...

IN MORE PLACES

°EUROPE'S FLOOD PART OF GLOBAL DELUGE
CHRISTIAN SCIENCE MONITOR
15 AUGUST 2002

...The Czech Republic is the latest country hit in a season of torrents and torment in Europe. Heavy rains and flooding have also paralyzed parts of Slovakia, Italy, Spain, Germany, Romania, Bulgaria, Croatia, Hungary, and Ukraine. Austria has been declared a disaster area. The floods have claimed 94 lives in the past week and a half, 59 of them in flash floods in southern Russia.

As they watch their homes and businesses drown, some in Europe worry that these floods are the result of climate changes induced by man. In Asia, torrential rains in China have killed 900 people this year. In North and South Korea, Vietnam, India, Nepal, and Bangladesh, floods have claimed another 700 lives in the past month.

"Rainfall is becoming more intense," says Prof. Phil Jones of the Climactic Research Unit at the University of East Anglia, a prominent European meteorological institution. "This is not a [natural] cycle."...

°FLOODS SUBMERGE MOST OF BANGLADESH, HIT MILLIONS
SYNDNEY MORNING HERALD
26 JULY 2004

Two-thirds of Bangladesh has been inundated by the worst monsoon floods in 15 years and water and sewerage systems in Dhaka have broken down completely, officials said yesterday. Severe flooding across Bangladesh and parts of India and Nepal has killed almost 570 people in the past month, forced millions from their homes and sparked an outbreak of diarrhoea and other water-borne diseases.

...In Dhaka, the diplomatic enclave and commercial centre were under waist-deep water today. Sewerage pipes in the city have burst, contaminating the water supply.

...The floods in Bangladesh have halted the impoverished country's oil and gas operations and closed thousands of garment factories, hitting a key export sector worth $US5 billion ($7.02 billion) a year. Tens of thousands of homes, mainly made of bamboo and straw, have been washed away in the three nations and rescue and relief services are stretched to breaking point.

Millions of people have no food or drinking water.

In the north-eastern Indian state of Assam, where roads and rail links have been cut for weeks, hundreds of soldiers handed out food, water, medicine and clothing by boat to some of the estimated 12 million people left homeless.

...The floods have so far killed 351 people and affected over 20 million others in Bihar...

°GLOBAL WARMING BLAMED AS AUSTRALIA'S BIGGEST CITY GETS WATER CURBS
AGENCE FRANCE-PRESSE
11 SEPTEMBER 2003

Residents of Australia's biggest city, Sydney, were ordered to stop sprinkling their lawns or hosing clean their cars Thursday under strict water curbs local officials blamed on global warming.

The premier of New South Wales state imposed the mandatory water restrictions on the city and its surrounding areas for the first time in nine years because of the country's worst drought on record and stubbornly rising domestic water use.

"The reason we're doing this is that the dam levels behind me are way lower than they should be at this time of year," [Primier Bob] Carr said.

..."This is the ninth consecutive year, speaking nationally, when rainfalls have been lower than average and average temperatures are climbing," he said.

"Those people who are sceptical about global warming ought to think again because this is the first very practical intimation of global warming being upon us," he said.

"Years from now, you might recall this announcement as the first time global warming affected our way of life," he said.

°SEA ENGULFING ALASKAN VILLAGE
BBC NEWS ONLINE
30 JULY 2004

It is thought to be the most extreme example of global warming on the planet. The village of Shishmaref lies on a tiny island on the edge of the arctic circle - and it is literally being swallowed by the sea.

Houses the Eskimos have occupied for generations are now wilting and buckled. Some have fallen into the sea. Not only is the earth crumbling underfoot, but the waves are rising ominously all around.

As we walked across the narrow strip of beach that was his playground as a kid, village elder Tony Weyiouanna pointed to a series of barricades that have been erected over the years in the hope of stemming the tide....Tony estimates the tide moves an average of 10 feet (three metres) closer to the land every year. When he was growing up, it was roughly 300 feet (91 metres) from where it is now.

...Because temperatures in Alaska have increased by as much as 4.4C over the last 30 years, glaciers are starting to melt, causing the sea levels to rise.

The increased temperature is also thawing the frozen ground, which is known as permafrost, on which the arctic communities such as Shishmaref were built. It is this thawing that is causing the ground to crumble like sand...

°GLOBAL WARMING TO AFFECT WATER SUPPLY: MORE RAINFALL MEANS SMALLER SIERRA SNOWPACK
SAN FRANCISCO CHRONICLE
15 JUNE 2001

California water planners face a problem they never thought they'd encounter: global warming is hitting the High Sierra snowpack. And just how the planners cope with it could affect every city-dweller, every farmer and every water-using industry in the state for years to come.

Scientists are in broad agreement that the world's climate is steadily warming -- whether due to "greenhouse gas" emissions from industry and automobiles, or to natural variability. And there is evidence that it is already altering the annual ebb and flow of the state's water supplies.

It's a matter of "more rain, less snow," says Dan Cayan, director of the climate research division at the Scripps Institution of Oceanography in La Jolla (San Diego County). And that's bad, he says, because California's water supply largely depends on the winter snowpack in the high mountains that must feed the state's lowlands the rest of the year.

A major change is already evident in the decreasing depths of the mountain snows that pile up each winter, in the unseasonal winter rainfalls that drench the mountains instead of snow, and in the speed of the snowmelt during spring.

Total precipitation over California hasn't changed significantly on average over the years, but seasonal variations between rain and snow show that a significant warming trend is under way, according to Cayan...

°GLOBAL WARMING TO COST $300 BILLION A YEAR
REUTERS
3 FEBRUARY 2001

An increase in natural disasters as a result of global warming could cost the world over $300 billion annually by the year 2050, a new United Nations commissioned report says.

According to the report from leading German re-insurers Munich Re, the losses would result from more frequent tropical cyclones, loss of land as a result of rising sea levels and consequent damage to agriculture and fishing stock.

°KILLER SNOWSTORM TRAPS 1.35 MILLION PEOPLE
AGENCE FRANCE-PRESSE
17 JANUARY 2001

Rescue workers are unable to reach northern areas where heavy snow has trapped 1.35 million people and three million livestock after a severe blizzard.

The Sina.com website said the New Year's Eve snowstorm in Inner Mongolia, which lasted more than 70 hours and left snow two metres high in some places, killed at least 27 people, with 14 others missing and feared buried in snow.

The Xinhua News Agency reported last week the harsh and unusual weather also killed more than 80,000 head of livestock. Direct economic losses are estimated at 910 million yuan (HK$855.4 million).

Much of the accumulated grazing for the animals has been destroyed and blocked roads have prevented rescue workers from delivering emergency food, blankets, fuel and animal feed to the disaster area...

"The herders had made preparations, but the preparations were not enough to handle such unusual circumstances," Mr Gao said...

°EFFECTS OF CLIMATE WARMING ARE HERE AND NOW
SCRIPPS HOWARD NEWS SERVICE
5 MAY 2004

Just thinking of wintertime in Northern Michigan can make some people shiver, but the region's days as one of the nation's premier iceboxes may be numbered.

Warmer temperatures, less snow and less lake ice in recent decades have made Michigan's winter tourism industry one of the early victims of a changing climate.

Ski-resort operators say the average length of the winter season has shrunk one to two weeks. "We're fortunate we can make snow," said Jim MacInnes, president of the Crystal Mountain ski resort near Cadillac. "Whether this is just an anomaly or a long-term trend it's hard for me to know, but I do believe global warming is a real issue."

Michigan is not alone. Rather than a distant theoretical possibility, most scientists agree that global warming in the United States is a here-and-now reality. From the drought-parched Rocky Mountains to Florida's beaches to the hardwood forests of New England, early signs of climate change are virtually everywhere.

Evidence that the impact of global warming is already being felt "is becoming overwhelming," said John Magnuson, a professor emeritus at the University of Wisconsin who authored a landmark study of lake ice. It found a 150-year warming trend throughout the Northern Hemisphere. "I think we're just beginning to recognize that it's happening everywhere and the effects are different in the different regions of the country and different parts of the world."...

ALL OF THIS WILL AFFECT THE STABILITY AND SECURITY OF OUR GROWING SETTLEMENTS

°DISASTERS WILL OUTSTRIP AID EFFORT AS WORLD HEATS UP
THE GUARDIAN
29 JUNE 2001

International aid will not be able to keep up with the impact of global warming, the Red Cross said yesterday, after reporting a sharp increase in the late 1990s in the number of weather-induced disasters

In its annual World Disasters Report the International Federation of Red Cross and Red Crescent Societies says that floods, storms, landslides and droughts, which numbered about 200 a year before 1996, rose sharply and steadily to 392 in 2000.

"Recurrent disasters from floods in Asia to drought in the Horn of Africa, to windstorms in Latin America, are sweeping away development gains and calling into question the possibility of recovery," the report says.

Blaming the trend on global warming, Roger Bracke, its head of disaster relief operations, said: "These are also the most deadly events; it is probable that these kind of disasters will increase even more spectacularly...

°WORLD FORUM SCIENTISTS: 'CLIMATE CHANGE WILL BECOME A SECURITY ISSUE'
REUTERS
2 FEBRUARY 2002

Scientists at this week's World Economic Forum have predicted a grim future replete with unprecedented biological threats, global warming and the possible takeover of humans by robots...

...At a session on climate change on Thursday, Robert Watson who chairs an international panel on the issue said the earth's climate would warm by at least 1.4 degrees centigrade in the next century even if urgent action was taken right now to stem emissions of carbon dioxide. If insufficient action was taken, warming could be as great as 5.8 degrees. He predicted more droughts in some areas and floods in others, more intense cyclones and massive social and economic disruption especially in poor countries.

Peter Wadhams of Cambridge University in England said that while rich nations could and would protect themselves against flooding by building sea defenses, a nation like Bangladesh could expect ever more frequent and severe flood disasters. Howard Ris, President of the Union of Concerned Scientists, said climate change could well lead to future conflict as nations found themselves confronted with unmanageable new challenges.

"Climate change will become a security issue," he said. "Hundreds of millions of people will find themselves fighting new threats to survival"...

°CLIMATE 'WILL LEAD TO HUNGRY CENTURY'
BBC NEWS ONLINE
19 FEBRUARY 2001

Scientists say rising global temperatures will condemn millions to hunger this century. In a United Nations report, they say agricultural production will decline in Asia and Africa, while Australia and New Zealand will become short of water. Europe will face a higher flooding risk, and the eastern seaboard of the US may expect more storm surges and coastal erosion. And the harder climate change bites, the likelier it is that profound and possibly irreversible changes will occur...

°GLOBAL WARMING TRIGGERS PUBLIC HEALTH WARNING
ENVIRONMENTAL NEWS NETWORK
8 MAY 2001

Climate change could have a far-reaching impact on health patterns in the United States, according to a recent assessment by a broad coalition of scientists from academia, government and private industry.

"This assessment is not one of doom and gloom, but does warrant concern within the public health community," said co-chairmen of the report Jonathan Patz, an assistant professor of environmental health sciences at the Bloomberg School of Public Health...

°EUROPE TOLD THERE IS NO CHOICE BUT TO ADAPT
THE GUARDIAN
2 NOVEMBER 2000

Europe must adapt to extremes of climate which will cause new deserts in the south and floods and wind storms in the north, according to a new report before the European commission...Agriculture in the south will suffer as underground water is exhausted and already sparse summer rain disappears. There will not be enough water to grow the fruit and vegetables that fill northern supermarkets. Hot summers will double in frequency by 2020 and be five times as likely in Spain. Forest fires will increase across the whole region. Species of wading birds which live on the Mediterranean wetlands will become extinct as sea levels rise, and environmental refugees will invade from Africa as local people move north in search of a gentler climate...

°CIVILISATIONS 'DESTROYED BY CLIMATE CHANGE'
THE TELEGRAPH
26 JANUARY 2001

Scientits warn today of "unprecedented social disruptions" that could result from global warming, after linking the collapse of societies throughout history to climate change.

There is "mounting evidence" that the demise of some civilisations was climate-driven, report Prof Harvey Weiss of Yale University and Prof Raymond Bradley of the University of Massachusetts, Amherst.

Scientists are now able to link the rise and fall of societies recorded in the archaeological record with evidence of the timing and magnitude of climate change held in ice cores, corals and sediments. Prof Weiss said: "We find a very precise coincidence between the abrupt climate changes and the archaeological record of collapse."

...The professors suggest that modern societies, faced with prospects of global warming, may not be immune to social disruptions triggered by abrupt climate change. Most of the world's people will continue to be subsistence or small-scale market farmers. But unlike ancient societies, who could migrate to where cultivation of crops was possible, the world is now too crowded for "habitat tracking."

The authors say: "We do, however, have distinct advantages over societies in the past because we can anticipate the future using computers," say the authors. We must use this information to design strategies that minimise the impact of climate change on societies that are at greatest risk. This will require substantial international co-operation without which the 21st century will likely witness unprecedented social disruptions.

°CLIMATE RELATED PERILS COULD BANKRUPT INSURERS
ENVIRONMENT NEWS SERVICE
7 OCTOBER 2002

Climate change is causing natural disasters that the financial services industry must address, a group of the world's biggest banks, insurers and re-insurers warned Monday. They estimated the cost of financial losses from events such as this summer's devastating floods in central Europe at $150 billion over the next 10 years.

"Climate Change and the Financial Services Industry," a report supported by 295 banks and insurance and investment companies, was launched Monday at the Swiss Re Greenhouse Gas conference in Zurich.

A partnership between the United Nations Environment Programme (UNEP) and the financial institutions, known as UNEP Finance Initiatives commissioned the report. It shows that losses as a result of natural disasters appear to be doubling every decade and have reached $1 trillion in the past 15 years.

"The increasing frequency of severe climatic events, threatening the social stability or coupled with significant social costs, has the potential to stress insurers, reinsurers and banks to the point of impaired viability or even insolvency," the report concludes.

John Fitzpatrick, CFO and member of the Executive Board of Swiss Re, said, "Climate change and substantial emissions reductions - like any other strategic global business challenge - ultimately becomes a financial issue. The problems associated with environmental disasters quickly become measured in dollars and cents. Our industry needs to lead by developing financial solutions and risk mitigation techniques to assist our clients in achieving global emission reductions." ...

°REFUGEES, DISEASE, WATER AND FOOD SHORTAGES TO RESULT FROM GLOBAL WARMING
AGENCE FRANCE-PRESSE
2 FEBRUARY 2005

Global warming will boost outbreaks of infectious disease, worsen shortages of water and food in vulnerable countries and create an army of climate refugees fleeing uninhabitable regions, a conference here was told. The scale of these impacts -- the theme of the second day of the major scientific forum on global warming -- varies according to how quickly fossil fuel pollution is tackled, how fast the world's population grows and how well countries can adapt to climate shift. But a common expectation is that widespread misery is lurking, a few decades down the road.

According to a study quoted by Rajendra Pachauri, chairman of the UN's top scientific authority on climate change, by 2050 as many as 150 million "environmental refugees" may have fled coastlines vulnerable to rising sea levels, storms or floods, or agricultural land that became too arid to cultivate. In India alone, there could be 30 million people displaced by persistent flooding, while a sixth of Bangladesh could be permanently lost to sea level rise and land subsidence, according to the study. Pachauri's body, the Intergovernmental Panel on Climate Change (IPCC), estimated in 2001 that by 2100, temperatures would rise by between 1.4 C (2.5 F) and 5.8 C (10.4) compared to 1990 levels, driven by atmospheric carbon pollution which stokes up heat from the Sun. The mean global sea level would rise by between nine and 88 centimetres (four and 35 inches).

...Global warming will also add significantly to Earth's worrisome water problems...By the 2050s, water availability in these water-stressed regions -- but also in parts of central, north and south America -- may be further crimped because of changed rainfall patterns. Between 700 million and 2.8 billion people in such areas will be affected, depending on population growth and the pace of temperature rise...

AND LET US NOT FORGET THE REST
OF THE LIVING PLANET

AS THE ONSET OF SEASONS CHANGE
AS THE HABITATS CHANGE
AS THE GLOBAL TEMPERATURE CHANGES
AS THE CLIMATE AND WEATHER CHANGES

°GLOBAL WARMING HITS SPECIES ALL OVER WORLD-STUDY
REUTERS
27 MARCH 2002

From dying coral reefs to later autumns and endangered male painted turtles, global warming has started to affect plant and animal life across the planet, scientists said Wednesday. The world's mean temperature increased by around 0.6 degrees Celsius in the 20th century -- most of the rise came in the last 30 years -- and its impact is already being felt by flora and fauna from the equator to the poles. Some species are doomed as they battle ever-rising temperatures in an increasingly crowded planet that offers fewer escape routes, according to scientists writing in the journal *Nature*.

"Temperature has increased by no more than 0.6 degrees and already the signs are very obvious," said geobotanist Gian-Reto Walther from the University of Hanover in Germany, who collated the research from across the branches of the natural sciences. The study's conclusions highlight the seriousness of global climate change by showing parallel trends in plants, birds, animals and fish. "This is a major concern," Walther told Reuters, adding extinction for some species was inevitable.

"The big difference between now and previous periods of climate change, like the Ice Age, is that seven billion people live on Earth now and many migration corridors for species are blocked," Walther said...

°STUDY WARNS OF GLOBAL WARMING EXTINCTIONS
ASSOCIATED PRESS
7 JANUARY 2004

Hundreds of species of land plants and animals around the globe could vanish or be on the road to extinction over the next 50 years if global warming continues, scientists warn.

The researchers concede that there are many uncertainties in both climate forecasts and the computer models they used. But they said their prediction could come to pass if industrial nations do not curtail emissions of greenhouse gases that trap heat in the atmosphere.

"We're already seeing biological communities respond very rapidly to climate warming," said Chris Thomas, a conservation biologist at the University of Leeds in England, and the study's lead author. The findings by Thomas and 18 other researchers appear in Thursday's issue of the journal *Nature*

They found that more than one-third of the 1,103 native species they studied could disappear or approach extinction by 2050 as climate change turns plains into deserts or alters forests...

°MINUTE SHIFT IN TEMPERATURE HAS HAD A MAJOR EFFECT ON EARTH, STUDIES SHOW
LOS ANGELES TIMES
2 JANUARY 2003

Gradual warming over the last 100 years has forced a global movement of animals and plants northward, and it has sped up such perennial spring activities as flowering and egg hatching across the globe -- two signals that the Earth and its denizens are dramatically responding to a minute shift in temperature, according to two studies published today.

One study showed that animals have shifted north an average of nearly four miles per decade. Another study showed that animals are migrating, hatching eggs and bearing young an average of five days earlier than they did at the start of the 20th century, when the average global temperature was 1 degree cooler.

That 1 degree, according to the studies, has left "climatic fingerprints" -- pushing dozens of butterfly and songbird species into new territories, prompting birds and frogs to lay eggs earlier and causing tree lines to march up mountain slopes...

..."There is a consistent signal," said Terry L. Root, a biologist at Stanford University and lead author of one report. "Animals and plants are being strongly affected by the warming of the globe...It was really quite a shock, given such a small temperature change," she said...

THE LIVING PLANET WILL CHANGE

BY THE HUNDREDS

BY THE THOUSANDS

BY THE MILLIONS

°HUNDREDS OF SPECIES PRESSURED BY GLOBAL WARMING
ENVIRONMENT NEWS SERVICE
2 JANUARY 2003

Hundreds of plant and animal species around the world are feeling the impacts of global warming, although the most dramatic effects may not be felt for decades, according to new research from a Stanford University team.

They predict that a rapid temperature rise, together with other environmental pressures, "could easily disrupt the connectedness among species" and lead to numerous extinctions.

"Birds are laying eggs earlier than usual, plants are flowering earlier and mammals are breaking hibernation sooner," said Terry Root, a senior fellow with Stanford University's Institute for International Studies (IIS) and lead author of the article published in Thursday's issue of the journal *Nature*.

Root and her colleagues analyzed 143 scientific studies involving a total of 1,473 species of animals and plants for the article, "Fingerprints of global warming on wild animals and plants."

After analyzing all 143 studies, the Stanford team concluded that global warming is having a statistically significant impact on animal and plant populations around the world...

°OBLIVION THREAT TO 12,000 SPECIES
BBC NEWS ONLINE
18 NOVEMBER 2003

Another 2,000 species have been added to the annual Red List of the world's most endangered animals and plants.

The "official" catalogue produced by IUCN-The World Conservation Union now includes more than 12,000 entries.

This year, IUCN has highlighted the problems faced by many island habitats which it claims face a bleak future. It says many native animals and plants on the Seychelles and the Galapagos, for example, are being driven to extinction by invasive species.

Since AD 1500, IUCN says 762 plants and animals have vanished, with another 58 known only in cultivation or captivity.

Achim Steiner, the organisation's director-general, said: "While we are still only scratching the surface in assessing all known species, we are confident [the list of 12,259 species] is an indicator of what is happening to global biological diversity.

"Human activities may be the main threat to the world's species, but humans can also help them recover - the Chinese crested ibis, the Arabian oryx and the white rhino are just a few examples."...

°WARMING MAY THREATEN 37% OF SPECIES BY 2050
WASHINGTON POST
8 JANUARY 2004

In the first study of its kind, researchers in a range of habitats including northern Britain, the wet tropics of northeastern Australia and the Mexican desert said yesterday that global warming at currently predicted rates will drive 15 to 37 percent of living species toward extinction by mid-century.

Dismayed by their results, the researchers called for "rapid implementation of technologies" to reduce emissions of greenhouse gases and warned that the scale of extinctions could climb much higher because of mutually reinforcing interactions between climate change and habitat destruction caused by agriculture, invasive species and other factors.

"The midrange estimate is that 24 percent of plants and animals will be committed to extinction by 2050," said ecologist Chris Thomas of Britain's University of Leeds. "We're not talking about the occasional extinction -- we're talking about 1.25 million species. It's a massive number."...

AND AS HISTORY IS REPEATED

THE EARTH WILL EXPERIENCE ANOTHER EXTINCTION OF MASS PROPORTION IN A VERY SHORT PERIOD OF TIME

REMEMBER TO TELL STORIES ABOUT THE PENGUINS AND POLAR BEARS

as if **yOUr** pollutants in **mY** blood wasn't bad enough
thanks alot you stupid apes

°ANTARCTICA BECOMES TOO HOT FOR THE PENGUINS: DECLINE OF 'DINNER JACKET' SPECIES IS A WARNING TO THE WORLD
THE INDEPENDENT
3 FEBRUARY 2002

Penguins are starting to desert parts of Antarctica because the icy wastes are getting too hot. The numbers of adelie penguins on the Antarctic peninsula – the most northerly part of the frozen continent – are falling as global warming takes hold. And experts predict that, as the climate change continues, they may abandon much of the 900-mile-long promontory altogether.

The archetypal "tuxedoed" species like the cold even more than other penguins. And the peninsula has been warming up faster than almost anywhere else on earth, with temperatures increasing at least five times faster than the world average. Scientists believe this is disrupting their food supplies...

°POLAR BEAR 'EXTINCT WITHIN 100 YEARS'
BBC NEWS ONLINE
9 JANUARY 2003

The polar bear could be driven to extinction by global warming within 100 years, warns an ecology expert. The animal, which relies on sea ice to catch seals, is already starting to suffer the effects of climate changes in areas such as Hudson Bay in Canada. Scientists say Arctic sea ice is melting at a rate of up to 9% per decade. Arctic summers could be ice-free by mid-century.

Dr Andrew Derocher, of the University of Alberta, Edmonton, has used the data to assess the impact on the Arctic's top predator. He believes the polar bear could disappear in the wild by the end of the century unless the pace of global warming slows.

He told BBC News Online: "Polar bears are a species whose whole life history is dependent on having sea ice.

"As the sea ice changes in distribution and pattern we can expect this to have fundamental changes on the ecology of polar bears.

"As the sea ice disappears, so will the polar bears."...

BECAUSE THEY WILL SOON DISAPPEAR WITH THE POLAR FRINGE

INTERMISSION

extra sensory perception

That was quite a bit to chew in one bite.

Facing the consequences of global temperature change during this century will take some clearheadedness, so breathe in...breathe out...

We still have some time to plan, some time to change, some time to adapt, some time to live.

Put this book down and go outside.

Look at the sky, feel the raindrops on your lips, stare at the clouds, grasp the grass, listen to the cars pass by, or the birds chirping, or the river flowing, or the squirrels squawking, or the people talking.

Think about your family, your friends, your work, your leisure, your past, your present, your future.

Think about the people and everything else you don't know, but share the planet with.

Think about how lucky you are to experience this life, or how lucky you might be in the next if you abide by the natural and divine laws in this one.

Think about the mountain.

Think about the sea.

Think about the grains of sand.

come be with me

Think about nothing and just be.

Think about the predictions coming from the international
scientific community. But do not worry too much about them.
Their predictions are not a death sentence, they are more of a life
sentence.

"YOU ARE HEREBY SENTENCED TO A LIFE ON PLANET EARTH AND
ARE SEMI-FREE TO LIVE THIS LIFE AS YOU SEE FIT. SO MAKE IT
THE BEST LIFE YOU CAN WITH THE CIRCUMSTANCES YOU ARE
PROVIDED, AND MAKE IT BETTER FOR THOSE AROUND YOU, THOSE
WHO WILL FOLLOW, AND THE PLANET YOU ALL INHABIT."

What does it mean to be fit as a human being at
the beginning of the 21st century?

What does it mean to be fit in the face of imminent
changes to the foundation of modern settlements?

What does it mean to live in a benign way within the
fabric of the Earth and still have a high quality of
life?

Is it possible with the way our settlements function?

How can we keep our settlements robust in the
face of global temperature change?

Is it possible to change the way our
settlements function without disrupting our
daily lives?

Would these changes be less disruptive than
the predicted consequences of global temperature
change?

Ponder these things as you finish this book.

But for now, go outside.

Go be with the planet that gives you life.

You will both appreciate it.

7

READING THE RINGS OF TIME

°TRIGGERS TO CLIMATE CATASTROPHE STILL POORLY UNDERSTOOD
AGENCE FRANCE-PRESSE
2 FEBRUARY 2005

Scientists at a global warming conference say they see potential triggers for runaway climate change but admit that when and how these notional doomsdays may be unleashed are debatable or quite unknown. The theoretical triggers are the apocalyptic side to global warming, giving the lie to the common perception of it as an incremental threat that will rise predictably, like a straight line on a graph.

A widespread view of climate change is that the Earth's surface temperature will gently rise as more and more carbon gas is spewed out by fossil fuels, trapping heat from the Sun. The change would be progressive, which means humans would have enough time to respond to the crisis and plants and animals have a better chance of adapting to its effects. But scientists at a conference here on global warming say there is also the risk of sudden, catastrophic, irreversible and uncontrollable climate change that could be triggered in as-yet unknown conditions.

"There's still a great deal we don't know about these rapid non-linear events," British scientist Sir John Houghton, a leading member of the UN's top panel on global warming, said on Tuesday...

During the past 30 years, and especially since the beginning of the 21st century, the world's scientific community has laid out some pretty convincing evidence that the Earth is warming at an accelerated rate compared to the past few thousand years. Evidence that shows the present trend is beyond natural variability. Evidence that shows the present trend is a direct result from the industrializing activities of a global civilization during the past few hundred years. Evidence that shows the present trend will continue during this century and beyond.

How do we know their findings are credible? Maybe they are just fabricating data to pad their own pockets with research funding? Why should we believe that glaciers are melting and oceans are warming? Why should we believe that satellite readings and computer models can tell us anything about what is happening now and what might happen in the near future?

The answer is fairly straightforward: science and technology are at a very advanced level at the beginning of the 21st century. There are thousands of people doing research in dozens of fields who are supported by billions of dollars to investigate the mysteries of global dynamics related to global temperature change.

ECOLOGISTS say global warming and human activities are damaging the integrity of ecosystem biodiversity throughout the planet.

BIOGEOCHEMISTS say biodiversity plays a key role in global greenhouse gas cycle regulation and global temperature regulation.

METEOROLOGISTS say the last decade of the 1900 s was the warmest of the century and the century was the warmest of the past 1000 years.

GEOLOGISTS say the world's ice is melting.

OCEANOGRAPHERS say the thermohaline pumps in the North Atlantic are weakening.

All of their discoveries point in the same direction.

As scientists, these people have dedicated their lives to objectively understanding the ways of nature for the betterment of humankind. They are not hired by a wealthy and powerful lobby, they are not in conspiracy with one another, trying to corrupt and destroy civilization.

These are just people like you and me.

They look at bubbles through a microscope, analyze it, and present what they find.

There are dozens of satellites and weather balloons observing the Earth's present activities from above, measuring changes in ozone concentration, greenhouse gas levels, the greenhouse effect, global surface and atmospheric temperature, land-use changes, water circulation, and more. There are hundreds of ice cores from deep below the surface that take us back in time, showing us the gases and particles of past atmospheres which allows us to decipher global temperatures and climates of the ancient world, and compare them with our present situation. There are super-computers that recreate the past and project the future conditions with ever-increasing levels of complexity, speed, and accuracy.

All of this is thrown into a collective kettle that continues to brew an ever clearer picture of Earth-System dynamics. Scientists do not necessarily do their research with the purpose of having it considered within a larger context, but when it concerns global temperature change and the fate of modern settlements, it has to be taken into account.

The Earth is a big place and constantly changing, it is not an isolated object in a neutralized laboratory. So it is natural that discrepancies in measurements appear occasionally. One of the biggest sources of controversy has been the discrepancy in average global temperature measurements taken at ground level, because the measurements taken at the surface show a higher temperature than those taken in the upper atmosphere.

But this discrepancy can be expected. As the greenhouse effect intensifies, the upper atmosphere cools, because a higher concentration of greenhouse gases in the atmosphere reflects more solar radiation back into space. At the same time, the higher concentration of greenhouse gases allow less heat energy to escape from the surface, reflecting it back to the surface. Anyway, we do not live in the clouds, we live on the surface, which is undeniably warming.

No scientist will ever say they are 100% certain until an experiment has been repeated ad nauseam and produced the same results. A global experiment like the one we are presently a part of cannot be replicated in a laboratory, it can not be reproduced. We can only use correlations of evidence from similar times in the past and look at present trends.

There is 99% certainty on the reality of a present global warming trend. There is 99% certainty that the present global warming trend is a result of human activity since the industrial revolution.

Is 99% certainty enough for you?

Nevertheless, there will be skeptics who use the 1% of uncertainty to confuse the issues and support a 'wait and see' attitude. An attitude that allows us to follow our present path without any concern of where it might lead; even if it leads us into the rocky coast.

Skeptics and skeptical inquiry are of course important for society. Skepticism keeps us from being complacent or too agreeable with the conventional framework. Skepticism often promotes ingenuity and progress as well.

°NO DOUBTS GLOBAL WARMING IS REAL, U.S. EXPERTS SAY
REUTERS
3 DECEMBER 2003

There can be no doubt that global warming is real and is being caused by people, two top U.S. government climate experts said.

Industrial emissions are a leading cause, they say -- contradicting critics, already in the minority, who argue that climate change could be caused by mostly natural forces.

"There is no doubt that the composition of the atmosphere is changing because of human activities, and today greenhouse gases are the largest human influence on global climate," wrote Thomas Karl, director of the National Oceanic and Atmospheric Administration's National Climatic Data Center, and Kevin Trenberth, head of the Climate Analysis Section at the National Center for Atmospheric Research.

"The likely result is more frequent heat waves, droughts, extreme precipitation events, and related impacts, e.g., wildfires, heat stress, vegetation changes, and sea-level rise," they added in a commentary to be published in Friday's issue of the journal *Science*...

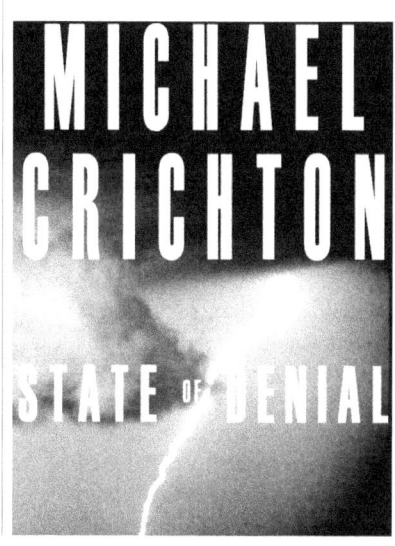

°U.S. STUDY LINKS HUMAN ACTIVITY TO GLOBAL WARMING
REUTERS
26 AUGUST 2004

Warmer temperatures in North America since 1950 were likely caused in part by human activities, the Bush administration said in a report that seems to contradict the White House position there was no clear scientific proof on the causes of global warming...

°ACCEPT IT: KYOTO WON'T STOP GLOBAL WARMING
THE GLOBE AND MAIL ONLINE
9 SEPTEMBER 2002

So Parliament will vote on the Kyoto Protocol before year's end. If Parliament assents, will it matter? Are efforts to reduce greenhouse-gas emissions the best way to spend our money in the face of global warming? What's the logical, rather than ideological, landscape surrounding the issue?

...When climatic conditions change on Earth, they often do so abruptly. The natural reasons for this can be highly varied, ranging from volcanoes to the state of the oceans and icecaps, Sun activity and external shocks, such as asteroids. The complexity of the environment suggests there are many "tipping points" where straws break camels' backs, and change comes rapidly. Even in the short history of humanity, whole societies have been decimated by relatively modest climatic change.

...Why are we furiously debating the best strategies to change the course of events marginally, rather than developing strategies to adapt to the body of what might happen?...

But, it is one thing to be skeptical about the predicted consequences of future climate changes. It is quite another thing to ignore the fact that global greenhouse gas levels rise and fall in tandem with the average global surface temperature. It is like the difference between being skeptical of an accurate weather forecast and being skeptical of whether or not gravity exists.

Nobody knows exactly what will happen in 100 years, in 50 years, in 20 years, in 5 years. But, as the level of technological achievement has advanced, the degree of uncertainty has been considerably reduced compared to 30 years ago. The degree of uncertainty continues to be reduced as we simply open our eyes and see the observable changes in the Earth.

Is the Earth experiencing global changes? Yes.

Is the Earth going to go through a great contortion next year?

Probably not, but maybe.

Is the Earth changing as a result of increased greenhouse gas levels and increased average global surface temperature? Yes.

Will next year bring more hurricanes than this year?

Perhaps.

Will the annual average number of hurricanes during the next 50 years be greater than the annual average number of hurricanes during the past 50 years?

Most likely.

Why do we have weather forecasters? So we can prepare.

Why do we have tornado sirens? So we can prepare.

Why do we have stock market analysts? So we can prepare.

Why do we save money for retirement? So we can prepare.

Who are you going to trust with your future?

Are you going to believe the few skeptics who believe the Earth is not experiencing global changes that are already disruptive. Those who believe there is nothing to worry about and base their arguments on anomalous events and selective research inconsistencies?

Or are you going to believe the other 99% of the world's scientists, organizations, insurance agencies, emergency workers, politicians, economists who are doing the research, taking the measurements, deciphering the results, and taking action?

This is the difference between the skeptics and the consensus.

Just as Galileo was a skeptic about the motion of the heavens, scientists in the field of global temperature change were once the skeptics of convention, believing that global changes were occurring. But what was once a fringe theory is now the established fact.

The skeptics have become the convention. Just like Galileo.

We now know the Earth moves around the Sun.

We now know the Earth is changing in ways we never thought imaginable as a result of human and natural activities.

It is one thing to be skeptical of a select piece of research that does not fit into the puzzle of climate change. It is another thing entirely to be skeptical of the other thousands of pieces that do fit.

It is also one thing to be skeptical of the policies designed to reduce greenhouse gases and 'stop' climate change deserve their skeptics as well. I am skeptical of the policies now being pursued to 'combat climate change,' but capitalizing on the 1% of uncertainty to do absolutely nothing is another thing altogether.

It is also understandable for those to be skeptical who are not yet familiar with the evidence and issues surrounding the topic.

But, for those who have seen the evidence, can understand the situation, and are still 'Skeptics,' a more appropriate term might be 'Denialists.'

These people are not heretics for their beliefs, they just have a distorted perception of reality, which everyone is entitled to.

Perhaps in 25 years I will be the one to look back at the beginning of this century and wonder what was I thinking. I really doubt it, but I would be joined by good company if that were the case.

Free will allows us to believe whatever we want about the existence of God, gravity, or global warming. But free will does not guarantee that one will do the best thing to promote one's survival.

There is plenty of evidence lifting the fog of uncertainty and revealing the rocky coast on the horizon.

Whether we believe in the rocky coast or not, the Earth could care less. Just because we do not believe the rocky coast exists, does not mean we will not slam into it.

Choices are choices.

The choices we make every day always affect the opportunities offered to us tomorrow.

We might not be able to choose to avoid the rocky coast, but at least we have the opportunity to prepare for impact.

°SCIENTISTS CLAIM FINAL PROOF OF GLOBAL WARMING
THE TIMES ONLINE
6 MAY 2004

Powerful evidence for global warming has been discovered by scientists funded by the US Government, demolishing the chief argument of sceptics who deny that the phenomenon is real. A new analysis of satellite data has revealed that temperatures in a critical part of the atmosphere are rising much faster than previously thought, strengthening the scientific consensus that the world is warming at an unnatural rate.

The discovery resolves one of the most contentious anomalies in climate science, which has often been invoked by the Bush Administration to question whether man-made global warming is happening.

While it is generally accepted that surface temperatures are increasing by an average of 0.17C (0.31F) per decade, satellites have been unable to detect a parallel trend in the troposphere the lowest level of the atmosphere, extending 7.5 miles above the ground, in which most weather occurs.

This lack of tropospheric warming has long puzzled scientists, as it is predicted by all the major models of climate change. It has also been seized on by a small but vocal minority of scientists, who have used it to raise doubts about whether global temperatures are rising at all. The enigma, however, has been explained by a team led by Qiang Fu, of the University of Washington in Seattle. His research reveals that the troposphere is warming almost precisely as the models predict it should: by about 0.2C (0.4F) per decade. Satellites have not previously detected the trend as they have been confused by colder temperatures in the atmospheric layer above.

The findings, details of which are published today in the journal *Nature*, provide one of the final pieces of proof that global warming is taking place, and that it is a human-induced phenomenon. Sceptics have often argued that if temperatures are rising at all, this is down to natural variation in the climate as the world emerges from a "little Ice Age". The tropospheric trend, however, is precisely what scientists would expect to see if man-made emissions of greenhouse gases were causing it to heat up. "I think this could convince not just scientists but the public as well," Dr Fu said.

Mike Hulme, director of the Tyndall Centre for Climate Change Research in Norwich, said: "It will become that much harder for people to claim that the world isn't warming and that the warming isn't caused by greenhouse-gas emissions."

..."Because of ozone depletion and the increase of greenhouse gases, the stratosphere is cooling about five times faster than the troposphere is warming, so the channel 2 measurement by itself provided us with little information on the temperature trend in the lower atmosphere," Dr. Fu said...Dr. Hulme said that while the results further confirm the overwhelming scientific consensus that man-made global warming is a proven phenomenon, he would be surprised if it were accepted by critics.

"Im under no illusions that it will knock down the critics altogether," he said. "In some quarters, people hold almost fundamentalist beliefs that are immune to carefully reasoned argument. A new paper that seems to take the legs away from one of their critiques may unfortunately not make much difference to their arguments.

"It is the totality of the evidence that has convinced the vast majority of experts that the planet is warming: surface temperature recordings, rises in sea level, retreating glaciers, shifting species domains. "The compendium of evidence from all these different sources means the overwhelming majority of scientists feel justified in warning society about this."

DATA DESIGNS

°CLIMATE MODELING
MUST CONSIDER ALL
'GREENHOUSE' GASES
TERRADAILY.COM
6 DECEMBER 2000

Global warming is a lot more complicated than is generally assumed, and we may have to rethink how we deal with the issue says University of Ottawa and Ruhr University (Germany) geologist Jan Veizer and his colleagues from Belgium who have assembled a very different picture of the physical and chemical conditions that have contributed to the past warming or cooling of our planet.

...Our own industrial contribution may enhance a given natural trend but not reverse it. The release and consumption of carbon dioxide goes hand in hand with other greenhouse gases.

In order to mitigate our impact on this balance, we should take into account the cumulative role of all greenhouse gases, particularly of water vapour as the leading agent that modulates our climate...

atmosphere = air

biosphere = life

cryosphere = ice

geosphere = land

hydrosphere = water

Scientists generate future scenarios of global temperature change with modern day supercomputers. Several decades ago, computers were generating rough estimates of how the global temperature and climates would look today and they turned out to be fairly accurate. At that time, a computer which now fits in a briefcase was the size of a garage. As computing capacity continues to increase in size, speed, and ability, the virtual reality it produces becomes more accurate and more reliable.

Today, GENERAL CIRCULATION MODELS or GLOBAL CLIMATE MODELS (GCMs) are incredibly complex and produce incredible scenarios about the future using incredibly powerful and fast computing technology.

Incredible.

Modeling the future global climate is a bit different than weather forecasting, but not totally different. Weather forecasting looks at the initial state of an event and infers how this event will evolve, looking at what will happen in a near future of hours or days. When a meteorologist looks at information on cloud formations, a prediction is generated about the most likely path those clouds will take and how they will develop. Weather forecasts are shorter in time frame and localized in scope. They are also not as sensitive to small changes in the variables used.

On the other hand, climate modeling does not infer results from an initial state. Climate modelers first establish a foundation which incorporates as many of the interrelated systems and activities of the Earth as possible, things that change over a longer period of time. Information about different variables are inserted and the forecast produced shows the likely outcome in a more distant future.

GCM's look decades into the future and are global in scope. They look at long term problems, probabilities and trends. They are more sensitive to the many variables involved and produce different results when the input of a single variable is different. Hence, there is a level of skepticism surrounding global climate models.

The five Earthly systems included in climate modeling are the ATMOSPHERE, the BIOSPHERE, the CRYOSPHERE, the GEOSPHERE, and the HYDROSPHERE. These systems all function in different time frames, have different sensitivities and react differently to different stimuli. All the information known about how these systems function and how they react over time to certain stimuli are poured into supercomputers, mixed with mathematical equations and stirred together with billions of computations per second to generate a future landscape that we and our descendants will inhabit.

GCM forecasts show how the systems of the Earth will behave when exposed to different variables and different circumstances at different points in time. GCM forecasts include information about the increase

supercruncher

or decrease in average surface temperature, snow and ice loss, precipitation changes, weather extremes, seasonal shifts, biodiversity dynamics, and greenhouse gas levels.

The word 'prediction' has two meanings. One is 'foretelling the future' like a fortune teller. The other is 'describing in advance what the outcome of an experiment will be.' Since global temperature change during the 21st century is a global experiment, the second definition is not that far off.

But forecasting the future is also an attempt to describe in advance the outcome of global temperature and climate changes, in order to foretell how settlements will be affected. So in this way, global climate models are used to foretell the future so that we may adequately prepare for the consequences predicted.

A climate model cannot predict with absolute certainty exactly where, when, and who will be affected with what problem or benefit, just as a weather forecast cannot predict with absolute certainty exactly where, when, and who will be affected by the approaching storm.

A climate model can be accurate in making a prediction that natural disasters on average will increase in frequency and severity; winters will become milder on average in some places and stormier on average in others; the onset of Spring will arrive earlier on average in some places; droughts will be more persistent in places.

Do we require 100% certainty from a weather forecast before we take cover from an approaching tornado? Do we wait and see if the tornado will rip off the roof?

Should we require 100% certainty from climate change forecasts before we prepare? Should we wait and see if global temperature change will disrupt our settlements before we prepare?

The future can never be predicted with 100% accuracy. But 99% sure is close.

And we keep getting closer.

°NASA STAKES OUT WORLD LEADERSHIP WITH NEW CLIMATE SUPERCOMPUTER
ENVIRONMENTAL NEWS NETWORK
23 JULY 2001

Questions about the effects of global warming will soon be a hallmark of the past. A new supercomputer at the National Aeronautics and Space Administration (NASA) is crunching climate data at warp speed. What used to take a year to calculate might be done in less than a day on the new machine.

NASA scientists are using the most powerful parallel supercomputer of its kind, 10 times more powerful than today's supercomputers, to evaluate the global impact of natural and human induced activities on the Earth's climate.

Developers say the new 512 supercomputer is 10 times more powerful than anything scientists have had until now. "This substantial increase in performance allows us to complete Earth climate simulations in days, rather than months," said Dr. Ghassem Asrar, associate administrator for earth science at NASA Headquarters in Washington, D.C...

°COMPUTER MODELS GENERATE EXTREME CLIMATE EVENTS
TERRADAILY.COM
16 JANUARY 2001

When scientists run computer models to simulate climate events, they often add elements, such as effect of volcanic eruptions, adding ice sheets, or altering the concentrations of greenhouse gases in the model's atmosphere. One NOAA scientist and his colleague discovered a large, abrupt climate event without the additions.

"When I first saw the results, I thought that I had bad data," said Ronald Stouffer, a meteorologist at NOAA's Geophysical Fluid Dynamics Laboratory in Princeton, N.J. "I would never would have guessed the 'bad data' was a very interesting event."

The "interesting event" is a severe and abrupt cooling of the North Atlantic Ocean near Greenland. The event will be described by Stouffer and Alex Hall of the Lamont-Doherty Earth Observatory in Palisades, N.Y. in the Jan. 11 issue of the science journal *Nature*...

°NCAR RELEASES NEW VERSION OF PREMIER GLOBAL CLIMATE MODEL

NCAR-UCAR NEWS RELEASE
22 JUNE 2004

The National Center for Atmospheric Research (NCAR) on June 23 is unveiling a powerful new version of a supercomputer-based system to model Earth's climate and project global temperature rise in coming decades. Scientists will contribute results from the new system to the next assessment by the Intergovernmental Panel on Climate Change, an international research body that advises policymakers on the likely impacts of climate change.

The system, known as the Community Climate System Model, version 3 (CCSM3), indicates in a preliminary finding that global temperatures may rise more than the previous version had projected if societies continue to emit large quantities of carbon dioxide into the atmosphere.

CCSM3 shows global temperatures could rise by 2.6 degrees Celsius (4.7 degrees Fahrenheit) in a hypothetical scenario in which atmospheric levels of carbon dioxide are suddenly doubled. That is significantly more than the 2 degree Celsius (3.6 degree Fahrenheit) increase that had been indicated by the previous version of the model.

NCAR developed the model in collaboration with researchers at universities and laboratories across the country, with funding from the National Science Foundation, the Department of Energy, the National Oceanic and Atmospheric Administration, and the National Aeronautics and Space Administration. NCAR is releasing the model results and the underlying computer codes to atmospheric researchers and other users worldwide.

William Collins, a NCAR scientist who oversaw the development of CCSM3, says researchers have yet to pin down exactly what is making the model more sensitive to an increased level of carbon dioxide. But he says the model overall is significantly more accurate than its predecessor.

"This model makes substantial improvements in simulating atmospheric, oceanic, and terrestrial processes," Collins says. "It has done remarkably well in reproducing the climate of the last century, and we're now ready to begin using it to study the climate of the next century."

...CCSM3 is one of the world's leading general-circulation climate models, which are extraordinarily sophisticated computer tools that incorporate phenomena ranging from the effect that volcanic eruptions have on temperature patterns to the impact of shifting sea ice on sunlight absorbed by the oceans.

Climate models work by solving mathematical formulas, which represent the chemical and physical processes that drive Earth's climate, for thousands of points in the atmosphere, oceans, sea ice, and land surface. CCSM3 is so complex that it requires about 3 trillion computer calculations to simulate a single day of global climate...

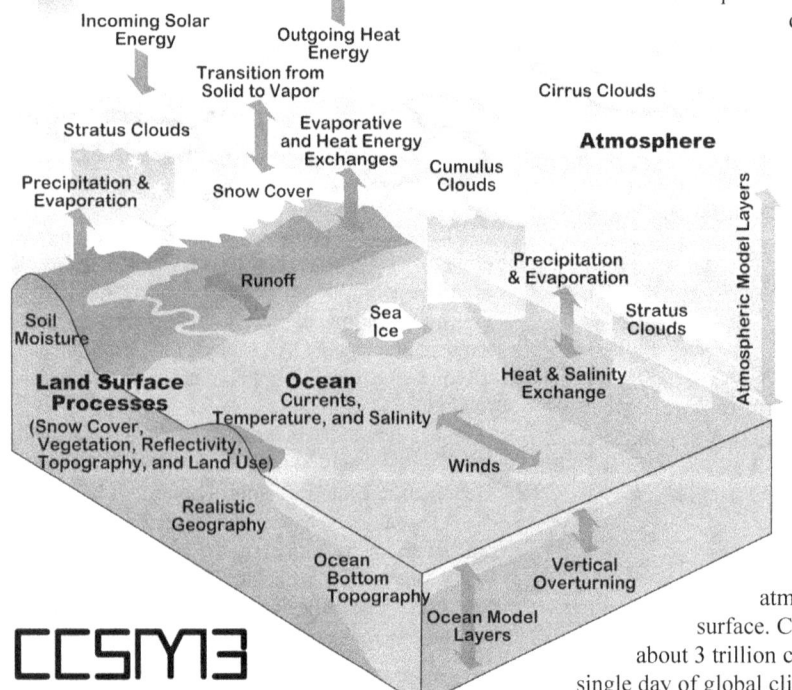

Incoming Solar Energy

Outgoing Heat Energy

Transition from Solid to Vapor

Stratus Clouds

Evaporative and Heat Energy Exchanges

Cirrus Clouds

Atmosphere

Precipitation & Evaporation

Snow Cover

Cumulus Clouds

Soil Moisture

Land Surface Processes
(Snow Cover, Vegetation, Reflectivity, Topography, and Land Use)

Runoff

Sea Ice

Precipitation & Evaporation

Stratus Clouds

Atmospheric Model Layers

Ocean
Currents, Temperature, and Salinity

Heat & Salinity Exchange

Realistic Geography

Winds

Ocean Bottom Topography

Vertical Overturning

Ocean Model Layers

CCSM3

A-PROXY-MATION

All of the scientists involved with researching the present trends of global warming and climate change can tell us what is happening now and we use that knowledge to project a future using supercomputers.

There is another very important discipline that we rely on to help us decipher what may lie ahead. This is where PALEOCLIMATOLOGY enters the stage from deep below the Earth's skin.

The Earth is a giant time capsule with evidence of its dynamic past buried all over the world. There are ancient atmospheres trapped in gas bubbles, ancient fires recorded in tree rings, meteor fragments from ancient collisions, and pollen from plants long gone.

Gas bubbles found in ice cores are used to decipher greenhouse gas concentrations and corresponding temperatures of a certain place from a given time. Sediment layers show what types of dust were circulating and can tell us about eruptions, impacts, and century-long droughts. Boreholes can show us temperature differences when compared to historical records.

From these and many other pieces of evidence, paleoclimatology can tell us when global and local changes in temperature and climate occurred, and sometimes even how they were initiated.

We call these pieces of evidence PROXIES.

By definition, a proxy-data is 'the data derived from deposits, which allow us to infer numerical values of meteorological and other parameters in the past.' Human ingenuity coupled with technology offers us very advanced methods to gather, process and derive results from proxies. Using these methods, scientists are able to recreate past scenarios, compare them to the present situation, and predict future possibilities.

When studying the past to learn about the future, it is not only important to look at the individual pieces of evidence, but also to look at the associations between pieces of evidence and observe the entire mosaic. All of this information gives us a very good understanding of how things have been and how they might become.

There are of course anomalies, uncertainties, conflicting data sets, and curiosities, but the overwhelming majority of proxy evidence tells us clearly that the Earth changes its global temperature in tandem with greenhouse gas levels; that global temperature and climate changes can be both gradual and abrupt; and that any change whatsoever is destabilizing to the life forms and settlements dependent on the way things are before changes occur.

Where do we find ourselves at the beginning of the 21st century?
Where will we find ourselves at the beginning of the 22nd century?
Let us once again dig into the evidence of our planet's past.
Let us once again scan the planet for its present trends.
The planet is not only a time capsule. It is a crystal ball.
Go ahead, take a look.

-The farther backward you can look, the farther forward you are likely to see.-

-Winston Churchill

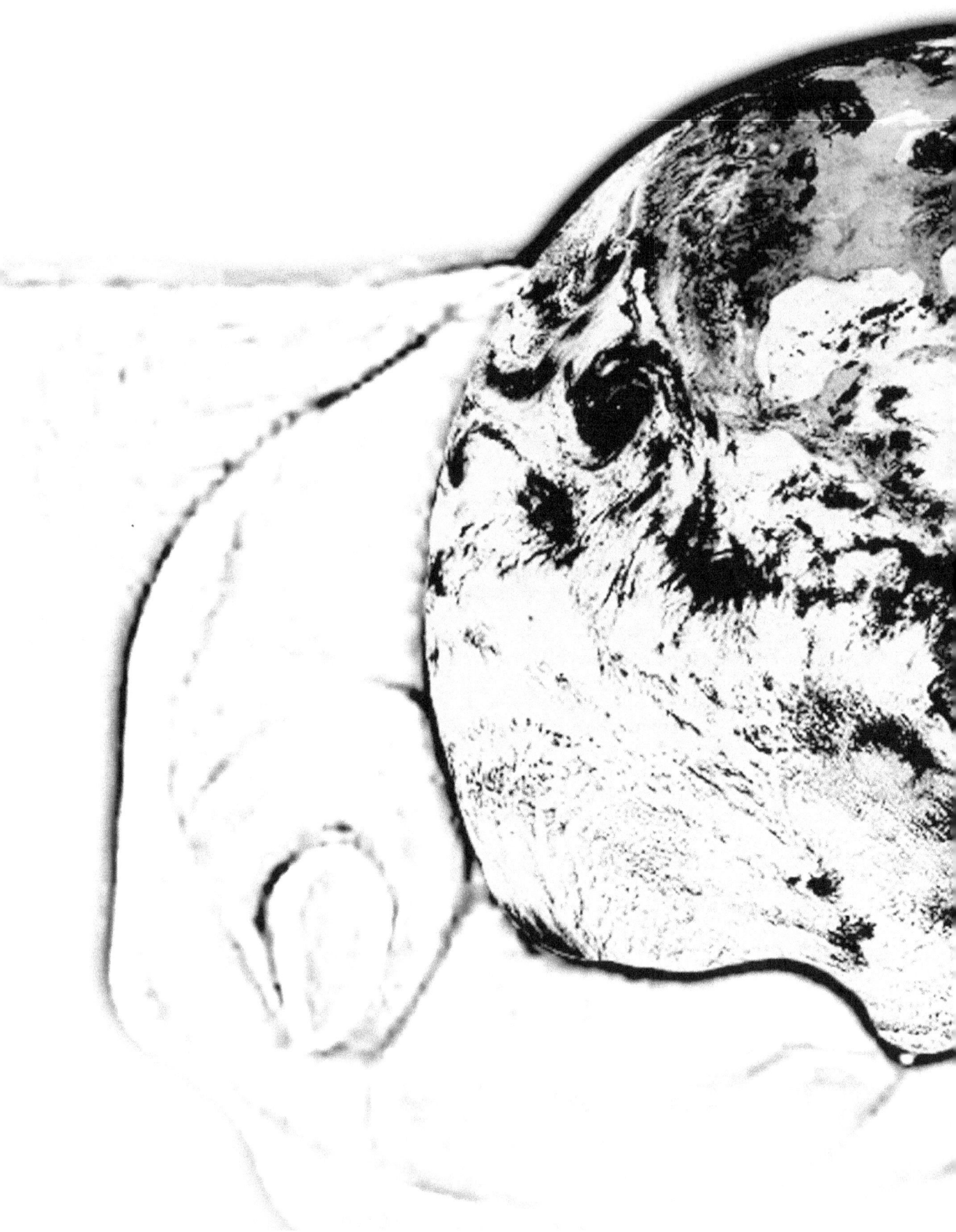

WHAT SHOULD WE LOOK AT?

HOW ABOUT
BOREHOLES

°BOREHOLE TEMPERATURES CONFIRM GLOBAL WARMING
ENVIRONMENTAL NEWS NETWORK
17 FEBRUARY 2000

Temperature readings taken from more than 600 holes drilled into Earth's surface confirm that a 500-year warming trend accelerated in the latter half of the 20th century.

"Some 80 percent of that warming corresponds with the growth of industrialization," said Henry Pollack, a geology professor at the University of Michigan in Ann Arbor and co-author of the study in today's issue of *Nature*...

°BOREHOLE TEMPERATURES CONFIRM GLOBAL WARMING PATTERN
UNISCI.COM
27 FEBRUARY 2001

Measuring temperatures inside holes in the ground is an accurate way of showing that Earth's Northern Hemisphere has warmed about 2 degrees Fahrenheit (1.1 degrees Celsius) since the Industrial Revolution began, University of Utah scientists have found.

"This is another piece of independent evidence that says global warming is real, and that it is proceeding at a rate faster than we have observed in recent geologic history," said David S. Chapman, graduate school dean and professor of geology and geophysics at the University of Utah.

"The warming we found implies a link between global warming and greenhouse gas emissions from industrialization" that began in about the 1750s, said Robert N. "Rob" Harris, an assistant professor of geology and geophysics.

"The warming is real and significant."...

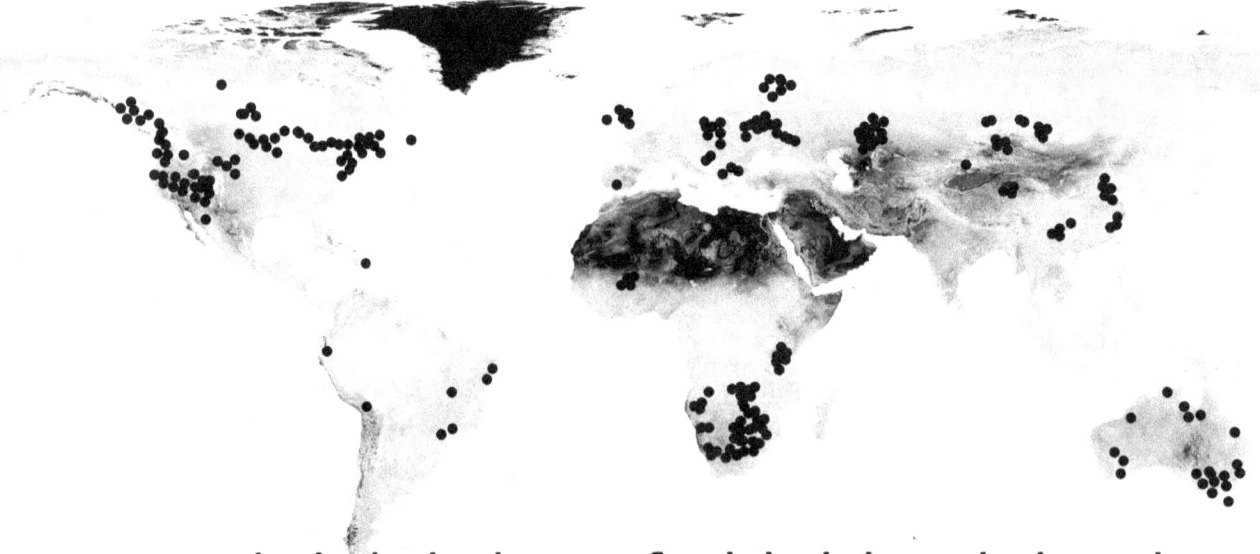

● general distribution of global borehole sites

HOW ABOUT ICE CORES

hard core bore

°ICE CORE REVEALS A WORRYING TRUTH ABOUT EARTH'S CLIMATE
THE INDEPENDENT
10 JUNE 2004

The deepest and oldest ice core yet drilled in the Antarctic suggests that the world's climate is headed for an unprecedented period of turmoil brought about by man-made greenhouse gases.

Chemical analysis of the ice within the core - nearly 2 miles long - has revealed details of the eight previous ice ages that have affected the Earth during the past 740,000 years. Scientists said yesterday that the present climate most closely resembled the warm "interglacial" period about 470,000 years ago, but with the difference that this time temperatures were set to spiral upwards as a result of global warming.

In a study published today in the journal *Nature*, the international team of scientists from 10 European countries warns that the Earth's climate would now be in a highly stable period if it were not for the extra carbon dioxide being pumped into the atmosphere from human activities. "Given the similarities between this earlier warm period and today, our results may imply that without human intervention, a climate similar to the present one would extend well into the future," the scientists say.

Eric Wolff, a senior member of the team from the British Antarctic Survey in Cambridge, said that anyone who suggested human- induced global warming was beneficial because it would avert the next ice age was misguided. "If the climate is left to its own devices, we have about another 15,000 years to go before the next ice age. If people say global warming is good because it stops us going into another ice age, they are wrong because we are not about to go into another ice age," Dr Wolff said.

The deepest ice cores were drilled at a site known as Dome C, where the East Antarctic ice sheet is about 3.4km (2 miles) thick. It is one of two sites being drilled on the frozen continent as part of the European Project for Ice Coring in Antarctica (Epica), which began field work in 1996.

Tiny bubbles of air were trapped in the ice when it formed from snow falling on the Antarctic ice sheet. That ancient air is being analysed to see how much carbon dioxide and other greenhouse gases such as methane were present in the atmosphere over many hundreds of thousands of years.

The ice cores retrieved from the Epica study will double the length of the record of greenhouse gases in the atmosphere, making it possible to judge just how unusual are today's high concentrations of carbon dioxide - the principal greenhouse gas behind global warming. "We've never seen greenhouse gas concentrations anything like as high as that we're seeing today," Dr Wolff said...

HOW ABOUT FOSSILS

°FOSSILS PROVIDE EVIDENCE OF A WARM ARCTIC
ENVIRONEMNTAL NEWS NETWORK
6 JANUARY 1999

Fossilized bones from several crocodile-like beasts known as champsosaurs have provided evidence of Arctic warming during the late Cretaceous period, scientist report recently in the journal *Science*.

The fossils indicate that at least once in Earth's history high amounts of the greenhouse gas carbon dioxide warmed Earth to much higher temperatures than usual, according to scientists at the University of Rochester in Rochester, N.Y.

The carbon dioxide was spewed by massive volcanic eruptions about 90 million years ago and as a result, Arctic temperatures were as warm as present-day Florida...

HOW ABOUT METEORS

°SCIENTIST SAYS ICE METEORS A SIGN
OF CLIMATE CHANGE
REUTERS
27 SEPTEMBER 2002

A Spanish scientist says global warming may be to blame for giant blocks of ice which fall from clear skies and rip gaping holes in cars and houses. Jesus Martinez-Frias has spent the last two-and-a-half years investigating so-called megacryometeors -- ice meteors -- which tend to weigh more than 22 lb and have been known to leave five feet holes in houses.

He fears the formation of these hailstone-like blocks on clear days could be a worrying symptom of climate change. "I'm not worried that a block of ice might fall on your head ... but that great blocks of ice are forming where they shouldn't exist," said Martinez-Frias, director of planetary geography at Spain's Astrobiology Center in Madrid. "Components of the atmosphere, like ozone and water, are changing in different levels of the atmosphere.

...We think these signs could be evidence of climate change," he said in a telephone interview with Reuters...

HOW ABOUT ROCKS

°EARTH'S ROCKS SHOW SIGNS OF
GLOBAL WARMING
CBC ONLINE
12 APR 2002

The rocks on the Earth's continental crust have warmed significantly over the past 50 years, a team of U.S. and Canadian researchers have found. The finding adds to the body of evidence collected for global warming in the Earth's oceans, atmosphere and ice.

"Our findings remove any last doubt that this is anything other than a global phenomenon," said University of Michigan geology Prof. Henry Pollack, who collaborated on the study appearing in the April 15 issue of the *Geophysical Research Letters*.

Until now, most signs of global warming have all been based on surface measurements. About a year ago, another group of researchers determined how much heat had been gained during the last half of the 20th century throughout the Earth's atmosphere, oceans and frozen or snow covered surfaces. But their analysis didn't included continental rock, which covers almost 30 per cent of the planet's surface...

HOW ABOUT CORAL

°CORAL PROVIDES CLUES TO CLIMATE CHANGE
ENVIRONMENTAL NEWS NETWORK
6 FEBRUARY 2000

Like tree rings that reveal the age of centuries-old cedar, growth rings in Indian Ocean coral tell how El Niño, the warm-water phenomenon of the tropical Pacific Ocean, influences marine temperatures a continent away.

"We found a record of climate change that reflected the influence of the tropical Pacific (El Niño) on [the Indian Ocean]. This record tells us a story that has broad climate implications," said Julia Cole of the University of Arizona and co-author of a report on coral growth rings. "ENSO (El Niño/Southern Oscillation) has a very long reach. El Niño brings more rain to East Africa just as it does to Arizona, but it also warms the oceans there.

..."One new piece of information the long-term perspective gives us is that the temperature in the region over the last decade was warmer than during the past two centuries," said Cole. "There are not many records of actual ocean temperatures, so adding ocean measurements to the pool, that's one more piece of information that makes a strong case something unusual is going on."

The unusual occurrence is global warming...

°ANTARCTIC MUD REVEALS ANCIENT EVIDENCE OF GLOBAL CLIMATE CHANGE
STANFORD REPORT
16 JANUARY 2002

Scientists concerned about global warming are especially troubled by dramatic signs of climate change in Antarctica -- from rapidly melting glaciers to unexplained declines in penguin populations.

Records show that average winter temperatures today are 10 degrees higher in parts of Antarctica than 50 years ago. If that warming trend continues, say many climate experts, the vast Antarctic ice sheets could melt, causing catastrophic coastal flooding as the world's oceans rise. Ironically, say researchers, the most pristine continent on Earth is heating up primarily because of increased greenhouse gas emissions from cars, power plants and other human endeavors elsewhere on the planet.

But new geologic evidence unearthed from deep-sea mud deposits strongly suggests that Antarctica experienced periods of extreme warming and cooling long before the invention of the automobile.

"We've got a sedimentary record that reveals very significant changes in water temperature and ice melt during the past 7,000 years," said Robert Dunbar, professor of geological and environmental sciences. "The cause of these highly variable climate changes is still a mystery."

The researchers based their study on a biogeochemical analysis of sediments obtained during a recent cruise of the JOIDES Resolution, a research vessel operated by the Ocean Drilling Program (ODP) -- an international project dedicated to exploring the geological history and evolution of the Earth. ODP is principally funded by the National Science Foundation with additional support from institutions representing nearly two dozen other countries, including Germany, Japan and Australia.

In 1998, ODP scientists extracted a 150-foot-long sediment core from the muddy bottom of the Palmer Deep -- a submerged section of the continental shelf along the west Antarctic Peninsula about 3,000 feet below sea level. The sediment sample was loaded with the shells of microscopic creatures called diatoms dating back some 10,000 years to the beginning of the Holocene -- the most recent geologic epoch.

"The Antarctic Peninsula is an ideal region to investigate climate change at decadal to millennial time scales due to its location in one of the Earth's most dynamic climate systems," noted Dunbar. "The ODP sample gives us the first continuous, high-resolution Holocene sediment record from the Antarctic continental margin."...

°MAJOR TEMPERATURE RISE RECORDED IN ARCTIC THIS YEAR: GERMAN SCIENTISTS
AGENCE FRANCE-PRESSE
28 AUGUST 2004

German scientists probing global warming said they had detected a major temperature rise this year in the Arctic Ocean and linked this to a progressive shrinking of the region's sea ice.

Temperatures recorded this year in the upper 500 metres (1,625 feet) of sea in the Fram Strait -- the gap between Greenland and the Norwegian island of Spitsbergen -- were up to 0.6 C (1.08 F) higher than in 2003, they said in a press release received here.

The rise was detectable to a water depth of 2,000 metres (6,500 feet), "representing an exceptionally strong signal by ocean standards," it said.

The experts, from the Alfred Wegener Institute for Polar and Marine Research in Bremerhaven, have been recording temperatures aboard a specialised vessel, Polarstern (Pole Star), for the past six weeks. The sampling has been taking place in the West Spitsbergen Current, which carries warm water from the Atlantic into the Arctic Ocean.

The institute said water in the Fram Strait has been warming steadily since 1990 and over the past three years, satellite images had documented "a clear recession" of sea ice edges, both in the strait and the Barents Sea...

HOW
ABOUT
MUD

HOW
ABOUT
WATER

HOW ABOUT MORE ROCKS

°EARTH'S ANCIENT ATMOSPHERE TRAPPED IN ROCKS
ENVIRONMENTAL NEWS NETWORK
13 JULY 2000

Scientists on a quest to characterize the long-term chemical evolution of Earth's atmosphere need to understand what the air was like millions of years ago. To do this, they've come to realize they can leave no stone unturned.

Uncovering the signature of so-called "fossil air" in terrestrial rocks and sediment is reported for the first time in today's issue of *Nature*...

HOW ABOUT TREES

°CLUES TO BRONZE AGE COMET STRIKE
BBC NEWS ONLINE
25 MAY 1998

Evidence is growing that a huge comet smashed into the Earth about 4,000 years ago. Scientists are pointing to studies of tree-rings in Ireland which have revealed that about 2,354-2,345 BC there was an abrupt change to a colder climate. They have also highlighted discoveries by archaeologists in northern Syria of a catastrophic environmental event at about the same time. This is also about the time that Bronze Age civilisations collapsed...

HOW ABOUT THE SHAPE OF THE ENTIRE PLANET

°RESEARCH OFFERS EXPLANATION FOR EARTH'S BULGING WAISTLINE
NASA - JPL NEWS RELEASE
9 DECEMBER 2002

A team of researchers from NASA's Jet Propulsion Laboratory, Pasadena, Calif., and the Royal Observatory of Belgium has apparently solved a recently observed mystery regarding changes to the physical shape of Earth and its gravity field. The answer, they found, appears to lie in the melting of sub-polar glaciers and mass shifts in the Southern, Pacific and Indian Oceans associated with global-scale climate changes.

The team of researchers sought to find a climatic reason for the dramatic changes in Earth's gravity field observed since 1997. These changes have resulted from large redistributions of mass around the globe and are characterized by an increased bulge in Earth's equator and mass movement away from the poles - an occurrence known as oblateness, which can be thought of as the difference between a football and a soccer ball; the football has a larger radius at the equator

Their results are published in the December 6 issue of *Science*...

i'm feeling waterlogged

WE HAVE SEEN
THIS ALL BEFORE

STILL

BLIND?

WHERE SHOULD WE LOOK?

°SCIENTISTS MINE CLUES OF EARLY GLOBAL WARMINGS
ENVIRONMENTAL NEWS NETWORK
10 NOVEMBER 1997

Drilling into the sea floor south of Haiti has uncovered conclusive ashy evidence that multiple massive volcanic eruptions occurred roughly 55 million years ago in the Caribbean Basin. Those cataclysmic events appear to have caused abrupt inversion of ocean waters, triggering one of the most dramatic climatic changes ever, according to a geologist on the team that discovered the eruptions.

The inversion caused release of massive amounts of sea floor methane into the atmosphere, leading to global warming and possibly speeding evolution of countless new plant and animal species, including many primates and carnivores, the scientist believes. At the same time, close to half of all deep-sea animals went extinct, asphyxiated in the suddenly warmer and stagnant deep waters...

HOW ABOUT AT THE BOTTOM OF THE OCEAN

°OLDEST ICE CORE FROM THE TROPICS RECOVERED, NEW ICE AGE EVIDENCE
OHIO STATE UNIVERSITY PRESS RELEASE
3 DECEMBER 1998

An analysis of ice cores drilled from a glacier atop a Bolivian volcano is painting a vivid picture of climate conditions in the tropics over the past 25,000 years. The ice at the bottom of the cores was formed during the last glacial maximum -- the coldest part of the last ice age -- making it the oldest core recovered from the tropics.

In a paper in the journal *Science*, the research team describes a climate in the tropics that was different from what many researchers have thought. The findings are the latest result from a 20-year effort to build a global climate record that reaches from the North to the South Pole...

HOW ABOUT IN BOLIVIAN VOLCANOES

°ICE CORES SHOW RECORD OF CLIMATE DATING BACK 20,000 YEARS
OHIO STATE UNIVERSITY PRESS RELEASE
26 JULY 1995

A new analysis of ice cores retrieved from the heart of a Peruvian glacier has unlocked a record of past climate dating back at least 20,000 years. That record is painting a new picture of the climate in the tropics during the last ice age. The Amazon River basin environment was probably drier at the time, with less extensive rain forests and more savanna vegetation such as exists in Oklahoma today.

The report by researchers from Ohio State University and Louisiana State University was published in the July 7 issue of the journal *Science*. These findings are the latest from nearly two decades of study of ice cores from glaciers in the South American Andes, Greenland, Antarctica, China and the former Soviet Union and they provide what may arguably be the best view of how world climate varied over the last 20,000 years...

HOW ABOUT IN PERUVIAN GLACIERS

HOW ABOUT IN THE HIMALAYAS

°RESEARCHERS IN HIMALAYAS RETRIEVE HIGHEST ICE CORE EVER DRILLED
OHIO STATE UNIVERSITY PRESS RELEASE
21 NOVEMBER 1997

An expedition to a glacier atop the world's 14th highest peak in the Himalayan Mountains has returned with ice cores containing climate records that could reach back into the last glacial stage -- some 12,000 years ago. The international expedition, led by Ohio State University researchers, included American, Chinese, Peruvian, Russian and Nepalese members and capped perhaps their most successful season in two decades of retrieving frozen climate records...

°HIMALAYAN ICE REVEALS CLIMATE WARMING, CATASTROPHIC DROUGHT
OHIO STATE UNIVERSITY PRESS RELEASE
14 SEPTEMBER 2000

Ice cores drilled through a glacier more than four miles up in the Himalayan Mountains have yielded a highly detailed record of the last 1,000 years of earth's climate in the high Tibetan Plateau. Based on an analysis of the ice, both the last decade and the last 50 years were the warmest in 1,000 years. The core also showed a clear record of at least eight major droughts caused by a failure of the South Asian Monsoon, the worst of these a catastrophic seven-year-long dry spell that cost the lives of more than 600,000 people...

°RESEARCHERS DATE CHINESE ICE CORE TO 500,000 YEARS
OHIO STATE UNIVERSITY PRESS RELEASE
12 JUNE 1997

A team of scientists has reconstructed a detailed climate record for the last 130,000 years from a thousand-foot-long ice core they drilled into a glacier on the Tibetan Plateau in 1992. Analyses suggest that the record in the core actually may go back more than 500,000 years. The ice core draws a vivid picture of a climate that changed frequently and dramatically in that region throughout the last glacial sequence -- an interval that reached back 110,000 years through the last glacial period and into the warm stage, the Eemian, that preceded it. Their report was carried in this week's issue of the journal *Science*.

...With support from the National Science Foundation, the National Geographic Society, Ohio State and the Chinese National Science Foundation, an international expedition retrieved the core five years ago from the Guliya Ice Cap, a 77-square-mile glacier sitting 22,014 feet high in the Kunlun Shan Mountains of western China. Although Guliya is in the sub-tropics the ice is very cold, making it a valuable reservoir of ancient climate records.

The team, headed by Lonnie Thompson, professor of geological sciences at Ohio State, used mechanical and then thermal drills to remove a 1,012-foot (308.6-meters) core from the ice cap. The core, which reached through the ice to bedrock, was split and divided between Chinese and American researchers. Ohio State researchers cut their half of the core into 34,800 individual samples that were then tested for oxygen isotope ratios, dust, pollen, and nitrate, chloride and sulfate ions...

the himalayas as viewed from the international space station

HOW ABOUT RIVERS AND LAKES

°EARLY 'ICE-OFF' MAY BE SIGN OF GLOBAL WARMING
ENVIRONMENTAL NEWS NETWORK
22 APRIL 1999

Lakes in the Northeast are freezing over and thawing earlier, according to a professor of biological sciences at the University of Buffalo in New York. "Thus far, the records from a few selected lakes show, over time, an earlier spring ice-off date and shorter duration of ice cover," said Kenton Stewart. Stewart, who studies the freeze and thaw dates of lake ice as an indicator of global climate change, has data dating back to 1970 on 250 lakes in New York and hundreds of others in Maine, Wisconsin and Minnesota. "Because the shortened duration of lake ice cover seems to be associated with known increases in global temperature over the same time period, there is increased interest in lake ice dates as proxy indicators of climate change," he said...

°ICE RECORDS HOLD COLD TRUTH ABOUT CLIMATE CHANGE
ENVIRONMENTAL NEWS NETWORK
9 SEPTEMBER 2000

They combed newspaper archives. They pored over transportation ledgers. They even noted religious observances.

And when they were done amassing lake and river ice records spanning the Northern Hemisphere, the scientists made "a very robust observation," in the words of researcher John Magnuson. They saw a steady, 150-year warming trend.

"It is clearly getting warmer in the Northern Hemisphere," says Magnuson. "The importance of these records is that they come from very simple, direct human observations, making them very difficult to refute in any general way."

The study, which includes 39 records of freeze dates or breakup dates from 1846 to 1995, represents one of the largest and longest records of observable climate data ever assembled...

HOW ABOUT THE PERMAFROST AND THE TUNDRA

°PERMAFROST STUDIED FOR GLOBAL WARMING CLUES
ENVIRONMENTAL NEWS NETWORK
15 DECEMBER 1998

Scientists are monitoring temperature changes in frozen regions of the world in an effort to have a better and more realistic understanding of climate change.
Headed by Ken Hinkel, a geographer at the University of Cincinnati, the multi- national monitoring program tracks temperatures in permafrost, the mixture of permanently frozen water, soil, organic matter and pockets of unfrozen brine found covering 20 percent of the Earth's land area.

"Permafrost regions may serve as a harbinger of global change as temperatures rise in response to increased atmospheric greenhouse gases," said Hinkel. Hinkel has rounded up colleagues in more than 11 countries to collect and track temperature data and changes in permafrost. Among the nations participating are Austria, Canada, China, Denmark, Greenland, Kazakstan, Poland, Russia, Sweden, Switzerland, Tibet and the United States.

Hinkel would like more sites in Russia and Canada, two countries that encompass much of the Northern Hemisphere. Antarctica will be added to the list in 1999. The more than two dozen researchers already participating have placed precise temperature sensors called thermisters at about 70 permafrost sites to measure and record air, surface and underground temperatures every hour...

°TRACKING CLIMATE CHANGE ON THE TUNDRA
NASA EARTH OBSERVATORY NEWS
15 DECEMBER 2001

If global warming occurs as many scientists predict, changes in Arctic thaw lakes and their basins could make it worse. A team from the University of Cincinnati is now developing ways to monitor these basins, which make up the majority of the landscape on the Arctic Coastal Plain.

"The Arctic plays a crucial role in global change because it responds sensitively to changes in climate," said University of Cincinnati (UC) assistant professor of geography Wendy Eisner.

By 2100, the global temperature will rise 2.5 to 10.4 degrees Fahrenheit, according to predictions made earlier this year in a report from the Intergovernmental Panel on Climate Change.

Lakes form on top of frozen Arctic tundra when just the top layer of the tundra thaws in summer. These thaw lake basins contain a lot of peat in the frozen Arctic subsoil. If global warming occurs, the permafrost could melt and peat would start to decompose, releasing greenhouse gases into the atmosphere.

"This could accelerate global warming," said Eisner. "If this scenario turns out to be true, we would probably want to do everything we could to keep these basins intact and storing, not releasing, greenhouse gases."...

WE CAN LISTEN TO THOSE WHO LIVE AND TRAVEL UP NORTH

OR MAYBE WE CANNOT

°EXPLORER SAYS ARCTIC ICE THINNING NOTICEABLY
REUTERS
27 MAY 2001

The ice sheets covering the Arctic seas have thinned noticeably over the last seven years, most likely as a result of global warming, said a Norwegian explorer who has just skied alone across the top of the world. Boerge Ousland, speaking after an 82-day trip in which he traveled 1,300 miles from the northern tip of Russia to the North Pole and then down to Canada, said on Sunday he had seen other evidence which hinted strongly at the effects of climate change.

...Earlier studies showed the Arctic sea ice had thinned over the last 30 years or so to six feet from 10 feet and had shrunk by around six percent since 1975.

..." I saw between 50 and 60 polar bear tracks on the Russian side. In 1994 I saw two tracks, so that's a big, big change," he said. One explanation could be that thinning ice meant the bears needed to travel further to hunt seals, he added...

°AS ICE THAWS, ARCTIC PEOPLES AT LOSS FOR WORDS
REUTERS
18 NOVEMBER 2004

What are the words used by indigenous peoples in the Arctic for "hornet," "robin," "elk," "barn owl" or "salmon?" If you don't know, you're not alone. Many indigenous languages have no words for legions of new animals, insects and plants advancing north as global warming thaws the polar ice and lets forests creep over tundra.

"We can't even describe what we're seeing," said Sheila Watt-Cloutier, chair of the Inuit Circumpolar Conference which says it represents 155,000 people in Canada, Alaska, Greenland and Russia. In the Inuit language Inuktitut, robins are known just as the "bird with the red breast," she said. Inuit hunters in north Canada recently saw some ducks but have not figured out what species they were, in Inuktitut or any other language.

An eight-nation report this month says the Arctic is warming twice as fast as the rest of the planet and that the North Pole could be ice-free in northern hemisphere summer by 2100, threatening indigenous cultures and perhaps wiping out creatures like polar bears.

The report, by 250 scientists and funded by the United States, Canada, Russia, Sweden, Finland, Norway, Denmark and Iceland, puts most of the blame on a build-up of heat-trapping gases from human use of fossil fuels like coal and oil...

°SIGNS OF THAW IN A DESERT OF SNOW: SCIENTISTS BEGIN TO HEED INUIT WARNINGS OF CLIMATE CHANGE IN ARCTIC
WASHINGTON POST
28 MAY 2002

And so it has come to be, the elders say, a time when icebergs are melting, tides have changed, polar bears have thinned and there is no meaning left in a ring around the moon. Scattered clouds blowing in a wind no longer speak to elders and hunters. Daily weather markers are becoming less predictable in the fragile Arctic as its climate changes. Inuit elders and hunters who depend on the land say they are disturbed by what they are seeing swept in by the changes: deformed fish, caribou with bad livers, baby seals left by their mothers to starve. Just the other year, a robin appeared where no robin had been seen before. There is no word for robin in Inuktitut, the Inuit language.

...There is increasing evidence that the Arctic, this desert of snow, ice and killing cold wind, one of the most hostile and fragile places on Earth, is thawing. Glaciers are receding. Coastlines are eroding. Lakes are disappearing. Fall freezes are coming later. The winters are not as cold. Mosquitoes and beetles never seen before are appearing. The sky seems to be clapping as thunderstorms roll where it was once too cold for them.

"The Inuit always observed the Sun and astrology for direction and for weather," Jayko Pitseolak, an Inuit elder here, said through an interpreter. "We were taught that one day the world will change, and it has...

°WARMING CLIMATE DISRUPTS ALASKA NATIVES' LIVES
REUTERS
16 APRIL 2004

Anyone who doubts the gravity of global warming should ask Alaska's Eskimo, Indian and Aleut elders about the dramatic changes to their land and the animals on which they depend. Native leaders say that salmon are increasingly susceptible to warm-water parasites and suffer from lesions and strange behavior. Salmon and moose meat have developed odd tastes and the marrow in moose bones is weirdly runny, they say. Arctic pack ice is disappearing, making food scarce for sea animals and causing difficulties for the Natives who hunt them. It is feared that polar bears, to name one species, may disappear from the Northern hemisphere by mid-century...

WE HAVE HEARD
THIS ALL BEFORE

STILL

DEAF?

WHAT SHOULD WE LOOK WITH?

HOW ABOUT WITH HIGH PRECISION EQUIPMENT

°GREENLAND ICE YIELDS CLIMATE CHANGE CLUES
ENVIRONMENTAL NEWS NETWORK
13 OCTOBER 1998

Using instruments developed at the U.S. Geological Survey, scientists have for the first time accurately determined how much temperatures have changed at a Northern Hemisphere site in central Greenland during the last 50,000 years, through the end of the last ice age. Previous studies using plant pollens stored in lake sediments, chemical isotope ratios stored in glaciers and various other climate indicators, have shown that past climates have been both warmer and colder than the present.

...USGS geophysicist Gary Clow used high-precision equipment to measure small temperature variations within the Greenland Ice Sheet resulting from past climatic changes. These measurements were made in a 10,000-foot access hole drilled through the ice sheet by the European Science Foundation, using equipment that can measure temperature to within 0.0004 F. With these data, researchers from the University of Copenhagen, Denmark, and the USGS reconstructed the record of past temperature (climate) changes...

HOW ABOUT WITH AN OCEAN THERMOMETER

°ANTARCTICA'S CLIMATE CLUES
BBC NEWS ONLINE
26 DECEMBER 2001

If humans really are interfering with the Earth's climate and pushing up world temperatures, some of the best evidence could come from Antarctica. Changes in the environment on and around the White Continent can have far-reaching effects.

...The researchers on board were engaged in their annual survey of the currents here. They want to know how the temperature and movement of the water is changing. The Antarctic Ocean has the world's largest ocean current, carrying water clockwise around the continent and interacting with the Atlantic, the Pacific and the Indian oceans.

Mike Meredith of the British Antarctic Survey explained: "The waters that form around the Antarctic spread out and move northwards into the Atlantic, reaching the latitudes of the UK, so any changes in temperature for example will have a long-term effect on global climate. "Exactly how that works is something that nobody really knows, which is why the measurements we're making at the moment are so important."

The team is using a CTD (conductivity - which relates to the amount of salt in the water- temperature and depth) instrument to monitor the current. The instrument is like a giant thermometer and has to be lowered overboard using a crane-like device...

HOW ABOUT WITH A SPACE THERMOMETER

°SATELLITES ACT AS THERMOMETERS IN SPACE, SHOW EARTH HAS A FEVER
NASA-GSFC NEWS RELEASE
21 APRIL 2004

Like thermometers in space, satellites are taking the temperature of the Earth's surface or skin. According to scientists, the satellite data confirm the Earth has had an increasing "fever" for decades.

For the first time, satellites have been used to develop an 18- year record (1981-1998) of global land surface temperatures. The record provides additional proof that Earth's snow-free land surfaces have, on average, warmed during this time period, according to a NASA study appearing in the March issue of the Bulletin of the American Meteorological Society.

The satellite record is more detailed and comprehensive than previously available ground measurements. The satellite data will be necessary to improve climate analyses and computer modeling...

NASA AQUA

°NASA LAUNCHES WEATHER SATELLITE
ASSOCIATED PRESS
4 MAY 2002

NASA launched a satellite early Saturday on a $952 million mission to improve weather forecasts and track changes in the global climate. The Aqua satellite left Vandenberg Air Force Base shortly before 3 a.m. and separated from its rocket about an hour later at an altitude of 425 miles, said Alan Buis, a spokesman with NASA's Jet Propulsion Laboratory in Pasadena.

...Currently, forecasters send about 4,000 sensor-equipped balloons aloft every day to collect that data. Aqua will make 400,000 equivalent measurements each day, said Moustafa Chahine, an Aqua scientist...

°RUSSIA-NASA SHARE EARTH DATA
SPACE.COM
20 DECEMBER 2001

A Russian satellite hurled into orbit carries NASA gear to monitor global warming trends and keep an eye on ozone destruction in Earth's stratosphere.

The Stratospheric Aerosol and Gas Experiment (SAGE III) instrument was successfully launched December 10 aboard Russia's METEOR-3M satellite.

...SAGE III is part of NASA's Earth Observing System (EOS). The different SAGE III components will measure atmospheric gases including oxygen, water vapor and ozone; and aerosols such as volcanic ash and exhaust from smokestacks.

...NASA and the Russian Aviation and Space Agency (RASA) will share data from the first SAGE III instrument. Scientists will use this data to study human-derived and natural processes that contribute to environmental changes, primarily global warming and the destruction of ozone in the stratosphere.

°NEW WAY UPCOMING TO TEST CLIMATE CHANGE PREDICTIONS
UNISCI.COM
12 JUNE 2002

A team led by UK Royal Holloway geologist Dr. Michal Kucera will map sea-surface temperature of the Mediterranean over past millennia. The data will provide a new target to test the computer models on which our predictions of climate change are based.

We currently make climate change predictions using mathematical models developed using climate records of the last two centuries. The trouble is, the global climate was fairly static during this period, so it is difficult to know how accurate these models are in predicting climates very different from the current state.

The Earth has experienced large shifts in climate in the past and if a snapshot of the Earth surface at one of these intervals was available, we could use it to test the reliability of our mathematical models...

HOW ABOUT WITH GLOBAL PARTNERSHIPS

°MEET EUROPE'S GIGANTIC ENVIRONMENT SATELLITE
EUROPEAN SPACE AGENCY PRESS RELEASE
13 NOVEMBER 2001

Early in 2002 an Ariane 5 rocket will launch the largest and most advanced Earth observation satellite ever built in Europe from the European Spaceport at Kourou in French Guiana...This gigantic Earth observation satellite, the size of an articulated lorry, was developed under the supervision of ESA experts from the European space research and technology centre (ESTEC) in Noordwijk, the Netherlands, at a cost of 2.3 billion euros.

oom pah oom pah
oom pah dee doo,
we've got a rid-dle
waiting for you.

oom pah oom pah
oom pah dee dee,
can you i-ma-gine
what it could be?

WE HAVE USED
THIS ALL BEFORE

STILL

DUMB?

THOUSANDS OF PEOPLE
LOOKING AT THOUSANDS
OF PIECES OF EVIDENCE

IMPACTS FROM AFAR
IMPACTS FROM WITHIN

CHANGES THAT TAKE MILLENNIA
CHANGES THAT TAKE CENTURIES
CHANGES THAT TAKE DECADES
CHANGES THAT TAKE YEARS

THE EARTH PRESENTS A LEGEND

HUMANKIND PROVIDES A COMPASS

THE TIME HAS PASSED TO BE
BLIND, DEAF, AND DUMB

IT IS TIME TO WAKE UP...

...AND SMELL THE DRYAS

8

PREDICTING APOCALYPSE NOW...
OR MAYBE LATER

CATASTROPHE HAPPENS

Apocalypse, doomsday, catastrophe.

These are favorite, albeit frightening topics for the human to ponder.

Look at the popular movies, video reality games, even the evening news: terror, fear, aggression, destruction, loss, and mayhem. If it isn't action packed, smash up, blow up, kill, destroy, explode, annihilate, detonate, bash and blast, then it isn't interesting.

The End of the World is a popular topic, be it through alien invasion, asteroid strike, biblical prophecy, or nuclear war. The United States has the 'Terror Alert Level' and even the weather forecasters are getting in on the action with "The first to warn you about severe weather!" There is even a 'Severe Weather Awareness Week' around here that is dedicated to practicing preparations for severe weather events.

Nature the Terrible has even made it to the Silver Screen.

This is exciting stuff!

Stories about cataclysm have been with us for a long time, as long as modern civilization has been around and most likely as long as humans have been thinking about the meaning or non-meaning of life on planet Earth. Perhaps we are interested in catastrophic events, because they are dramatic and thrilling. Perhaps it is an innate, unconscious, animal-like hunger for destruction. Perhaps it is a desire for action, because the daily routine is so mundane. Perhaps it is an expression of discontent with the world and the only way for to feel better is to think how it could be worse or imagine it could all just disappear in the blink of an eye and be oh so painless.

Perhaps there are no answers, bad things happen to good people and good things happen to bad people. There will always be winners and there will always be losers. There will always be a balance of the Yin and the Yang. There will always be creation and always destruction. There will always be a beginning and always be an end. There will always be life and always be death.

We are told this the way the world works.

But as sentient beings we can choose what will win and what will lose; what will live and what will die. With this choice comes great power and great responsibility.

Catastrophe is good for us, unless of course, it is you or someone you know who is washed away by the tsunami or vaporized under the rubble of a skyscraper.

Catastrophe helps desensitize us to the chaos of an uncertain world. When we are exposed to catastrophe, it prepares us for when it happens in our own back yard. This preparation can usually reduce the severity of the consequences and can even save one's life.

Catastrophes perpetuated by humans on humans are despicable, deplorable, tragic, depraved, and usually preventable.

Catastrophes perpetuated by things beyond human control are tragic, but sometimes the pain and suffering can be reduced if one is well prepared.

Catastrophe shows us just how precarious it is to be alive. Sometimes it takes catastrophe to wake us up from a dream and shock us into reality. Sometimes it takes catastrophe to make us change our way of thinking and behaving. Tragedy of such a degree makes us realize what we have, what we can lose, what we take for granted, and what we live for.

Catastrophe makes us stop for a moment and reflect upon who we are, what we are doing, and where we are going.

For these reasons alone, change of such magnitude can be beneficial, even if it seems horrendous at the time.

In Europe during the late 1700's, there was a great interest by science to explain the features of the Earth through the occurrence of dramatic, catastrophic events. This theory was termed CATASTROPHISM and was developed when science was still guided by The Church. The idea of Catastrophism was good because it could conform to and confirm a literal interpretation of the Bible's story of creation.

According to the beliefs of the The Church, the Earth was only a few thousand years old and was created in a matter of days, so there must have been tremendous events of divine intervention that gave us the features we see. But, catastrophism slowly became discredited as advances in scientific dating began stretching out the history of the Earth and stretching out the time it took for events to happen.

When the theory of evolution came around, science was lulled to sleep by the creeping pace of Earthly activities. Evolution told us that the Earth and its life forms developed at a snail's pace, over thousands and millions of years, steadily crawling along, absorbing indiscernible changes in the surrounding environment and making small adjustments to diversify and adapt.

It turned out that the history of the planet was one big, long yawn.

Then in the 1950's, a Russian pyschoanalyst/doctor/scientist named IMMANUEL VELIKOVSKY sparked a wave of thought that was again oriented around the theory that dramatic, catastrophic, split-second events had indeed shaped the planet and the Universe.

Velikovsky wrote several books during his life, the first and most well known was *Worlds in Collision*, which described events that happened throughout the Solar System and Universe. After being ridiculed by the academic community for using ancient legends as proof for his theories, he used a more conventional approach when he wrote *Earth in Upheaval*, which described events that shaped the face of the Earth.

Velikovsky used an interdisciplinary method which teetered on the boundaries of conventional scientific analysis for his time. Because some of his theories were outlandish and his methods unconventional, he was believed to be a pioneer by some and a quack by others. It is interesting to note that the interdisciplinary approach is encouraged in the academic halls of science today. It is also interesting to note that legends from ancient civilizations are given much more credibility today as modern scientific methods discover hard evidence of the events recorded.

In his book *Worlds in Collision*, Velikovsky explains how the planet Venus came hurling by the Earth a few thousand years ago. According to Velikovsky, the massively active and giant planet Jupiter went through a dramatic heaving convulsion, expunging enough material into space to create the protoplanet Venus. Perhaps this convulsion was generated internally, perhaps it resulted from an external collision.

In either case, the energy involved with such an event is unimaginable. It would have been far greater than the impact of the Shoemaker-Levy comet that struck Jupiter in July of 1994, an impact that produced scars the size of planet Earth and shot plumes of debris thousands of kilometers above the surface. It would also have been far greater than the energy contained in The Great Red Spot of Jupiter, which is the size of six Earths and twists and churns like a salt-water taffy machine, producing lightning the strength of a few hundred atomic bombs.

go big red

Velikovsky's theory suggested that after this massive burst, the protoplanet Venus was catapulted on its course toward the Sun. It and its cometary tail sailed by the Earth and showered the planet with searing meteors and cosmic dust for decades. As the Earth was captured in the comet's gravitational pull, the Earth convulsed and shifted on its axis, tearing the crust, leveling some mountain ranges and creating others, changing river flow, toppling forests, burying civilizations under ocean waves, and sentencing the living planet to centuries of darkness and demise.

Velikovsky claimed this event took place 3,500 years ago and that the event, as well as the immediate and long term consequences of the event were detailed in ancient stories dating back to that time.

He noted that the birth of Venus and fiery showers from space are found in many ancient texts. He also noted that ancient names for Venus include the 'Long Haired Star' and 'Bearded Star,' and that ancient symbols which mean 'Planet Venus' also carry the meaning 'Comet.'

Judging from the legends of our ancestors and the pock-marked impressions of impacts dating back to this time, big rocks did indeed hit the Earth and cataclysms did embrace human settlements; around 4,000-5,000 years ago. Velikovsky's timing may have been a little off and these events have since been attributed to the breakup and passage of Comet Encke, nevertheless, he seemed to be on the right track.

Velikovsky made several other predictions about celestial activities. At the time when the first satellites were sent into space to measure the activities of Venus and Jupiter, he made predictions about the age and composition of both planets. As data from the orbiting satellites came back, it turned out his predictions about certain planetary features and dynamics were correct.

Velikovsky also theorized the Universe was filled with magnetic fields, that the Sun and Earth are magnetically charged bodies, and that electromagnetism is a driving force behind this charge.

This we also know today to be correct.

The true value of Velikovsky is not whether his theories were correct or not, but rather the jolt he gave to mainstream conventional thought. A catastrophic jolt which went against the popular theory of a stable, slowly changing Earth, Solar System and Universe. A jolt that again shocked the world of science and the public into imagining the possibilities of cataclysmic events shaping the world in which we live.

It is interesting to note that Albert Einstein disputed Velikovsky's claims about planetary magnetism and the electromagnetic properties of celestial bodies. When Einstein passed from this life into the unknown, *Worlds in Collision* was open on his desk.

It is also interesting to note that in the 45 million kilometer area of space between the Earth and Venus, there is a 45 million kilometer long electrically charged plasma-based magnetotail emitted from our sister planet that almost reaches the surface of the Earth. Scientists say this is a result of solar wind blowing over a planet that has no magnetic field of its own. It also means Venus has acquired its present position only recently, is a young orbiting planet, and that it has not yet achieved a magnetic charge equilibrium with its environment.

Venus touches the Earth. There is nothing better than sibling love.

If this is the case, maybe Venus did indeed pass by the Earth before landing in its present orbit. Maybe not 3,500 years ago, but maybe a few million years ago, or maybe a few hundred thousand years ago.

Some believe the existence of Venus' tail is part of a larger trend of increased energy coursing through the entire Solar System. Since the discovery of the tail in the 1970's, it has grown 60,000 percent larger.

Other events happening in our Solar System are also used as proof of this elevated energy buzz. On Mars, between June and September in 2001, the planet went from totally clear to being covered in a planet-wide dust storm. Three months, that is quick.

The Sun has recently experienced increased activity in solar flares and the strength of its magnetic field. In the past 60 years, the Sun has been more active than in the 1150 years before combined, solar flares are more powerful than ever recorded, and its magnetic field has increased in strength by 230% since 1901.

These are indeed massive events happening in the Solar System, but they are nothing new to celestial life in the Universe.

Maybe we are noticing these things because we have better instruments for measuring.

Maybe Galactic Marbles is still being played.

Maybe our own planet is far more capable of producing and experiencing mind-boggling events than we ever imagined.

Looking at the past, it seems this may be the case.

There are a number of mysteries that have been discovered.

There are a number of mysteries still out there.

It is awe inspiring.

It is awesome.

°PLANET'S TAIL OF THE UNEXPECTED
NEW SCIENTIST
31 MAY 1997

One of our neighbouring planets can still pack a few surprises, it seems. Using satellite data, an international team of researchers has found that Venus sports a giant, ion-packed tail that stretches almost far enough to tickle the Earth when the two planets are in line with the Sun.

"I didn't expect to find it," says team member Marcia Neugebauer of the Jet Propulsion Laboratory in Pasadena, California. "It's a really strong signal, and there's no doubt it's real."

NASA's Pioneer Venus Orbiter first found the tail in the late 1970s. Around 70,000 kilometres from the planet, the spacecraft detected bursts of hot, energetic ions, or plasma.

...But now Europe's Solar and Heliospheric Observatory (SOHO), a project partly sponsored by NASA, has shown that the tail stretches some 45 million kilometres into space, more than 600 times as far as anyone realised. This satellite, which sits about 1·5 million kilometres away from the Earth, passed through the tail last July, when it was roughly in line with Venus and the Sun...

THE SHEPHERD BOY IS NOT LYING

°GLOBAL WARMTH
UP, OZONE HOLE AT
RECORD, ARCTIC ICE
DOWN FOR 2003 - UN
UNITED NATIONS PRESS
RELEASE
17 DECEMBER 2003

This year is on track to be the third warmest over the past century and a half, the size of the Antarctic ozone hole matched an all-time high and the extent of Arctic sea ice neared a record low, according to new figures from the United Nations World Meteorological Organization (WMO).

Global surface temperature for 2003 is expected to be 0.45 C (centigrade) above the 1961-90 annual average, according to records maintained by WMO members, making the year the third warmest, just behind 2002 with 0.48 C. The warmest year was 1998 with 0.55 C.

...It noted that the rate of change since 1976 was roughly three times that for the past 100 years as a whole, and analyses of proxy data for the Northern Hemisphere indicate that late twentieth century warmth was unprecedented for at least the past millennium...

Not too long ago, some individuals in the Western World began thinking about the possibility of large-scale impacts to the Earth and our settlements as a result of the 'progress' and 'development' being made by a rapidly industrializing civilization.

By the end of the 18th century, the concern in London was that a growing human population would outstrip the ability of nature to keep this population alive. In the 19th century, the worry was over a growing industrial beast that would consume the Earth and all of its resources, that at some point would collapse under its own obesity.

In 1970, **THE LIMITS TO GROWTH** by the **CLUB OF ROME** provided the first analytic projections using computer scenarios and interdisciplinary research methods to report the planetary impact created by humankind's cumulative activities within the body of the Earth. Their theory was that the limits of the Earth's ability to replenish itself had already been or was going to be breached soon, and this would cause modern civilization to fall apart.

Since then, a flood of global catastrophe scenarios have emerged. Some address over-population, some address global pollution, some address acid rain, a looming energy crisis, water scarcity, food shortages, asteroid impacts, nuclear war, biotechnology, ozone layer disappearance, mass extinctions, accelerated global warming, natural disasters, and a drop into the next ice age.

The extreme scenarios of the past have not really worked out to the proportion that was suggested at the time of their writing, but some of the scenarios made in the recent past are not really that far off from how the world looks at the beginning of the 21st century.

Civilization has not collapsed under its own pressure, but people are ever more concerned about overpopulation and widening economic disparities; every corner of the Earth is polluted with artificial chemicals; the seas are being fished clean; resources are being consumed ever more rapidly; the sixth mass extinction is occurring; the ozone layer is thinning; the Earth is warming; climates are changing.

Fortunately for us, the Earth appears to be far more resilient than we once believed.

Unfortunately for us, the Earth appears to be far more sensitive and prone to abrupt changes than we once believed.

As far as limits are concerned, the big worry today is not whether human settlements can sustain themselves with their present or future resource needs, nor whether the Earth can replenish itself from what we take. The big worry is whether a threshold has already been reached or will soon be reached that will alter the entire functioning of the Earth's systems in a way so dramatic that it will destabilize the foundation of all 21st century settlements within a couple of generations.

Besides the doomsday reports, there are also reports to the contrary. Reports that quantitatively say nothing is wrong with the planet and everything is peachy keen. There are reports that say global warming is not really happening, that it is a scam by the scientists to keep them employed. If it is happening, it is completely natural and will be a good thing by extending growing seasons and opening oil fields at the North Pole, as if the Earth were giving us a gift because we are so good.

There are other reports that start off by saying, 'We will never be able to know if greenhouse gases cause global warming,' and 'Climate change science is uncertain.'

It is also uncertain to drive a car, but most of us do not wait to be sent through the windshield before we put on our seatbelt.

The key is not to wait for 100% certainty, see what happens, and then do something about it. The key is to recognize the knowledge we have, take a baby-step of faith to fill in the blanks, and figure out how to use our knowledge to prepare in advance, without making things worse than they already are and are predicted to become.

Since the scenarios of the past have not been fully realized, it is understandable for some to be leery of those who claim the sky is falling and discredit them as an overreacting, irrational menace to progress. But those who adamantly deny the scientific realities are just as radically wrong, just as much of a menace, stunting growth, innovation, even evolution with their complacency.

Those who discredit the reports do not seem to realize that the scientists are not trying to disturb our way of life, they are not trying to promote a political or economic agenda.

They are simply presenting the objective results of research.

If anything, they are trying to protect our way of life by telling us what is happening, so that we can prepare.

Living in a world with greater natural disruptions will cause greater disruptions to our settlements and our way of life.

That is a conservative statement at most.

There have of course been improvements in environmental problems at the local level. Water has become cleaner in some places, air has become easier to breathe in some places, people have become healthier in some places, more energy is being produced by renewable and benign sources, materials are being recycled more frequently. Although local improvements are critical to the individuals and settlements affected, these improvements do not reflect the quality or stability of entire global systems.

The state of the planet may not be in dire straits, but it also is not as rosy as many claim and are led to believe.

As more evidence surfaces, maybe Chicken Little was right. The sky is falling. Just a little slower than we once thought. And the boy who cried 'Wolf'? What is all the fuss about?

°IT'S A WARMER WORLD, BUT DOES THAT MEAN ARMAGEDDON?
REUTERS
11 FEBRUARY 2005

When bears wake early from hibernation, Australia suffers its worst drought in 100 years and multiple hurricanes hammer Florida should we believe The End is nigh? That's the nub of a debate over the human impact on global warming that pits scientists who say such anomalies are signs of impending doom against those who say they are evidence that the earth's climate has always been chaotic.

Amid those signs of warming, for instance, Algeria had its worst snow in 50 years last month. This month 141 countries will attempt the best effort to arrest a forecasted continued rise of global temperatures by bringing into force the Kyoto protocol. The treaty is an agreement aimed at curbing emissions of gases from cars and industry, blamed for trapping the earth's heat. "Dealing with (global warming) will not be easy. Ignoring it will be worse," the United Nations says.

At issue is how humanity should deal with global warming, the risks of which are not yet fully understood despite broad consensus among scientists that people are heating the planet with the emission of such heat-trapping gases as carbon dioxide.

Not everyone is convinced of Kyoto's importance. President Bush pulled the United States out of Kyoto in 2001, reckoning it will be too costly and that it wrongly excludes developing countries from cuts in emissions until 2012.

..."We're talking about spending perhaps $150 billion a year on Kyoto with fairly little benefit," said Bjorn Lomborg, Danish author of "*The Skeptical Environmentalist.*" Lomborg said that money would be better spent on combating AIDS and malaria, malnutrition and promoting fair global trade.

Many climate scientists say that floods, storms and droughts will become more frequent and that climate change is the most severe long-term threat to the planet's life support systems. Rising temperatures could force up ocean levels, swamping coasts and low-lying Pacific islands and drive thousands of species to extinction by 2100. But full proof is elusive...

—There once was a young shepherd's boy...

°GREENHOUSE GASES BUILDING QUICKLY
ENVIRONMENTAL NEWS NETWORK
25 JUNE 2001

Scientists who once disagreed about the amount of the greenhouse gas carbon dioxide is emitted by the United States have now reconciled their differences. The exact amount is important because it is a measure of how much the country contributes to global warming.

...In a new analysis, an international consortium of scientists led by Princeton University predicts that greenhouse gases will build up quickly at the same time as the uptake of carbon dioxide slows down. "The greenhouse problem is going to get worse faster than we expected," said the study's lead author Princeton professor Stephen Pacala...

°GLOBAL GROWTH OF CARBON DIOXIDE STILL RISING
CSIRO
28 MARCH 2004

CSIRO has measured above average growth in carbon dioxide levels in the global atmosphere, despite global attempts to reduce these emissions. The source of the increase is most likely from the burning of fossil fuels - coal, oil and gas.

"The results are concerning because carbon dioxide is the main driver of climate change," says CSIRO Atmospheric Division chief research scientist Dr Paul Fraser. "I am a little bit surprised that the level is so high without input from forest wildfires."

Measurements at Cape Grim in Tasmania, Cape Ferguson in Queensland, sub-Antarctic Macquarie Island, Mawson in Antarctica, and the South Pole, show that carbon dioxide over the last two years has increased at near-record levels. The persistent increases measured over such a large region of the Southern Hemisphere ensure that they closely reflect the total global emissions.

These results support findings from the National Oceanic and Atmospheric Administration (NOAA) in the United States. NOAA announced this week that independent data from Mauna Loa in Hawaii showed peak seasonal carbon dioxide levels last year...

°GREENHOUSE GAS JUMP SPURS GLOBAL WARMING FEARS
REUTERS
11 OCTOBER 2004

An unexplained jump in greenhouse gases since 2002 might herald a catastrophic acceleration of global warming if it becomes a trend, scientists said on Monday.

But they said the two-year leap might be an anomaly linked, for instance, to forest fires in Siberia or a freak hot summer in Europe in 2003 rather than a portent of runaway climate change linked to human disruption of the climate system.

"There have been two years where the rise of carbon dioxide (CO2) has been faster than average," said Richard Betts, Manager for Ecosystems and Climate Impacts at Britain's Hadley Center...

...Scientists said the figures were confirmed at sites including Mauna Loa, Hawaii, west Ireland or the Norwegian Arctic island of Svalbard, about 800 miles from the North Pole...

°INCREASE IN GREENHOUSE GASES SEEN FROM SPACE
REUTERS
14 MARCH 2001

Scientists dispelled any lingering doubts about the increase of greenhouse gases in the atmosphere Wednesday with new evidence from satellites orbiting the Earth.

...New sets of data taken 27 years apart from two satellites orbiting the Earth have now provided the first observational evidence from space of a rise in greenhouse gases...

°GLOBAL WARMING GAS SEEN INCREASING DRAMATICALLY
REUTERS
19 NOVEMBER 2003

Worldwide emissions of carbon dioxide, considered a culprit in global warming are expected to increase by 3.5 billion tonnes, or 50 percent, annually by the year 2020, an executive for ExxonMobil Corp said on Wednesday.

At the same time, global demand for energy will rise by 40 percent as the world population increases and economies grow, said Randy Broiles, global planning manager for Exxon's oil and gas production unit.

"Between now and 2020 we estimate increases of some 3.5 billion tonnes per year of additional carbon emissions, so it's definitely increasing," Broiles said at an energy conference sponsored by accounting and consulting firm Deloitte.

He said about 7 billion tonnes of carbon dioxide, which is a byproduct of burning fossil fuels, go into the earth's atmosphere each year from power plants, cars and other sources.

Experts say the United States, which has the world's largest economy and 4 percent of its population, is responsible for about 25 percent of so-called "greenhouse" gases now produced, but Broiles said most future growth in output will come from developing countries...

...Who tended his sheep at the foot of a mountain near a dark forest...

°FURTHER EVIDENCE FOUND OF GLOBAL WARMING
ENVIRONMENTAL NEWS NETWORK
23 APRIL 1998

Three years in the decade of the 90s have been found to be warmest on record of all years dating back to 1400 A.D. This finding bolsters the argument that the Earth is warming at least in part due to human emissions of so-called greenhouse gases, according to a study published in current issue of the journal *Nature*.

Climatologists at the University of Massachusetts and Amherst, working on a National Science Foundation-funded study, have reconstructed global temperature over the past 600 years and determined that 1997, 1995 and 1990 were the warmest...

°20TH CENTURY GLOBAL WARMING IS UNPRECEDENTED
ENVIRONMENTAL NEWS NETWORK
22 DECEMBER 1998

Paleoclimatologists, studying a variety of natural sources from around the Northern Hemisphere, have confirmed that 20th century global warming is unprecedented relative to the last 1,200 years...

°TWENTIETH CENTURY 'WARMEST IN 500 YEARS'
BBC NEWS ONLINE
16 FEBRUARY 2000

Studies of temperature records preserved deep in underground rocks show that the Earth has been gradually warming over at least the last 500 years. And the studies, by scientists in the US and Canada, show that the trend accelerated markedly during the 20th Century, which was the warmest of the past five centuries. Since 1500, the Earth's temperature has increased by about one degree Celsius, with half of that increase occurring in the last century...

°EARTH HITS '2,000-YEAR WARMING PEAK'
BBC NEWS ONLINE
1 SEPTEMBER 2003

The Earth appears to have been warmer since 1980 than at any time in the last 18 centuries, scientists say. They reconstructed the global climate from data derived from ice cores, vegetation and other records.

They believe their research provides unequivocal confirmation that humans are affecting the climate...

°EARTH GETS A WARM FEELING ALL OVER
NASA-GISS NEWS RELEASE
8 FEBRUARY 2005

Last year was the fourth warmest year on average for our planet since the late 1800s, according to NASA scientists.

To determine if the Earth is warming or cooling, scientists look at average temperatures. To get an "average" temperature, scientists take the warmest and the coolest temperatures in a day, and calculate the temperature that is exactly in the middle of those high and low values. This provides an average temperature for a day. These average temperatures are then calculated for spots all over the Earth, over an entire year...

°SATELLITE DATA SHOWS 1998 WARMEST THIS CENTURY
ENVIRONMENTAL NEWS NETWORK
18 JANUARY 1999

°1999 CLOSES WARMEST DECADE AND WARMEST CENTURY OF THE MILLENNIUM
ENVIRONMENT NEWS SERVICE
22 DECEMBER 1999

°WORLD TEMPERATURE SECOND HIGHEST ON RECORD
REUTERS
18 DECEMBER 2001

°SCIENTISTS: '02 SECOND HOTTEST AS WARMING SPEEDS
REUTERS
17 DECEMBER 2002

°2004 WAS FOURTH WARMEST YEAR EVER RECORDED
NEW YORK TIMES
10 FEBRUARY 2005

°2005 COULD BE WARMEST YEAR RECORDED-NASA
REUTERS
10 FEBRUARY 2005

...One day, after becoming rather lonely and bored, he thought up a trick to play on the villagers to have a little fun and conversation...

°PACE OF GLOBAL WARMING 'COULD DOUBLE'
THE TELEGRAPH
23 JANUARY 2001

Global warming could happen twice as quickly as previously forecast over the next 100 years, the most authoritative report yet produced on the science of climate change said yesterday. Global average temperatures could rise by between 1.4C and 5.8C by the end of the century, according to the latest report of the Intergovermental Panel on Climate Change, made up of scientists from 100 countries, and sponsored by the UN Environment Programme and the World Meteorological Organisation. The panel's previous forecast in 1995 was that the greatest likely temperature rise over the next 100 years was 3C.

Scientists have revised their forecasts of rising temperatures upwards in the report because of the gradual removal from the atmosphere of sulphate aerosols - pollution produced by industry - which reduce global warming by blocking sunlight. The likely temperature rise is expressed as a range because of uncertainties over man's ability to rein back the use of fossil fuels which produce carbon dioxide, the gas mainly responsible for global warming. In the report published yesterday at a meeting in Shanghai, scientists say there is now a greater degree of certainty about ascribing the warming seen since 1861 to human influences. The panel's third assessment report concludes: "There is new and stronger evidence that most of the warming observed over the past 50 years is attributable to human activities."

...Since 1750, before the Industrial Revolution, the atmospheric concentration of carbon dioxide has increased by 31 per cent, from 289 parts per million to 367 ppm today. This concentration has not been exceeded during the past 420,000 years, and probably during the past 20 million years...

°SCIENTISTS ALARMED BY CONTINUED WARMING TREND
LOS ANGELES TIMES
12 DECEMBER 2002

The year 2002 is the second-warmest in recorded history, according to NASA scientists who monitor global air temperatures. A record-breaking stretch of warmth in recent years -- with 2001 now going down as the third-warmest year on record and 1998 still holding the all-time record -- has scientists and climate experts concerned that greenhouse gases are warming the planet more quickly than previously expected.

"Studying these annual temperature data, one gets the unmistakable feeling that temperature is rising and that the rise is gaining momentum," said Lester R. Brown, an economist and president of the Earth Policy Institute in Washington.

...The string of warmer years provides strong evidence that humans are in large part to blame for changing the climate, said Peter Frumhoff, an ecologist and senior scientist with the Union of Concerned Scientists in Cambridge, Mass. "It's important we pay attention to this drumbeat of evidence as the signal of human impact starts to emerge from the noise of natural climate patterns," he said...

°US EXPERTS SAY GLOBAL WARMING FASTER THAN THOUGHT
AGENCE FRANCE-PRESSE
24 JUNE 2004

A new US supercomputer has shown that global temperatures could be rising more than scientists had thought, experts said.

The computer at the National Center for Atmospheric Research projects that temperatures could rise by 2.6 degrees Celsius (4.7 degrees Fahrenheit) if countries continue to emit large amounts of carbon dioxide. The previous estimates were a rise of about two degrees Celsius (3.6 degrees Fahrenheit)...

°STUDIES: GLOBAL WARMING TO WORSEN
CNN NEWS ONLINE
18 APRIL 2002

Two new climate studies predict that global warming by the end of the century will be even more dramatic than a United Nations group has predicted. But more important than that long-range outlook, said one climate expert, is the data that both teams of scientists show for the years 2020-2030.

"These very different approaches both tell us that two to three decades from now, it will be warmer than it is now," said Francis Zwiers, a statistics and climate expert with the Meteorological Service of Canada.

A British study says that in those years the Earth will be 0.5 to 2.3 degrees Fahrenheit warmer than the period between 1990-2000. A Swiss study predicts a temperature increase ranging from 0.9 to 1.9 degrees Fahrenheit. "In that time scale, we can do planning for changing fuel sources, so there are less greenhouse gases," Zwiers said.

The two studies are detailed in the latest issue of the journal *Nature*. The studies are consistent with findings from the U.N. Intergovernmental Panel on Climate Change. That group has predicted world temperatures could rise as much as 10.5 degrees or as little as 3 degrees Fahrenheit by the end of the century…

...He rushed into town calling "Wolf Wolf" and was met by villagers who stopped to ask what all the fuss was about...

°HUMANS COULD TRIGGER DRAMATIC CLIMATE CHANGE
ENVIRONMENTAL NEWS NETWORK
28 MAY 1997

"We are playing Russian roulette with our climate," one scientist told the opening session of GW8, the Eighth International Global Warming Conference and Exposition being held this week at Columbia University.

Annual precipitation shown in ice cores and pollen distribution from seafloor sediments show that Earth's climate is subject to extremely abrupt and dramatic changes, said Wallace Broecker, Newberry Professor of Earth and Environmental Sciences at Columbia and a well-known paleoclimatologist. Patterns of ocean currents can change, and climate temperatures can rise or drop by 15 degrees Fahrenheit or more.

Human activity -- such as our dumping 6 billion tons of carbon dioxide into the atmosphere annually -- could precipitate such a dramatic change, Professor Broecker said, and the fact that our climate is currently between major periods of glaciation means the climate could as easily turn colder as warmer.

"The Earth's climate system is an angry beast subject to unpredictable responses, and by adding carbon dioxide to the atmosphere we may be provoking the beast," he said...

°GLOBAL WARMING WILL PERSIST AT LEAST A CENTURY EVEN IF EMISSIONS CURBED NOW
NASA EARTH OBSERVATORY MEDIA ALERT
17 FEBRUARY 2002

Though significant uncertainty remains regarding the amount of global warming that will occur over the next century or two, scientists agree that the trend will continue for the next hundred years even if fossil fuel consumption is dramatically reduced Scientists predict significant increases in global temperature and sea level this century. And related changes in weather patterns are expected to affect agricultural production. Global warming is likely to have the greatest human impact in poor countries unable to adequately respond to the changes.

Professor Robert Dickinson of the Georgia Institute of Technology's School of Earth and Atmospheric Sciences will present the evidence behind this assessment at the annual meeting of the American Association for the Advancement of Science (AAAS) on Feb. 17 in Boston. Dickinson's presentation, titled "Predicting Climate Change," is part of the symposium "Climate Change: Integrating Science, Economics and Policy."

"Current climate models can indicate the general nature of climate change for the next 100 to 200 years," Dickinson says. "But the effects of carbon dioxide (CO2) that have been released into the atmosphere from the burning of fossil fuels last for at least 100 years. That means that any reductions in CO2 that are expected to be possible over this period will not result in a cleaner atmosphere and less global warming than we see today for at least a century."...

°CLIMATE CHANGE 'INEVITABLE': CSIRO CHIEF
CSIRO
13 OCTOBER 2000

Increases in atmospheric concentrations of carbon dioxide and other greenhouse gases are inevitable during the coming century and it is inevitable that we will experience climate change.

This is the message from Dr Graeme Pearman, Chief of CSIRO Atmospheric Research. Dr Pearman was speaking at a national gathering of leaders from industry, research and government today in Melbourne to discuss the greenhouse effect, technical and policy options for Australia, and possible Government responses.

"There is no question that greenhouse gas concentrations are rising," said Dr Pearman.

"The levels we are now experiencing are certainly higher than for the past 400,000 years and quite possibly higher than at any time in the past 20 million years."...

°CLIMATE OF CHANGE IS HERE FOR GOOD
THE INDEPENDENT
20 MARCH 2002

Glaciologists from the British Antarctic Survey in Cambridge said yesterday that the 500-billion-ton Larsen B ice shelf is no more. Another team of Antarctic scientists in America said that they had spotted a huge iceberg half the size of Cyprus floating majestically away from the frozen continent.

Meanwhile, closer to home, climate researchers from East Anglia University have found that the British growing season – as defined by the period between two cold snaps lasting five consecutive days – is now longer than ever.

"If the trend continues, it is possible that we will have a year-round growing season within a generation," predicted Tim Mitchell of East Anglia's Tyndall Centre for Climate Change Research.

At the same time, their colleagues at University College London have calculated that we are heading for a balmy spring. They forecast that temperatures will be as much as 1.3C warmer than the 30-year seasonal norm...

...This pleased him greatly, so a few days later, he played the trick again and again the villagers came to his calling...

°STUDY HINTS AT EXTREME CLIMATE CHANGE
ENVIRONMENTAL NEWS NETWORK
28 OCTOBER 1999

Abrupt climate change could be in the future for the Earth if a recently discovered pattern repeats itself. After analyzing sediment from the subtropical Atlantic Ocean deposited during the last ice age, scientists discovered extreme temperature fluctuations occurred during and at the end of the period. The scientists found that even during an ice age, warm oceans can heat up.

"What is new here is clear evidence that the warm Atlantic, like the polar Atlantic, was undergoing very large and very rapid temperature changes during the last glacial period," said Scott Lehman, a research associate at the University of Colorado Institute of Arctic and Alpine Research.

..."The temperature of the warm ocean realm regulates the water vapor content of the atmosphere and its greenhouse capacity," said Lehman. Past temperature records and climate models suggest ocean circulation changes, like those in the last glacial period, can be triggered by human activity, showing that "the impact of possible future circulation changes may be more dramatic and widespread than suspected," he said...

°CATASTROPHIC CLIMATE CHANGE '90% CERTAIN'
THE INDEPENDENT
20 JULY 2001

The world is 90 per cent certain to experience a potentially catastrophic global warming over the next century caused by man-made emissions of greenhouse gases, scientists say.

New estimates suggest that there is a nine out of ten chance that the Earth will warm by between 1.7C and 4.9C by 2100, generating serious disturbances to the climate and a rise in sea levels.

The fresh probability estimates are made by Tom Wigley of the National Center for Atmospheric Research in Boulder, Colorado and Sarah Raper of the University of East Anglia in Norwich in a study published today in the journal *Science*...

...But shortly after, a wolf really did come out of the woods and began to worry the sheep...

...The boy returned to find his sheep in much distress...

°THE PAST SAYS ABRUPT CLIMATE CHANGE IN OUR FUTURE
PENNSYLVANIA STATE UNIVERSITY PRESS RELEASE
12 DECEMBER 2001

Past climates changed abruptly, suggesting that abrupt changes in the future will also occur, according to a Penn State geoscientist. "When we look at records of the past, climate often changed abruptly rather than smoothly," says Dr. Richard B. Alley, the Evan Pugh professor of geosciences at Penn State. "This is true wherever and whenever you look."

Alley, who is currently chairing the National Academy of Science Committee on Abrupt Climate Change: Science and Public Policy, told attendees today (Dec. 13) at the fall meeting of the American Geophysical Association in San Francisco, that while studies of ice cores, sediments and other relics of the past indicate these abrupt changes, the models currently used by those predicting the future of climate change do not do a good job of simulating abrupt changes in the past.

"If we look at what we know about climate, there is much we don't understand," says Alley. "However, we do know that abrupt change occurred in the past." The abrupt changes are especially notable in temperature near the north and south poles and in precipitation away from the poles. In the near term, nature sometimes changes smoothly, sometimes remains the same and sometimes changes all at once. In the long term, abrupt change appears to be the norm. Current models all tend to change smoothly and do not capture abruptness.

"It is possible that climate change in the future will include abruptness, even though the current models do not show this," says Alley. The Penn State geoscientist suggests that climate change includes a process of approaching and crossing a series of thresholds. Climate forcing factors are like a tower of blocks. Building the tower, blocks can be added, and the tower remains stable, but eventually the block height crosses the threshold of stability and the tower abruptly topples. With climate, the thresholds in the past have sometimes been reached in as few as 10 years...

°CLIMATE CHANGE COULD COME FAST AND FURIOUS
ENVIRONMENT NEWS SERVICE
9 DECEMBER 2002

The effects of global climate change could be more abrupt and more catastrophic than many scientists have predicted, warns a Penn State climatologist.

Debate in the U.S. over climate change often focuses on whether things will be as bad as some scientists say they will be. Dr. Richard Alley of Penn State says the more important question may be whether researchers are confident that things will be as good as they are predicting…

…Given that the future could be quite challenging, it would be wise for policy makers to start looking for ways that people can adapt when climate changes, Alley said. He noted that there is ample historic evidence of human groups who refused or were unable to adapt to climatic changes, and their societies collapsed or failed, while other groups adapted to the new environment and coped and sometimes thrived.

Congress, federal agencies and even local governments who must deal with these changes when they happen should look at ways to plan for changes in water supply, crop production, heating oil demand, flood control and other things likely to be affected by climate change, Alley said. These groups should establish contingencies to meet problems with scarcity of resources before there is competition for these resources, he advised...

…The boy ran back into the village crying "Wolf Wolf" much louder than before…but this time, the villagers being fooled twice before, ignored the boy….

°EARTH WARNED ON 'TIPPING POINTS'
BBC NEWS ONLINE
26 AUGUST 2004

The world has barely begun to recognise the danger of setting off rapid and irreversible changes in some crucial natural systems, a scientist says. Professor John Schellnhuber says the most important environmental issues for humans are among the least understood.

He told a briefing in Sweden that the Asian monsoon was one of the "tipping points" that could change very quickly. He said a better understanding of the risks was as important as the programme to prevent collisions with asteroids.

Professor Schellnhuber is research director of the UK's Tyndall Centre for Climate Change Research. He was speaking at the EuroScience Forum in Stockholm, at a briefing by the International Geosphere-Biosphere Programme entitled Beyond Global Warming: Where On Earth Are We Going?

Professor Schellnhuber said 12 "hotspots" had been identified so far, areas which acted like massive regulators of the Earth's environment. If these critical regions were subjected to stress, they could trigger large-scale, rapid changes across the entire planet. But not enough was known about them to be able to predict when the limits of tolerance were reached.

"We have so far completely underestimated the importance of these locations," he said. "What we do know is that going beyond critical thresholds in these regions could have dramatic consequences for humans and other life forms."...

°GLOBAL WARMING COULD TRIGGER CASCADE OF CLIMATIC CHANGES
OREGON STATE UNIVERSITY NEWS RELEASE
13 MARCH 2003

Global warming and the partial melting of polar ice sheets can dramatically affect not only sea levels but also Earth's climate, in ways that may be complex, rapid and difficult to adjust to, scientists say in a new study to be published Friday in the journal *Science*.

Sea level and climatic changes in Earth's distant past, near the end of the last Ice Age about 14,600 years ago, offer significant clues to some phenomena that Earth may experience in the near future, possibly in coming decades or centuries, the study found.

The research was done by scientists at the University of Victoria, Oregon State University, and the University of Toronto. It revealed changes in global temperature, sea level and ocean currents that can occur with surprising rapidity.

"With the advent of global warming, we're trying to identify the climatic surprises that may be in store for us, the events that we really aren't expecting," said Peter Clark, a professor of geosciences at OSU and a co-author of the study. "The more we look at this, the more it appears there have been large and abrupt changes in climate and sea level that are interconnected. If these changes were to happen in the future, they could cause huge societal disruptions."...

…And the wolf made a good meal out of his flock…

°STUDY HIGHLIGHTS GLOBAL DECLINE
BBC NEWS ONLINE
30 MARCH 2005

The most comprehensive survey ever into the state of the planet concludes that human activities threaten the Earth's ability to sustain future generations. The report says the way society obtains its resources has caused irreversible changes that are degrading the natural processes that support life on Earth. This will compromise efforts to address hunger, poverty and improve healthcare.

The Millennium Ecosystem Assessment was drawn up by 1,300 researchers from 95 nations over a period of four years. It reports that humans have changed most ecosystems beyond recognition in a dramatically short space of time. The way society has sourced its food, fresh water, timber, fibre and fuel over the past 50 years has seriously degraded the environment, the assessment (MA) concludes. And the current state of affairs is likely to be a road block to the Millennium Development Goals agreed to by world leaders at the United Nations in 2000, it says.

"Any progress achieved in addressing the goals of poverty and hunger eradication, improved health, and environmental protection is unlikely to be sustained if most of the ecosystem 'services' on which humanity relies continue to be degraded," the report states. "This report is essentially an audit of nature's economy, and the audit shows we've driven most of the accounts into the red," commented Jonathan Lash, the president of the World Resources Institute. "If you drive the economy into the red, ultimately there are significant consequences for our capacity to achieve our dreams in terms of poverty reduction and prosperity."

The MA is slightly different to all previous environmental reports in that it defines ecosystems in terms of the "services", or benefits, that people get from them - timber for building; clean air to breathe; fish for food; fibres to make clothes. The study finds the requirements of a burgeoning world population after WW II drove an unsustainable rush for these natural resources. Although humanity has made considerable gains in the process - economies and food production have continued to grow - the way these successes have been achieved puts at risk global prosperity in the future.

"When we look at the drivers of change affecting ecosystems, we see that, across the board, the drivers are either staying steady or increasing in severity - habitat change, climate change, invasive species, overexploitation of resources; and pollution, such as nitrogen and phosphorus," said Dr William Reid, the director of the MA. More land was converted to agriculture since 1945 than in the 18th and 19th Centuries combined. More than half of all the synthetic nitrogen fertilisers - first made in 1913 - ever used on the planet were deployed after 1985.

The MA authors say the pressure for resources has resulted in a substantial and largely irreversible loss in the diversity of life on Earth, with some 10-30% of the mammal, bird and amphibian species currently threatened with extinction.

The report says only four ecosystem services have been enhanced in the last 50 years: increases in crop, livestock and aquaculture production, and increased carbon sequestration for global climate regulation (which has come from new forests planted in the Northern Hemisphere). Two services - fisheries and fresh water - are said now to be well beyond levels that can sustain current, much less future, demands.

..."The MA is a very powerful consensus about the unsustainable trajectory that most of the world's ecosystems are now on." "There will undoubtedly be gainsayers, as there are with the IPCC; but I put them in the same box as the flat-Earthers and the people who believe smoking doesn't cause cancer," said Professor Sir John Lawton, former chief executive of the UK's Natural Environment Research Council...

°GLOBAL WARMING ALREADY DISRUPTING CLIMATE: SCIENTISTS TELL CONFERENCE
AGENCE FRANCE-PRESSE
1 FEBRUARY 2005

Evidence is growing that global warming is already starting to disrupt the world's delicately-balanced climate system, and the damage will reverberate for generations, a top science conference was told.

"There is no longer any doubt that the Earth's climate is changing," conference chairman Dennis Tirpak said Tuesday. "Globally, nine of the past 10 years have been the warmest since records began in 1861," he said. "Rising greenhouse gases are affecting rainfall patterns and the global water cycle."

Tirpak singled out the heatwave that gripped western Europe in 2003 as an example. Europe's worst natural disaster in 50 years killed as many as 30,000 people and inflicted an estimated 30 billion dollars (23 billion euros) in damage. "Since the 1970s, climatic warming has increased the extent and frequency of droughts over land," he said. "(...) Terrestrial ecological systems are shifting and marine systems are changing, all with outcomes that are difficult to predict."

...One problem is that even if pollution were slashed immediately, temperatures would continue to rise because of the gas which has already been spewed into the atmosphere. "The inertia can carry the impacts, especially on sea level rise, for centuries, if not for millennia," Pachauri said.

...When the boy complained...

244

°ALARM AT NEW CLIMATE WARNING
BBC NEWS ONLINE
26 JANUARY 2005

Temperatures around the world could rise by as much as 11C, according to one of the largest climate prediction projects ever run. This figure is twice the level that previous studies have suggested.

Scientists behind the project, called climateprediction.net, say it shows that a "safe" upper limit for carbon dioxide is impossible to define. The results of the study, which used PCs around the world to produce data, are published in the journal *Nature*.

Climateprediction.net is run from Oxford University, and is a distributed computing project; rather than using a supercomputer to run climate models, people can download software to their own PCs, which run the programs during downtime.

More than 95,000 people have registered, from more than 150 countries; their PCs have between them run more than 60,000 simulations of future climate...

...Overall, the project produces a picture of the possible range of outcomes given the present state of scientific knowledge. The lowest rise which climateprediction.net finds possible is 2C, ranging up to 11C. The timescale would depend on how quickly the doubling of CO2 was reached, but large rises would be on a scale of a century at least from now.

"I think these results suggest that our need to do something about climate change is perhaps even more urgent," the climateprediction.net chief scientist David Stainforth told BBC News...

...the wise man of the village said...

°CLIMATE CHANGE ALREADY AFFECTING GLOBAL ENVIRONMENT
THE CHIEF ENGINEER ONLINE
3 MARCH 2005

Global warming has had little noticeable impact in Washington, D.C. Politicians in the U.S. capital have been reluctant to set limits on the carbon dioxide pollution that is expected to warm the planet by two degrees to four degrees Celsius (four degrees to seven degrees Fahrenheit) during the next century, citing uncertainty about the severity of the threat.

But that uncertainty may have shrunk somewhat with the release of two scientific reports suggesting that global warming is not just a hypothetical possibility, but a real phenomenon that has already started transforming especially sensitive parts of the globe. Overall, the reports say, Earth's climate has warmed by about a half-degree Celsius (one degree Fahrenheit) since 1900. In the Arctic, where a number of processes amplify the warming effects of carbon dioxide, most regions have experienced a temperature rise of two to four degrees Celsius (four to seven degrees Fahrenheit) in the last 50 years.

That warmth has reduced the amount of snow that falls every winter, melted away mountain glaciers and shrunk the Arctic Ocean's summer sea ice cover to its smallest extent in millennia, according to satellite measurements. In Alaska, swaths of permafrost are thawing into soggy bogs, and trees are moving northward at the expense of the tundra that rings the Arctic Ocean.

..."Responses to climate change are being seen across the USA," said Camille Parmesan, a biologist at the University of Texas in Austin. She is the co-author, with Hector Galbraith of the University of Colorado in Boulder, of "Observed Impacts of Global Climate Change in the U.S." The report was released by the Pew Center on Global Climate Change, a non-partisan but not disinterested research organization dedicated to providing sound scientific information about global warming...

..."The canaries in the coal mine are squawking, and we should absolutely take that seriously," Galbraith said.

President George W. Bush's administration has argued that not enough is known about climate change to justify major efforts at forestalling or preventing future warming.

..."A LIAR WILL NOT BE
BELIEVED,

EVEN WHEN HE SPEAKS
THE TRUTH."–

—Aesop

BE WARY OF THOSE WHO CRY
DENY! DENY!

TO BELIEVE OR NOT TO BELIEVE...
IS THAT A QUESTION?

The Future.

We see the future every day with our local weather forecast.

We think about the future every day with our investments.

We worry about the future every day with our children.

We embrace the future every day with our plans, goals, and dreams.

The power to forecast future events is magnificent indeed, even if it is not always 100% accurate. The weather forecast where I live is accurate enough to be advertised as 'The Futurecast.'

Those meteorologists are pretty sure of themselves and their abilities.

The climate change scientists are no less certain.

Throughout the history of civilization, people have relied on those who have an ability to glimpse into the unknown and provide direction in an uncertain world. Whether this knowledge comes from seers, sages, or Doppler Radar, forecasting future events has played an important role in society.

Predicting future events is normal and expected these days.

From forecasting the weather to speculating the stock market, we try to sense and plan for the future. The unknown makes us uncomfortable and uncertainty makes us uneasy. Having an idea of the direction things will be gives us a feeling of control. If we have an idea of how things will be, we can alter the way things are, in order to maximize our comfort now and in the future, or simply improve our chances for survival.

Science, from the alchemists, through the catastrophists, to the modern analysts, has tried to decipher the riddles of nature to understand how it has behaved in the past, how it behaves in the present, and how it will behave in the future.

At the beginning of the 21st century, modern science has the power to very accurately predict future events, so powerful that modern science has been given the role and authority of an **ORACLE**.

oracle
A person or a place in which a person (as a priestess in ancient Greece) through whom a deity is believed to speak.

Our global family is using the most advanced tools and techniques, research, logic, intuition, inference, induction, and deduction to decipher the riddles wrapped into the fabric of the Earth.

Plain and simple, here is what scientists are telling us about the world in which we live and the world we will most likely experience in the near future:

° The Earth's average global surface temperature and the level of atmospheric global greenhouse gases are intertwined with one another, rising and falling in tandem.
° Atmospheric global greenhouse gas levels are increasing and this is causing an enhanced greenhouse effect.
° This increase can be attributed to human activities like fossil fuel burning and deforestation during the past few hundred years.

° As a result of an enhanced greenhouse effect, the average global surface temperature of the Earth is increasing.
° Disruptions to global systems and the amount of greenhouse gas emissions released from human activity during the past 350 years may be enough to tip the scale of the planet's equilibrium, in which it is beginning to enhance the greenhouse effect itself.
° The amount of greenhouse gas emissions released from human activity is negligible compared to the amount of greenhouse gases nature has stored in oceans, forests, and tundra.
° The longevity of greenhouse gases are 20-100 years, which means the observed trend will likely continue for this century, even if all anthropogenic sources of greenhouse gases were reduced to 0 today.
° A warmer ocean's expanding water, melting glaciers, and melting polar ice will raise ocean levels significantly during the 21st century, affecting coastal and island settlements.
° An infusion of freshwater from melting ice from the North Pole will most likely alter the circulation of ocean currents and effects of this change are already being seen in the Atlantic Ocean.
° The Earth's global temperature and climates have fluctuated throughout history and changes have sometimes been rapid, abrupt, unpredictable, and long-lasting.
° Positive feedbacks of greenhouse gases and ocean dynamics are perhaps the reason for rapid and abrupt shifts in global temperature and climates in the past and will most likely cause such events in the near future.
° A change in global climate systems will disrupt habitats and the species who live in them, and this disruption in ecosystems will affect global greenhouse gas levels.
° A change in global climate systems will cause a great change in the stable and predictable global weather patterns our settlements are dependent and based upon.
° Changing weather patterns will increase the frequency and severity of extreme weather events throughout the world, including droughts, floods, hurricanes, typhoons, warm and cold waves, and storms.
° Changing weather patterns will alter growing regions.
° Changing weather patterns will disrupt growing seasons.
° Changing weather patterns will increase freshwater stresses.
° Changing weather patterns will increase the costs of natural disasters.
° Changing weather patterns will cause larger disease outbreaks and disease migrations.
° All of these changes will create more insecurity and more instability within our settlements and between settlements.

Furthermore, increasing pressures from the natural environment, increasing pressures from a growing global population, increasing stresses on freshwater sources, increasing scarcity of a rapidly dwindling energy source, increasing conflicts of culture, and increasing conflicts over resources; all of this will stress modern settlements and jeopardize the well being of people and places around the world.

All of these predictions have been made with as much accuracy as your daily weather forecast.

These forecasts just happen to involve a lot more variables, are global in scale, and are long term.

Nevertheless, this is the future to be faced.

The rocky coast is in plain sight and it is too late to turn around.

Some settlements may crash their boat into splinters, some may sustain only a few dents in the hull; it all depends on how prepared the cast and crew are for an impact.

2 degrees Celsius = danger

°CLIMATE CHANGE: COUNTDOWN TO GLOBAL CATASTROPHE
THE INDEPENDENT
24 JANUARY 2005

The global warming danger threshold for the world is clearly marked for the first time in an international report to be published tomorrow - and the bad news is, the world has nearly reached it already.

The countdown to climate-change catastrophe is spelt out by a task force of senior politicians, business leaders and academics from around the world - and it is remarkably brief. In as little as 10 years, or even less, their report indicates, the point of no return with global warming may have been reached.

The report, Meeting The Climate Challenge, is aimed at policymakers in every country, from national leaders down. It has been timed to coincide with Tony Blair's promised efforts to advance climate change policy in 2005 as chairman of both the G8 group of rich countries and the European Union.

And it breaks new ground by putting a figure - for the first time in such a high-level document - on the danger point of global warming, that is, the temperature rise beyond which the world would be irretrievably committed to disastrous changes. These could include widespread agricultural failure, water shortages and major droughts, increased disease, sea-level rise and the death of forests - with the added possibility of abrupt catastrophic events such as "runaway" global warming, the melting of the Greenland ice sheet, or the switching-off of the Gulf Stream.

The report says this point will be two degrees centigrade above the average world temperature prevailing in 1750 before the industrial revolution, when human activities - mainly the production of waste gases such as carbon dioxide ($CO2$), which retain the Sun's heat in the atmosphere - first started to affect the climate. But it points out that global average temperature has already risen by 0.8 degrees since then, with more rises already in the pipeline - so the world has little more than a single degree of temperature latitude before the crucial point is reached.

More ominously still, it assesses the concentration of carbon dioxide in the atmosphere after which the two-degree rise will become inevitable, and says it will be 400 parts per million by volume (ppm) of $CO2$.

The current level is 379ppm, and rising by more than 2ppm annually - so it is likely that the vital 400ppm threshold will be crossed in just 10 years' time, or even less (although the two-degree temperature rise might take longer to come into effect).

...The report starkly spells out the likely consequences of exceeding the threshold. "Beyond the 2 degrees C level, the risks to human societies and ecosystems grow significantly," it says.

"It is likely, for example, that average-temperature increases larger than this will entail substantial agricultural losses, greatly increased numbers of people at risk of water shortages, and widespread adverse health impacts. [They] could also imperil a very high proportion of the world's coral reefs and cause irreversible damage to important terrestrial ecosystems, including the Amazon rainforest."

It goes on: "Above the 2 degrees level, the risks of abrupt, accelerated, or runaway climate change also increase. The possibilities include reaching climatic tipping points leading, for example, to the loss of the West Antarctic and Greenland ice sheets (which, between them, could raise sea level more than 10 meters over the space of a few centuries), the shutdown of the thermohaline ocean circulation (and, with it, the Gulf Stream), and the transformation of the planet's forests and soils from a net sink of carbon to a net source of carbon."

°REPORTS POINT TO PROOF OF GLOBAL WARMING
ASSOCIATED PRESS
13 NOVEMBER 2004

...Politicians in the nation's capital have been reluctant to set limits on the carbon dioxide pollution that is expected to warm the planet by 4 to 7 degrees Fahrenheit during the next century, citing uncertainty about the severity of the threat.

But that uncertainty may have shrunk somewhat with the release last week of two scientific reports suggesting that global warming is not just a hypothetical possibility, but a real phenomenon that has already started transforming especially sensitive parts of the globe.

Overall, the reports say, Earth's climate has warmed by about 1 degree Fahrenheit since 1900. In the Arctic, where a number of processes amplify the warming effects of carbon dioxide, most regions have experienced a temperature rise of 4 to 7 degrees in the last 50 years.

That warmth has reduced the amount of snow that falls every winter, melted away mountain glaciers and shrunk the Arctic Ocean's summer sea ice cover to its smallest extent in millennia, according to satellite measurements. Swaths of Alaskan permafrost are thawing into soggy bogs, and trees are moving northward at the expense of the tundra that rings the Arctic Ocean.

These changes seriously threaten animals such as polar bears, which live and hunt on the sea ice. The bears have already suffered a 15 percent decrease in their number of offspring and a similar decline in weight over the past 25 years. If the Arctic sea ice disappears altogether during the summer months, as some researchers expect it will by the end of the century, polar bears have little chance of survival.

..."The canaries in the coalmine are squawking, and we should absolutely take that seriously," (Hector) Galbraith said...

Changes may happen a bit sooner than predicted, they may happen a bit later. But changes, significant changes, will happen during this century, during the lives of our children, and during the lives of their children and their children's children.

The rocky coast is on the horizon and it will not go away.

The tundra is thawing, the oceans are warming.

The snowflakes are gathering very quickly.

And we are living under the avalanche without a care in the world.

The messages emanating from the Earth are open for all to hear.

It has all been laid right before our eyes by countless individual and collective discoveries in the fields of modern science.

So what are we going to do with this knowledge?

A good story about what a past civilization did with knowledge of future events comes from Greek mythology. The story of Cassandra, priestess, prophetess and daughter of the king and queen of Troy.

Cassandra was one of the most beautiful women in the world at the time and was fancied by Apollo, God of the Sun, who gave her the gift to see future events. However, Apollo's admiration for Cassandra turned into a desire for intimacy and Cassandra refused to consent.

Typical for Gods who do not get their way, Apollo still gave Cassandra the gift to see future events but with one condition: no one would believe her.

Those Gods, they are a clever bunch.

This was quite a predicament for Cassandra, who was now able to see the future, but was thought to be delirious. Cassandra told the people of Troy about their fate in the Trojan War, but nobody believed her. Cassandra told them during the war that the Trojan Horse was a deceptive trick by their enemy, the Greeks, but nobody believed her. She told them beforehand of several other attacks planned by the Greeks, but the people of Troy never believed anything she said, although she spoke the truth.

With the knowledge of future events screamed into their ears, the people of Troy could have known the true purpose of the Trojan Horse and prepared themselves for the invading Greeks. They could have prevented the collapse of their settlement if they would have listened. But they failed to heed her words and were all slaughtered in their sleep.

As for Cassandra, she was taken by the Greek Commander in Chief of the invasion, Agamemnon, who took her as his concubine. She later predicts Agamemnon's and her own death by the blade struck by Agamemnon's wife.

The morals to this story?

Gods like to have their way.

It is bad for your marriage to sleep with other women.

It is bad for your survival to ignore advice about future events.

The study of global temperature change has come a long way in 150 years and has been advancing with unprecedented speed, ingenuity, and understanding since the 1970's.

Earthly dynamics continue to be better understood, as are the interrelationships between human-human, Earth-human, human-Earth, and Earth-Earth activities. At the beginning of the 21st century, there are countless high-tech inventions, methods, theories, and proofs devoted to understanding Earthly systems and global environmental changes as they relate to global temperature change. The complexity and amount of research is almost as mind-boggling as the Earthly activities themselves.

It is clear what all of the different stories, proxies, measurements, trends, and the most advanced scientific and theoretical methods to date tell us. No matter what the initial cause, be it orbital variations, asteroid impacts, or a terrestrial initiator, the Earth will stay in relative equilibrium until it reaches a threshold. Once the threshold is passed, the planet in its present state will warm and release greenhouse gases which have been locked away, raising atmospheric greenhouse gas levels and enhancing the greenhouse effect. As the greenhouse effect is enhanced, the planet's large greenhouse gas sinks become greenhouse gas sources, less is absorbed and more stays in the atmosphere. The planet continues to warm, its oceans warm, its polar ice melts away and as larger amounts of land are exposed, the Earth absorbs more warmth. More greenhouse gases are released, the greenhouse effect is enhanced, etc.

During this time, short and long term weather patterns change, entire climate regimes change, and after a while, enough freshwater enters the planet's oceans at sensitive circulation points to breach a threshold and push an entire hemisphere or the entire planet into a different steady-state, one that is much cooler. As the planet cools, short and long term weather patterns change once again, entire climate systems change once again, and the planet begins locking greenhouse gases away again, reducing atmospheric greenhouse gas levels and diminishing the greenhouse effect. Poles begin to gather ice, glaciers begin to grow and the planet begins reflecting more of the Sun's energy which keep it cool.

Throughout eons of the past and eons into of the future, the planet moves in and out of cold and warm periods. The speed at which changes can occur is somewhat uncertain, but the map of the past tell us they can occur extremely fast after a threshold is breached. Sometimes shifts take millennia, sometimes centuries, sometimes decades.

GLOBAL TEMPERATURE AND CLIMATE CHANGES WILL OCCUR WHEN THE CONDITIONS ARE RIGHT. THOSE CONDITIONS ARE BEGINNING TO EXIST.

This is not an apocolyptic prediction. It is not doomsaying.

It is the forecast provided by 21st century science and legitimized by the international community of business, industry, insurance, and politics.

Black and white and all of the interdisciplinary colors in the rainbow.

This is your future on planet Earth.

This is our future on planet Earth.

I would say the proof is in the pudding.

Or rather, the proof is in the countless Dollars, Pounds, Yen, Euros, Rubles, and Pesos invested in the ingredients and the countless hours spent making the pudding over the past 150 years.

It is just too bad this pudding has an awfully bitter taste to it.

'GLOBAL WARMING REAL' SAY NEW STUDIES
THE FINANCIAL TIMES
18 FEBRUARY 2005

A leading US team of climate researchers on Friday released "the most compelling evidence yet" that human activities are responsible for global warming. They said their analysis should "wipe out" claims by sceptics that recent warming is due to non-human factors such as natural fluctuations in climate or variations in solar or volcanic activity.

Scientists from the Scripps Institution of Oceanography in California have been working for several years with colleagues at Lawrence Livermore National Laboratory to analyse the effects of global warming on the oceans. They combined computer modelling with millions of temperature and salinity readings, taken around the world at different depths over five decades.

The researchers released their conclusions on Friday at the American Association of the Advancement of Science meeting in Washington. They found that the "warming signals" in the oceans could only have been produced by the build-up of man-made carbon dioxide in the atmosphere. Non-human factors would have produced quite different effects.

The latest study to suggest that global warming is a real phenomenon, and one caused by human action, adds further weight to a body of scientific evidence that has been accumulating steadily.

Tim Barnett, the Scripps project leader, said previous attempts to show that human activities caused global warming had looked for evidence in the atmosphere. "But the atmosphere is the worst place to look for a global warming signal," he said. "Ninety per cent of the energy from global warming has gone into the oceans and the oceans show its fingerprint much better than the atmosphere."

Prof Barnett added: "The debate over whether there is a global warming signal is over now at least for rational people."...

EPILOGUE

B

INTERVIEW WITH THE AUTHOR PARTII

WHERE DO WE GO FROM HERE?

"Well, my friend, you've certainly laid it on thick in this thing."

I'm just a mouthpiece, I simply wanted to provide the information to give people a starting point. That's all, just "Here's the information, do what you want with it." Whether people do anything with the information I present doesn't really matter to me, it's their life.

I don't want people coming away from this thinking the world is going to end tomorrow. This isn't meant to be a fatalist piece, just a piece to get people thinking about things they might never have known about, but something that will affect them and their family, and maybe already is.

"But what you present for future scenarios is still just an estimated projection. Nothing is proven 100%."

You're right, no one can really see for certain what the future holds, and anyone can use selective evidence to prove a point and make a case. That's what I've done more or less. As you've seen, the evidence for the case of global temperature change is overwhelming and comes from a dizzying array of sources which corroborate and are legitimized by an overwhelming majority of their peers.

We can keep doing as many studies as we want, but there is no longer any doubt about what is happening with the planet in regards to global temperature change.

We are no longer blindfolded by uncertainty.

We can see the rocky coast in plain sight.

-...the balance of evidence suggest there is a discernible human influence on global climate.-

-Intergovernmental Panel on Climate Change:
Second Assessment Report
1995

-There is no debate among any statured scientists i.e. those currently engaged in relevant research and whose work has been published in the refereed scientific journals about what is happening.-

-James McCarthy
Scientific Committee for the International
Biosphere Programme
1998

-All the summaries of the IPCC have been agreed at the plenary meetings without dissent and none of us has received any subsequent letters of complaint from scientists regarding the final version. The process provides justification for the description of substantial scientific consensus.-

-Past IPCC Chair Bert Bolin
1997

Unfortunately, we're being pushed toward the rocky coast by the inevitable, something that is much stronger than all of our powers put together. A lot of people are screaming their heads off and the international community is listening.

The only uncertainties remaining are exactly when, how fast, and how much the global temperature will change, and who will be affected where with what consequences.

"Do you think the Earth is on the verge of a dramatic shift in functioning?"

The evidence certainly seems to point in that direction. This doesn't mean that everything will change tomorrow, but it could all change very quickly and that is why many people are concerned. It is fully possible that it take only a few years or decades for dramatic and abrupt changes to occur, depending on the Earth's dynamic systems and critical thresholds.

If the Earth didn't react as it has in the past, it would be very surprising, more surprising than if something did happen.

The Earth has entered what Earth Systems scientists call a NO-ANALOGUE STATE, which means modern civilization has no experience with the way the Earth is presently behaving. We know changes, big changes, could very well be on the horizon. But we don't know exactly how or when these changes are going to occur. But we know that when they come, they will seriously impact the stability of our settlements.

I don't think human civilization will collapse at once, but it's not out of the question. More probable is that little by little, the disruptions caused by global temperature change will severely strain 21st century settlements in many ways.

–There is new and stronger evidence that most of the warming observed over the last 50 years is attributable to human activities.–

–Intergovernmental Panel on Climate Change:
The Scientific Basis
2001

–It's clear that climate has both dimmer dials and switches.–

–Richard Alley
Pennsylvania State University
December 2001

°CLIMATE CHANGE 'ENTERING THE UNKNOWN'
THE BALTIMORE SUN
3 DECEMBER 2003

Two of the nation's top climate scientists say there's no longer any doubt that human activities are changing the Earth's atmosphere and its climate, and that our children and grandchildren will inherit the consequences.

Writing in tomorrow's edition of the journal *Science*, Thomas R. Karl and Kevin E. Trenberth say researchers remain uncertain about the precise course of climate change from here. That change has already "exceeded the bounds of natural variability. ... We are entering the unknown."

Karl is director of the National Climatic Data Center in Asheville, N.C. Trenberth heads the climate analysis section of the National Center for Atmospheric Research in Boulder, Colo. Their article in Science is part of the journal's "State of the Planet" series. A footnote states their conclusions are their own, and not those of the federal government.

Without more international cooperation to mitigate global climate change, the scientists say, "the likely result is more frequent heat waves, droughts, extreme precipitation events, wild fires, heat stress, vegetation changes and sea level rise."

...Over the past 50 years, the authors say, emissions of greenhouse gases - chiefly carbon dioxide released by the combustion of fossil fuels - have altered atmospheric chemistry. And because these gases remain in the atmosphere for decades or centuries, the effects have been cumulative, and will be long-lasting.

"Significant further change is guaranteed," Karl and Trenberth write. "The rate of change can be slowed, but it is unlikely to be stopped in the 21st century."

Since pre-industrial times, carbon dioxide in the atmosphere has increased 31 percent, while average global temperatures have risen in a parallel fashion. "There is no doubt that the composition of the atmosphere is changing from human activities, and today greenhouse gases are the largest human influence on global climate," Karl and Trenberth say.

The emission of greenhouse gases in the United States has increased between 0.5 percent and 1 percent per year in recent decades, studies have found.

..."Climate change is truly a global issue, one that may prove to be humanity's greatest challenge," Karl and Trenberth say.

–The Earth System is currently operating in a no-analogue state. Human activities are significantly altering the environment on a global scale.–

–International Geosphere Biosphere Programme:
Global Change and the Earth System
2004

°NASA: BUSH STIFLES GLOBAL WARMING EVIDENCE
ASSOCIATED PRESS
27 OCTOBER 2004

The Bush administration is trying to stifle scientific evidence of the dangers of global warming in an effort to keep the public uninformed, a NASA scientist said Tuesday night.

"In my more than three decades in government, I have never seen anything approaching the degree to which information flow from scientists to the public has been screened and controlled as it is now," James E. Hansen told a University of Iowa audience.

Hansen is director of the NASA Goddard Institute for Space Studies in New York and has twice briefed a task force headed by Vice President Dick Cheney on global warming.

Hansen said the administration wants to hear only scientific results that "fit predetermined, inflexible positions." Evidence that would raise concerns about the dangers of climate change is often dismissed as not being of sufficient interest to the public.

...Hansen said such warnings are consistently suppressed, while studies that cast doubt on such interpretations receive favorable treatment from the administration.

He also said reports that outline potential dangers of global warming are edited to make the problem appear less serious. "This process is in direct opposition to the most fundamental precepts of science," he said...

°CLIMATE 'BIGGER THREAT THAN TERROR'
THE AUSTRALIAN
27 OCTOBER 2004

Climate change posed a greater long-term threat to Australia than terrorism, South Australian Premier Mike Rann said today.

"The metabolism of the world's modern economy is on a collision course with the metabolism of our planet," Mr. Rann said.

"Action on climate change is not just about the environment, it is about economics and a future for our children.

"In the long term, climate change poses a greater threat to Australia than terrorism."...

°'U.S. CLIMATE POLICY BIGGER THREAT TO WORLD THAN TERRORISM'
THE INDEPENDENT
9 JANUARY 2004

Tony Blair's chief scientist has launched a withering attack on President George Bush for failing to tackle climate change, which he says is more serious than terrorism.

Sir David King, the Government's chief scientific adviser, says in an article today in the journal *Science* that America, the world's greatest polluter, must take the threat of global warming more seriously.

"In my view, climate change is the most severe problem that we are facing today, more serious even than the threat of terrorism," Sir David says.

..."Delaying action for decades, or even just years, is not a serious option. I am firmly convinced that if we do not begin now, more substantial, more disruptive, and more expensive change will be needed later on."...

°SCIENTIST 'GAGGED' BY NO 10 AFTER WARNING OF GLOBAL WARMING THREAT
THE INDEPENDENT
8 MARCH 2004

Downing Street tried to muzzle the Government's top scientific adviser after he warned that global warming was a more serious threat than international terrorism.

Ivan Rogers, Mr. Blair's principal private secretary, told Sir David King, the Prime Minister's chief scientist, to limit his contact with the media after he made outspoken comments about President George Bush's policy on climate change.

...Support for Sir David's view came yesterday from Hans Blix, the former United Nations chief weapons inspector, who said the environment was at least as important a threat as global terrorism...

°BUSH ATTACKS ENVIRONMENT 'SCARE STORIES' : SECRET EMAIL GIVES ADVICE ON DENYING CLIMATE CHANGE
THE OBSERVER
4 APRIL 2004

George W. Bush's campaign workers have hit on an age-old political tactic to deal with the tricky subject of global warming - deny, and deny aggressively.

The Observer has obtained a remarkable email sent to the press secretaries of all Republican congressmen advising them what to say when questioned on the environment in the run-up to November's election. The advice: tell them everything's rosy.

It tells them how global warming has not been proved, air quality is 'getting better', the world's forests are 'spreading, not deadening', oil reserves are 'increasing, not decreasing', and the 'world's water is cleaner and reaching more people'.

The email - sent on 4 February - warns that Democrats will 'hit us hard' on the environment. 'In an effort to help your members fight back, as well as be aggressive on the issue, we have prepared the following set of talking points on where the environment really stands today,' it states.

The memo - headed 'From medi-scare to air-scare' - goes on: 'From the heated debate on global warming to the hot air on forests; from the muddled talk on our nation's waters to the convolution on air pollution, we are fighting a battle of fact against fiction on the environment - Republicans can't stress enough that extremists are screaming "Doomsday!" when the environment is actually seeing a new and better day.'

Among the memo's assertions are 'global warming is not a fact', 'links between air quality and asthma in children remain cloudy', and the US Environment Protection Agency is exaggerating when it says that at least 40 per cent of streams, rivers and lakes are too polluted for drinking, fishing or swimming.

It gives a list of alleged facts taken from contentious sources. For instance, to back its claim that air quality is improving it cites a report from Pacific Research Institute - an organisation that has received $130,000 from Exxon Mobil since 1998.

The memo also lifts details from the controversial book *The Skeptical Environmentalist* by Bjorn Lomborg. On the Republicans' claims that deforestation is not a problem, it states: 'About a third of the world is still covered with forests, a level not changed much since World War II. The world's demand for paper can be permanently satisfied by the growth of trees in just five per cent of the world's forests.'

The memo's main source for the denial of global warming is Richard Lindzen, a climate-sceptic scientist who has consistently taken money from the fossil fuel industry. His opinion differs substantially from most climate scientists, who say that climate change is happening.

But probably the most influential voice behind the memo is Frank Luntz, a Republican Party strategist. In a leaked 2002 memo, Luntz said: 'The scientific debate is closing [against us] but not yet closed. There is still a window of opportunity to challenge the science.'...

COME NOW BOYS

HAVE SOME DIGNITY

WE MAY BE PLEBIANS

BUT WE AIN'T STUPID

°OIL FIRMS FUND CLIMATE CHANGE 'DENIAL'
THE GUARDIAN
27 JANUARY 2005

Lobby groups funded by the US oil industry are targeting Britain in a bid to play down the threat of climate change and derail action to cut greenhouse gas emissions, leading scientists have warned.

Bob May, president of the Royal Society, says that "a lobby of professional sceptics who opposed action to tackle climate change" is turning its attention to Britain because of its high profile in the debate.

Writing in the Life section of today's *Guardian*, Professor May says the government's decision to make global warming a focus of its G8 presidency has made it a target. So has the high profile of its chief scientific adviser, David King, who described climate change as a bigger threat than terrorism.

Prof. May's warning coincides with a meeting of climate change sceptics today at the Royal Institution in London organised by a British group, the Scientific Alliance, which has links to US oil company ExxonMobil through a collaboration with a US institute.

Last month the Scientific Alliance published a joint report with the George C Marshall Institute in Washington that claimed to "undermine" climate change claims. The Marshall institute received £51,000 from ExxonMobil for its "global climate change programme" in 2003 and an undisclosed sum this month...

°FORGET CLIMATE CHANGE, THAT'S THE LEAST OF OUR WORRIES, SAY NOBEL WINNERS
THE GUARDIAN
21 OCTOBER 2004

Climate change, predicted by the UN to change the way most people live over the next 100 years, is the least important of the world's immediate problems, says a group of economists, including three Nobel prize winners, who were asked to prioritise how money should be spent on helping the world's poor.

The team of six American and two other economists, brought together by controversial environmentalist Bjorn Lomborg, said it was not worth spending money on climate change because the effects were expected to be far in the future. They recommended that people became rich first and that money should be spent on HIV/Aids, water and free trade.

...The economists then considered the potential costs and benefits of spending money on problems like hunger, climate change, communicable diseases, sanitation, water, money laundering and financial instability. Rejecting spending anything on education, slums, terrorism, arms proliferation, deforestation, lack of energy or corruption, they narrowed the list to 10 areas, which they then divided and ranked into a number of initiatives.

..."This simplistic and rather banal ranking of these problems should not be taken too seriously," said Stephen Tindale, the director of Greenpeace. "It is an example of intellectual illiteracy. All these problems are linked."

...Mr. Lomborg, however, was unfazed. "The biggest problem is that we all die, yet no one is considering how to solve that."...

°RETIRING CLIMATE CHANGE
CENTER FOR AMERICAN PROGRESS
16 FEBRUARY 2005

Since his re-election, President Bush has spent much of his time, and political capital, on the security of seniors. Despite the fact that Social Security can pay full benefits at least until 2042, he has warned that "The cost of doing nothing – the cost of saying the current system is okay – far exceeds the cost of trying to make sure we save the system for our children."

While he creates a false crisis in Social Security, he ignores the real crisis of climate change...

°NO STOPPING GLOBAL WARMING, STUDIES PREDICT
REUTERS
17 MARCH 2005

Even if people stopped pumping out carbon dioxide and other pollutants tomorrow, global warming would still get worse, two teams of researchers reported on Thursday.

Sea levels will rise more than they have already risen, worsening the damage caused by extreme high tides and storm surges, and droughts, heat waves and storms will become more severe, the climate experts predicted.

That makes immediate action to slow global warming even more vital, the teams at the National Center for Atmospheric Research in Colorado report in the journal *Science*.

"Even if we stabilize greenhouse gas concentrations, the climate will continue to warm, and there will be proportionately even more sea level rise," said the NCAR's Gerald Meehl, who led one of the two studies.

"The longer we wait, the more climate change we are committed to in the future."...

-I want you to think about a Social Security system that will be flat bust, bankrupt, unless the United States Congress has got the willingness to act now.

And that's what we're here to talk about, a system that will be bankrupt.

Now, I readily concede some would say, well, it's not bankrupt yet, why don't we wait until it's bankrupt?

The problem with that notion is that the longer you wait the more difficult it is to fix.-

-United States President George W. Bush Jr. as quoted on social security 11 January 2005

- To say we'll worry about it later is dangerous and irresponsible, and I think, cowardly.-

-Nebraska Senator Chuck Hagel as quoted on social security 6 March 2005

ADDRESSING CLIMATE CHANGE IS NO DIFFERENT

°NOW THE PENTAGON TELLS BUSH: CLIMATE CHANGE WILL DESTROY US
THE OBSERVER
22 FEBRUARY 2004

Climate change over the next 20 years could result in a global catastrophe costing millions of lives in wars and natural disasters. A secret report, suppressed by US defence chiefs and obtained by The Observer, warns that major European cities will be sunk beneath rising seas as Britain is plunged into a 'Siberian' climate by 2020. Nuclear conflict, mega-droughts, famine and widespread rioting will erupt across the world.

The document predicts that abrupt climate change could bring the planet to the edge of anarchy as countries develop a nuclear threat to defend and secure dwindling food, water and energy supplies. The threat to global stability vastly eclipses that of terrorism, say the few experts privy to its contents.

'Disruption and conflict will be endemic features of life,' concludes the Pentagon analysis. 'Once again, warfare would define human life.'

The findings will prove humiliating to the Bush administration, which has repeatedly denied that climate change even exists. Experts said that they will also make unsettling reading for a President who has insisted national defence is a priority.

The report was commissioned by influential Pentagon defence adviser Andrew Marshall, who has held considerable sway on US military thinking over the past three decades. He was the man behind a sweeping recent review aimed at transforming the American military under Defence Secretary Donald Rumsfeld.

Climate change 'should be elevated beyond a scientific debate to a US national security concern', say the authors, Peter Schwartz, CIA consultant and former head of planning at Royal Dutch/Shell Group, and Doug Randall of the California-based Global Business Network.

An imminent scenario of catastrophic climate change is 'plausible and would challenge United States national security in ways that should be considered immediately', they conclude. As early as next year widespread flooding by a rise in sea levels will create major upheaval for millions.

Last week the Bush administration came under heavy fire from a large body of respected scientists who claimed that it cherry-picked science to suit its policy agenda and suppressed studies that it did not like. Jeremy Symons, a former whistleblower at the Environmental Protection Agency (EPA), said that suppression of the report for four months was a further example of the White House trying to bury the threat of climate change.

...Among those scientists present at the White House talks were Professor John Schellnhuber, former chief environmental adviser to the German government and head of the UK's leading group of climate scientists at the Tyndall Centre for Climate Change Research. He said that the Pentagon's internal fears should prove the 'tipping point' in persuading Bush to accept climatic change.

Sir John Houghton, former chief executive of the Meteorological Office - and the first senior figure to liken the threat of climate change to that of terrorism - said: 'If the Pentagon is sending out that sort of message, then this is an important document indeed.' Bob Watson, chief scientist for the World Bank and former chair of the Intergovernmental Panel on Climate Change, added that the Pentagon's dire warnings could no longer be ignored.

'Can Bush ignore the Pentagon? It's going be hard to blow off this sort of document. Its hugely embarrassing. After all, Bush's single highest priority is national defence. The Pentagon is no wacko, liberal group, generally speaking it is conservative. If climate change is a threat to national security and the economy, then he has to act. There are two groups the Bush Administration tend to listen to, the oil lobby and the Pentagon,' added Watson...

°AN ABRUPT CLIMATE CHANGE SCENARIO AND ITS IMPLICATIONS FOR UNITED STATES NATIONAL SECURITY
THE UNITED STATES PENTAGON -
PETER SCHWARTZ AND DOUG RANDALL
OCTOBER 2003

There is substantial evidence to indicate that significant global warming will occur during the 21st century. Because changes have been gradual so far, and are projected to be similarly gradual in the future, the effects of global warming have the potential to be manageable for most nations.

Recent research, however, suggests that there is possibility that this gradual global warming could lead to a relatively abrupt slowing of the ocean's thermohaline conveyor, which could lead to harsher winter weather conditions, sharply reduced soil moisture, and more intense winds in certain regions that currently provide a significant fraction of the world's food production. With inadequate preparation, the result could be a significant drop in the human carrying capacity of the Earth's environment.

The research suggests that once temperature rises above some threshold, adverse weather conditions could develop relatively abruptly, with persistent changes in the atmospheric circulation causing drops in some regions of 5-10 degrees Fahrenheit in a single decade. Paleoclimatic evidence suggests that altered climate patterns could last for as much as a century, as they did when the ocean conveyor collapsed 8,200 years ago, or, at the extreme, could last as long as 1,000 years as they did during the Younger Dryas, which began about 12,700 years ago...

KEY FINDINGS OF THE PENTAGON REPORT ON ABRUPT CLIMATE CHANGE

°A significant drop in the planet's ability to sustain its present population will become apparent over the next 20 years.

°By 2010 the US and Europe will experience a third more days with peak temperatures above 90F. Climate becomes an 'economic nuisance' as storms, droughts and hot spells create havoc for farmers.

°Between 2010 and 2020 Europe is hardest hit by climatic change with an average annual temperature drop of 6F. Climate in Britain becomes colder and drier as weather patterns begin to resemble Siberia.

°Bangladesh becomes nearly uninhabitable because of a rising sea level, which contaminates the inland water supplies.

°GLOBAL WARMING INEVITABLE FOR DECADES TO COME, SCIENCE CONFERENCE TOLD
AGENCE FRANCE-PRESSE
1 FEBRUARY 2005

A climate conference opened to renewed concern about the worsening threat of global warming and appeals from Britain to its ally, the United States, not to stand on the sidelines...

..."A significant impact (on the world's climate system) is already inevitable," she said.

"What is certainly clear is that temperatures will go on rising... most of the warming we are expecting over the next few decades is now virtually inevitable." Beckett warned: "No one country, not even one continent, can solve the problem by acting alone."

She hailed the Kyoto Protocol, the UN's pact on carbon pollution, which takes life on February 16 after more than seven years of haggling to complete its rulebook and secure its ratification. "Kyoto is very much a first step"," said Beckett, who also lobbied for clean technology and encouragement for developing countries not to follow rich nations down the path of fossil-fuel pollution...

°Future wars will be fought over the issue of survival rather than religion, ideology or national honour.

°Rich areas like the US and Europe would become virtual fortresses to prevent millions of migrants from entering after being forced from land drowned by sea-level rise or no longer able to grow crops. Waves of boatpeople pose significant problems.

°China's huge population and food demand make it particularly vulnerable.

°BLAIR WARNS OF CLIMATE CHANGE'S THREAT
ASSOCIATED PRESS
14 SEPTEMBER 2004

Prime Minister Tony Blair warned on Tuesday of the threat posed by climate change and urged support for the principles of the Kyoto accord on global warming, a treaty rejected by President Bush as unfair toward U.S. industry.

Blair promised to make global warming a focal point of Britain's presidency of the Group of Eight summit next year and said he will push for greater international commitment to cut greenhouse gases.

..."Climate change will be a top priority for our G-8 presidency next year," Blair said. "This remains an issue of high and fraught politics for many countries. But it is imperative we try.".

...The world needs a "new green industrial revolution" to tackle the crisis that is seeing temperatures rise, glaciers melt, and sea levels rise, he said...

°GLOBAL WARMING THREAT CENTRAL TO POLICY-BRITAIN
REUTERS
15 MARCH 2005

Britain told the world's biggest polluters including the United States Tuesday that only by placing the environment at the heart of economic policy could they prevent a crisis caused by global warming.

Britain hosted a two-day brainstorming on climate change by ministers and senior officials from 20 countries in the run-up to a July meeting of the eight most industrialized nations -- the G8 group -- currently led by London.

The need for action to avert a looming climate catastrophe was rammed home by graphic images of melting glaciers and makeshift sea defenses displayed at the venue of the meeting.

"We must make climate stability, energy investment and energy security central to economic policies," British Chancellor of the Exchequer Gordon Brown told the meeting. "International cooperation is again the only way forward."...

The changes described in future forecasts involve dramatic events, catastrophic events. Although they appear strange and unbelievable to us, dramatic changes like this are normal for the Earth.

"Do you think humankind does not believe the Earth will dramatically change, because we doubt it will change? Or do we not believe the Earth is capable of such changes?"

I think humankind both doubts the Earth will change to such a degree in our lifetime and also can't believe it will change. The scenarios are formidable and a bit unfathomable. The forecasts for the 21ˢᵗ century signal an abrupt end to the world as we know it. They forecast an end to the foundation of our settlements, an end to our way of life, an end to our lifestyles, and an end to all that we hold familiar and sacred.

But, I think an end to one way can also be a new beginning for another way, and I think our warring global civilization is due for a change. I don't want to be misconstrued as a misanthrope, I think humans are awesome, but maybe we deserve a swift spanking after the way we have been behaving toward one another and the planet that we depend on for survival.

It might help to set us straight and pointed in the right direction.

It may take dramatic events to change our worldview, change our paradigms, change our behaviors, and change the way we live with one another and with the Earth.

We will have to adapt to the changing Earth either before or after the changes have occurred. We will have to change, not for environmental, social or economic reasons, not to 'Save the Earth' and not for the betterment of our global family. We will have to change, because things will happen beyond our control that will force us to change for our own security, for our own survival.

But in doing so, we can make changes that will be economically, environmentally, and socially beneficial, as well as strengthen the ties within our global family and become a more benign actor within the fabric of the Earth.

"But global temperature change and climate changes, these things are really far off aren't they? I mean, there are so many pressing issues today, like developing country debt relief, or genocides around the world, or the AIDS epidemic, or the War on Terror. Couldn't we spend our money better on these issues which affect humans in the here and now?"

Let's face it, at the beginning of the 21ˢᵗ century there are many troubling issues to address within our species. There are great disparities between rich and poor, stresses on food, water and energy resources, cultural differences, religious differences, disease epidemics, and changing global climate patterns.

These problems often lead to violent conflicts within the species.

WE KNOW
WHAT TO
EXPECT

WHAT ON
EARTH
ARE WE
EXPECTED
TO DO?

intraspecies
Arising or occurring within a species, involving the members of one species, i.e. intraspecific competition.

—To me the question of the environment is more ominous than that of peace and war. We will have regional conflicts and use of force, but world conflicts I do not believe will happen any longer.

But the environment, that is a creeping danger. I'm more worried about global warming than I am of any major military conflict.—

—Hans Blix
U.N. Weapons Inspector
12 March 2003

There is an increase in terrorism and nuclear proliferation, mass migrations, and genocides. All of the above are sentencing millions to death and billions more to famine, hydrological poverty, economic impoverishment, and a general degraded circumstance for our fellow humans.

What happens to the Western World when the oil it is dependent on is being sucked up by other countries in the East?

What happens when countries go to war over fresh water supplies?

What happens when climate changes disrupt growing seasons and the global food supply?

What happens when millions of people are displaced as a result of rising sea levels?

What happens when 2 billion more people start driving cars?

All of our biological, chemical, and nuclear weapons, all of our issues with genetic modifications and cloning, all of our viral inventions, all of our North/South and East/West cultural clashes, all of our economic inequalities, all of our terrorists, all of our militarization of space, all of our bombs, guns, bullets, and mines, all of our crimes against humanity, all of our plundering of the Earth, all of our self-destructiveness; all of these sick manifestations and disgusting expressions of being 'civilized.'

All of these problems are INTRASPECIFIC.

All of the problems that revolve within the human sphere are created by humans, can be prevented by humans, and can be solved by humans.

We all came from the same place a long, long time ago. We are all brothers and sisters from the same ancestors in a big family of 6 billion and growing. Sure, we have our differences, which are as important as our similarities and give us our own unique understanding of the world around us.

Human beings have made great strides over the years, but the family feuds being perpetrated, whether it is military imperialism and a war for energy resources in the Fertile Crescent, religious feuds and terrorist attacks in the Holy Land, or genocide in the birthplace of humans, all of the conflicts within the species can be resolved by our species, no matter how great they may appear.

In regards to the general degradation of the planet we depend on for survival, we often hear people say, "Oh, the Earth is sick," or "Oh, we're killing the Earth," or "Oh, we have to save the Earth."

The Earth can get along by itself, it has been around a lot longer than we have and will be around a lot longer after we're gone. The Earth doesn't care what we do to it, the Earth has no feelings like we know feelings. For all we care, the Earth is dead matter and should be used and misused however we like. It may be an energetic living, breathing blob of metal, rock, water, gas, and complex carbon beings, but it is a living blob without a brain and it does not think when it reacts, it simply reacts.

The Earth doesn't care if we pollute it, bomb it, chop it down, dig it up, or cover it with our garbage. It is not the Earth that is sick, it

Support Oil Dependency

is the way in which humans live within the Earth that is sick and our relationship to one another is sick. We are only poisoning, trashing, killing, and impoverishing ourselves and our children; dumping our garbage in our own home and burning it down.

Is this any way to live on the only life bearing planet we know of?

We as a species are smarter than this.

We are better than this.

We can rise above this and make our lives on this planet better for everyone.

The Earth does not care about us, does not care what we do to it, and does not care what we do to each other. Humans may think they're the greatest thing to ever walk the Earth, but the Earth does not care how great we think we are.

Who cares if we are here or not?

We care if we are here or not.

We care what will happen to us and our family in the present and in the future.

We care about how healthy our surrounding natural and social environment is.

We care who we are, where we are from, and what we believe from our traditions and cultures.

We care about all of these things.

This is what makes us all human beings and is what being human is all about.

"How do you think humankind will react to the predicted consequences of global temperature change?"

Good question. Do you mean as a result of actual changes occurring this century or as a result of simply knowing about predicted changes, because we aren't changing much at all as a result of this knowledge, in fact, we seem to be going in the opposite direction when it comes to energy use, our infrastructures, and a globalizing consumer economy.

When the natural and human systems and structures that supply water, food, energy, security, employment, and safety break down, society can become unstable and civil unrest can ensue. When these pressures breach a threshold, even the most advanced civilization can be left totally helpless. Unpreparedness and intraspecific conflicts will only exacerbate the problems created by a changing global temperature, changing climates, and changing weather patterns.

If individuals, communities, towns, cities, states, and nations are unprepared, there will be a weakening of settlements to the point of breakdown in many areas of society. Communities and nations will struggle to supply basic necessities to their people, while dealing with larger disruptions caused by increased natural disasters. Furthermore, international insurance agencies will have difficulties providing funding for increased damages and disaster areas.

THE END OF OIL IS CLOSER THAN YOU THINK
THE GUARDIAN
21 APRIL 2005

The one thing that international bankers don't want to hear is that the second Great Depression may be round the corner. But last week, a group of ultra-conservative Swiss financiers asked a retired English petroleum geologist living in Ireland to tell them about the beginning of the end of the oil age.

They called Colin Campbell, who helped to found the London-based Oil Depletion Analysis Centre because he is an industry man through and through, has no financial agenda and has spent most of a lifetime on the front line of oil exploration on three continents. He was chief geologist for Amoco, a vice-president of Fina, and has worked for BP, Texaco, Shell, ChevronTexaco and Exxon in a dozen different countries.

..."The issue is the long downward slope that opens on the other side of peak production. Oil and gas dominate our lives, and their decline will change the world in radical and unpredictable ways," he says.

...If he is correct, then global oil production can be expected to decline steadily at about 2-3% a year, the cost of everything from travel, heating, agriculture, trade, and anything made of plastic rises. And the scramble to control oil resources intensifies. As one US analyst said this week: "Just kiss your lifestyle goodbye."

...Most serious of all, he and other oil depletion analysts and petroleum geologists, most of whom have been in the industry for years, accuse the US of using questionable statistical probability models to calculate global reserves and Opec countries of drastically revising upwards their reserves in the 1980s.

...In the wake of the Iraq war, the rapid economic rise of China, global warming and recent record oil prices, the debate has shifted from "if" there is a global peak to "when"...

On top of that, there will be huge numbers of people migrating away from places that have been rendered uninhabitable.

Where are they supposed to go?

But I am still optimistic, even a little naive, maybe even a bit idealistic. But I am still grounded in the reality of the situation.

I envision a landscape for our descendants in which the principles of **ANTICIPATORY ADAPTATION** are applied and based on sustainable development, self-reliance, and self-preservation. I believe the shift in this direction will put communities and nations on the offensive, allowing them to address predicted disruptions in a preventive manner, secure their way of life, and improve their quality of life and the community they inhabit.

The solutions for the issues you mentioned and the consequences of global temperature change are interlinked. One cannot simply prioritize issues, one must look at the relationships between issues and address the problems holistically.

Anticipatory adaptation measures can remedy many of the problems simultaneously. This allows us to not only hold on to a semblance of our way of life, but allow others to improve their way of life.

TIME TO START THINKING ABOUT SOLUTIONS THAT:

SECURE LIFESTYLES

IMPROVE QUALITY OF LIFE

ARE ECONOMICALLY ROBUST

REDUCE IMPACT ON EARTH

REDUCE GREENHOUSE GASES

Anticipatory adaptation allows us to secure necessities, become economically resilient, reduce our impact on the natural world, reduce greenhouse gases, and remedy our intraspecific conflicts.

Anticipatory adaptation is intended to mitigate the consequences brought by global temperature change and climate changes before they happen. As we confront the issue of global temperature change under the emerging worldview of global connectedness and responsibility, we will notice that our intraspecific problems and those resulting from a warming planet are interrelated.

There are already real solutions available today that can solve many problems facing people and the planet at the same time, while creating more robust settlements and improve livelihoods. It is time to start figuring out what to do and how to reconstruct our settlements before they are destabilized by forces beyond our control.

Settlements that act on this will be able to prepare for the worst case scenarios, which include energy instabilities, unstable weather for growing seasons, economic collapse of certain sectors and large scale unemployment, increasingly frequent and severe natural disasters, freshwater scarcity, coastal metropolis flooding, health problems, transportation disruptions, and many other fundamental problems that can tear a society apart.

Fortunately, millions of years of evolution have given us a brilliant brain to work through complex problems and this will help us reduce the threat we have created for ourselves. We have developed the most

-Necessity is the mother of invention.-

-Plato

advanced organ for thinking and the most manipulative abilities of anything ever known to exist. We are meant to think complexly, we are supposed to solve problems. Our ancestors may have gotten us into this predicament, but our contemporaries can figure a way out.

We just have to put our heads, hands, and hearts together.

I think our predicament may be a blessing in disguise.

"How do you think we'll get to this safe harbor?"

International coordination and global treaties are certainly helpful, and the **KYOTO PROTOCOL**, the international agreement to curb anthropogenic sources of greenhouse gas emissions, provides a fantastic framework for cooperative endeavors. But the goal is quite weak and doesn't really address people's security in a future with climate change. I don't think it has the appropriate focus of what needs to be accomplished.

The Kyoto Protocol focuses primarily on developed countries reducing their greenhouse gas emissions, but only by 5% from 1990 levels. The idea is that by reducing anthropogenic greenhouse gas emissions, we will be able to slow global warming and allow our settlements to continue functioning as they presently do, with less emissions.

Believing we can control the systems of the Earth in this way is what I call **ACUTE ANTHROPOCENTRIC ARROGANCE**, a prime example of the mentality that led us into our present predicament. It is this thinking that must be replaced if we hope to create adequate solutions.

We have to realize what humans can and cannot do; what humans can and cannot control. Realizing this is critical to using our time and energy effectively to create solutions which solve the problems and seize the benefits predicted as a result of global temperature and climate changes during this century, without making the situation any worse.

The trend of global warming, greenhouse gas emissions from melting permafrost, warming and rising oceans, melting glaciers and poles…this is all happening and predicted to continue. Even if 100% of anthropogenic emissions were eliminated today and stayed at 0 for the next 100 years, the trends would continue.

So the 5% reduction that the Kyoto Protocol enforces is fairly useless. Furthermore, it doesn't target developing countries, which are set to emit far more greenhouse gases during the coming years as they go through fossil fuel intensive industrialization.

I'm not against the idea of an international agreement, I'm just for targeting adaptation and mitigation strategies which focus on securing our way of life while reducing greenhouse gases and our impact on the planet, rather than only targeting emission reductions. It seems more logical to me to implement adaptation strategies that first secure livelihoods and simultaneously reduce emissions, improve economic resilience, and improve social and environmental aspects of a place.

International frameworks are good, but countries, regions, states,

5%

THE INTERNATIONAL CLIMATE CHANGE COMMUNITY IS TRYING TO STABILIZE GLOBAL GREENHOUSE GAS EMISSIONS

TO STABILIZE THE GLOBAL TEMPERATURE

TO STABILIZE CURRENT GLOBAL CLIMATE SYSTEMS

TO STABILIZE CURRENT GLOBAL WEATHER PATTERNS

TO PERPETUATE CURRENT HUMAN SYSTEMS AND STRUCTURES

WHICH ALL 21ST CENTURY SETTLEMENTS ARE BUILT AND TOTALLY DEPENDENT ON

IT'S WORTH A TRY

BUT IT WON'T GIT 'ER DUN

°CLIMATE TALKS SHIFT FOCUS TO HOW TO DEAL WITH CHANGES
NEW YORK TIMES
2 NOVEMBER 2002

The global climate is changing in big ways, probably because of human actions, and it is time to focus on adapting to the impacts instead of just fighting to limit the warming. That, in a nutshell, was the idea that dominated the latest round of international climate talks, which ended on Friday in New Delhi.

While many scientists have long held this view, it was a striking departure for the policy makers at the talks — the industry lobbyists, environmental activists and government officials. For more than a decade, their single focus had been the fight over whether to cut smokestack and tailpipe emissions of carbon dioxide and other heat-trapping greenhouse gases.

Many environmentalists had long avoided discussing adaptation for fear it would smack of defeatism.

...Although they conceded its importance, environmental campaigners said an approach that focused on adapting to climate change rather than preventing it would inevitably fail, because the impact of unfettered emissions would eventually exceed people's ability to adjust...

...Another impetus is the rising realization that many significant shifts have already been set in motion by a century-long accumulation of warming gases.

...The adaptation issue also got support from a new scientific analysis, published Friday, suggesting that the only way to safely stabilize greenhouse gases by midcentury was with a hugely ambitious Apollo-size research program on fusion, solar power, and other nonpolluting energy sources...

—Call on God, but row away from the rocks.—

—Hunter S. Thompson

°IT'S MUCH TOO LATE TO SWEAT GLOBAL WARMING
SAN FRANCISCO CHRONICLE
13 FEBRUARY 2005

Time to prepare for inevitable effects of our ill-fated future. At the core of the global warming dilemma is a fact neither side of the debate likes to talk about: It is already too late to prevent global warming and the climate change it sets off. Environmentalists won't say this for fear of sounding alarmist or defeatist. Politicians won't say it because then they'd have to do something about it. The world's top climate scientists have been sending this message, however, with increasing urgency for many years.

Since 1988, the United Nations Intergovernmental Panel on Climate Change, comprised of more than 2,000 scientific and technical experts from around the world, has conducted the most extensive peer-reviewed scientific inquiry in history. In its 2001 report, the panel said that human-caused global warming had already begun, and much sooner than expected. What's more, the problem is bound to get worse, perhaps a lot worse, before it gets better.

Last month, the climate change panel's chairman, Rajendra Pachauri, upped the ante. Although Pachauri was installed after the Bush administration forced out his predecessor, Robert Watson, for pushing too hard for action, the accumulation of evidence led Pachauri to embrace apocalyptic language: "We are risking the ability of the human race to survive," he said.

Until now, most public discussion about global warming has focused on how to prevent it - for example, by implementing the Kyoto Protocol, which comes into force internationally (but without U.S. participation) on Wednesday. But prevention is no longer a sufficient option. No matter how many "green" cars and solar panels Kyoto eventually calls into existence, the hard fact is that a certain amount of global warming is inevitable.

The world community therefore must make a strategic shift. It must expand its response to global warming to emphasize both long-term and short-term protection. Rising sea levels and more weather-related disasters will be a fact of life on this planet for decades to come, and we have to get ready for them."

...British Prime Minister Tony Blair regards climate change as "the single biggest long-term problem" of any kind facing his country. His government's top scientist, Sir David King, goes further, calling climate change "the biggest danger humanity has faced in 5,000 years of civilization."

Although the Bush White House continues to downplay the urgency of global warming, some parts of the Bush administration have recognized the gravity of the situation. A report released last year by the Pentagon's Office of Net Assessments said that by 2020, climate change could unleash a series of interlocking catastrophes including mega-droughts, mass starvation and even nuclear war as countries like China and India battle over river valleys and other sources of scarce food and water.

All of this underlines the urgency of revising the world's response to climate change. To be sure, it remains essential to reduce greenhouse gas emissions by strengthening the Kyoto Protocol and augmenting it with other measures. Otherwise, the amount of warming that civilization eventually will have to endure will prove too great to survive.

In the meantime, it is imperative to prepare against the climate change already on its way...

municipalities, and ultimately individuals have to accept that anticipatory adaptation has to be taken into our own hands, for their own survival.

"Is this something you think can be realized?"

Of course! It is already happening in places all over the world.

There are farmer-owned regional wind energy systems providing renewable electricity to economically and socially run-down rural areas in the United States. There are solar cooker projects that are providing thousands with cooking fuel through solar reflection in African villages. There are biodegradable 'plastics' made from corn starch that can replace all petroleum based plastics on the market. There are rain-water harvesting areas in South America. There are urban gardens all over the world and organic foods are becoming more prevalent than conventional foods.

From the highest of high-tech to traditional methods that go back thousands of years, these solutions are out there.

The future society I'm talking about is already emerging and it has to be nurtured and encouraged. But, it cannot be forced out.

As a collective species, we are finally taking to heart the repeated mistakes of the past. We are noticing that competition is no longer necessary and is in fact a detriment to survival in a modern world. We are beginning to see that cooperation is the only way we will be able to survive in a future that provides us with less suitable conditions for living in our modern settlements.

What I am talking about is a shift within our global civilization that puts traditional differences behind, while still holding on to cultural identities, and focuses on the security and safety of our shared species in the face of great changes.

This may be difficult for a lot of people to accept, because many of us are grounded with the mentality of 'us' and 'them.'

But in facing our global predicament, there can only be a 'we.'

Thinking about a future under different circumstances can be a daunting task, but it is useful to think about the future concerning global temperature change. It is as equally useful as thinking about retirement, investments, social security, or your child's education.

I remember when I was growing up my parents would tell me:

"You see son...You should always think about where you want to be five years from now. Don't give up living in the moment, but you have to prepare for the future you want and you have to begin preparing now to get yourself there. In order to secure the life you want in the future, you have to start thinking about it now and take the steps to realize it a few years down the road."

"Well G, this has been quite a ride. I'm looking forward to working with you again. Any parting words?"

Thanks for your interest, my friend. And thanks to you reader.

This book is only half of the story, it just framed our situation to give us a starting point.

The first step in preparation is understanding the situation. The second step is internalizing the understanding. The third step is actually doing something to secure a life for yourself, your family, and your neighbors in a future that offers different circumstances than today.

As a few parting words until next time............................

My fellow humans.

We face a common future of uncommon proportion.

An end to one way of life means the beginning of another way, which may be better and may be worse, probably a little of both, depending on how you look at it and the aspects you consider.

The future will be different than today.

The Earth will not function like it does today, because it will not be able to.

The course for the Earth is pretty well laid out in how it has functioned in the past. There are always unpredictabilities, but it is fairly straightforward chemistry and physics. If one drops a rock, it falls to the ground. If events occur which increase the level of greenhouse gases, the Earth warms. It has worked this way since the beginning of life on this planet and it will most likely work that way until the end.

However, the course for humankind is created everyday by the independent thinking and doing of members in our global family. The beauty of our global family lies in the fact that we can help provide for each other's needs, satisfy most people's wants, and prepare ourselves for impact with the rocky coast.

If we want to.

It all depends on the choices we make.

It is time to put our heads together.

It is time to put our hands together.

It is time to put our hearts together.

We've got a planet to give to our children.

There is a new game in town called 'adapting in anticipation to the consequences of 21st century global temperature change.'

I hope ya'll are ready to come outside and play.

Do you feel that?

It's not the buildup of artificial chemicals in your air, water, and food.

It is the truth aching to be set free.

Let it out and see where it takes you.

It's likely to give you a safe landing.

ACHTUNG BABIES

THIS IS YOUR CAPTAIN
SPEAKING

WE HAVE
ENCOUNTERED
SOME UNEXPECTED
TURBULENCE

PLEASE FASTEN
YOUR SEATBELTS
AND EXTINGUISH ALL
SMOKING DEVICES

WE HAVE PROVIDED
YOU WITH AN
EMERGENCY SCENARIO
FOR YOUR SAFETY

PLEASE TURN
YOUR ANTICIPATORY
ADAPTATION MANUAL
TO PAGE 2005

REMEMBER...

WE ONLY
BORROW THE
EARTH FROM
OUR CHILDREN

SOMEDAY...
THEY WILL
WANT IT BACK

TO BE CONTINUED............................

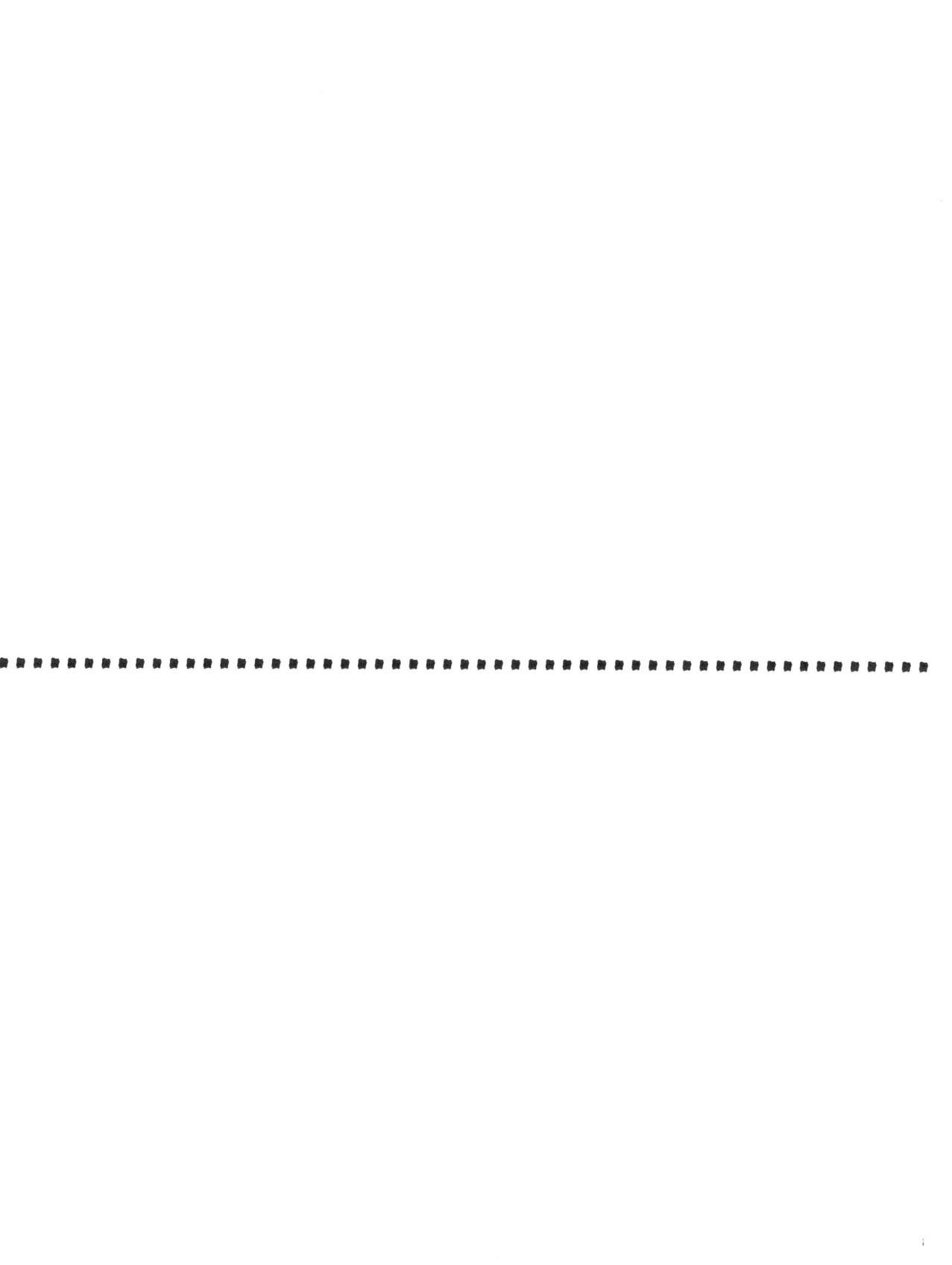

CREDITS

Concept, layout, and graphic design by G.E. Johnson.
All material is credited in order of appearance.

FRONT COVER

BACK COVER

SOMEDAY

North Atlantic ocean circulation system © 2011 The Estate of G.E. Johnson based on image of *The North Atlantic Ocean circulation system* from the NASA Goddard Space Flight Center Multimedia Design Studio. *Northern Hemisphere* courtesy of NASA Goddard Space Flight Center Multimedia Design Studio. Public Domain.

Evidence found of ancient climate swings © Environmental News Network 24 April 1998.

Earth enters big thaw by Alex Kirby BBC News Online 7 March 2001 © Author and/or Publisher.

Ocean circulation shut down by melting glaciers after last ice age © NASA-GSFC 19 November 2001.

Scientists warn of abrupt climate change by Andrew Quinn from Reuters 12 December 2001 © Author and/or Publisher.

Global warming will plunge Britain into new ice age 'within decades' by Geoffrey Lean from The Independent 25 January 2004 © Author and/or Publisher.

Satellites record weakening North Atlantic current © NASA-GSFC 15 April 2004.

"It would take no more..." quote by Vittorio Canuto from a personal communication of Peter Bunyard with Vittorio Canuto, May 1996. *How global warming could cause northern Europe to freeze* by Peter Bunyard. The Ecologist Vol. 29, No. 2. March/April 1999 © Author and/or Publisher.

The Springfield Interchange in suburban Washington, D.C. Source: http://www.whitehouse.gov Public Domain.

Chicago Skyline on Lake Michigan © 2011 The Estate of G.E. Johnson.

Catastrophic climate change '90% certain' by Steven Connor from The Independent 20 July 2001 © Author and/or Publisher.

Global warming will continue for the next century by Robert Dickenson from Unisci.com 20 February 2002 © Author and/or Publisher.

Climate change 'entering the unknown' by Frank Roylance from The Baltimore Sun 3 December 2003 © Author and/or Publisher.

Global warming spirals upwards by Geoffrey Lean from The Independent © Author and/or Publisher.

West warned on climate refugees by Alex Kirby from BBC News Online 24 January 2000 © Author and/or Publisher.

More hunger and poverty may be enduring impact of climate change © World Bank 14 June 2001.

World forum scientists :'climate change will become a security issue' © Reuters 2 February 2002.

A quarter of China's population at risk as glaciers start melting © Agence Farnce-Presse 13 May 2004.

The Golden Rule by Norman Rockwell. Printed by permission of the Norman Rockwell Family Agency © 1961 the Norman Rockwell Family Entities.

PROLOGUE-INTERVIEW WITH THE AUTHOR PART 1

"It's Global Warming Dude..." quote by Pito Robles, as quoted in *A Mittenless Autumn, For Better or Worse* by Pam Belluck and Andrew Revkin from New York Times 23 December 2001 © Author and/or Publisher.

Interview Pictures of Evan Green and G.E. Johnson taken at Museumplein and Rijksmuseum in Amsterdam, Netherlands © 2011 The Estate of G.E. Johnson.

Man has been changing climate for 8,000 years by Betsy Mason from Nature News Service 10 December 2003 © Author and/or Publisher.

Carbon at 20 million year high by Alex Kirby from BBC News Online 17 August 2000 © Author and/or Publisher.

Fear growing over a sharp climate shift by Usha Lee McFarling from Los Angeles Times 13 July 2001 © Author and/or Publisher.

Scientists warn of abrupt world climate change by Andrew Quinn from Reuters12 December 2001 © Author and/or Publisher.

Climate film 'flawed but useful' by Alex Kirby from BBC News Online 12 May 2004 © Author and/or Publisher.

Carbon reaching 'danger levels' by Alex Kirby from BBC News Online 13 October 2004 © Author and/or Publisher.

1 FEEL YOUR THOUGHTS

Der Trichter der Weisheit by Georg Philipp Harsdoerfer, Nuremberg 1648-1653 © Germanisches Nationalmuseum Nürnberg. Reprinted with Permission.

Feedback could warm climate fast: Holistic model hints next century could get even hotter than we thought by Tom Clark from Nature News Service 23 May 2003 © Author and/or Publisher.

Positive Feedback Example: High Latitude Warming © 2011 The Estate of G.E. Johnson. Image segment of *Methane in the Earth's atmosphere* courtesy of NASA. Public Domain.

Non-linear responses and surprises to global change by Ian Noble-Australian National University from the proceedings of the Challenges of a Changing Earth Conference held July 2001.

Nature plants doomsday devices by Fred Pearce from The Guardian Online 25 November 1998 © Author and/or Publisher.

2 A GRAIN OF SAND IN THE BEACH OF TIME

You are here © 2011 The Estate of G.E. Johnson. *Spiral Galaxy NGC 4414* courtesy of NASA and the Hubble Heritage Team (AURA/STScI/NASA). Public Domain.

Earth's offspring? The Collision Theory by Lee Siegel from Space.com 1 September 2000 © Author and/or Publisher.

And so the moon was born...© 2011 The Estate of G.E. Johnson. Pictures used in montage clockwise from top right: *Massive Terrestrial Strike* by Don Davis courtesy of NASA. Public Domain. *Graphic of the Morphology of Complex Craters* © Lunar and Planetary Institute. Reprinted with Permission. *The Earth-Moon System* courtesy of NASA. Public Domain. *Splash on a Glass* © Curtis Hurley and North Carolina School of Science and Mathematics. Reprinted with Permission. *Object Hitting Earth* by Don Davis courtesy of NASA. Public Domain. *Coastline Remodeling* by Don Davis courtesy of NASA. Public Domain.

Geographic Distribution of Major Craters on Earth © Lunar and Planetary Institute. Reprinted with permission. Font altered by G.E. Johnson.

K-2 Impact Event © Don Davis commissioned by NASA. Courtesy of NASA. Public Domain.

Space object that killed dinosaurs broke through Earth's crust by David Perlman from San Francisco Chronicle 18 December 2000 © Author and/or Publisher.

Giant crater found undersea by Irene Brown from Discovery.com 1 August 2002 © Author and/or Publisher.

Asteroid 253 Mathilde composite image constructed from four images acquired by the NEAR spacecraft on June 27, 1997 from a distance of 2,400 km (1,500 miles). Courtesy of NASA Near Earth Object Program. Public Domain.

Hunt for potentially deadly asteroids underfunded, panel says by Jason Bates from Space.com 10 July 2002 © Author and/or Publisher.

Unseen comets may raise impact risk for Earth by Mark Peplow from Nature News Service 18 October 2004 © Author and/or Publisher.

Don't tell public of doomsday asteroid by Mark Hendersen from The Times Online 15 February 2003 © Author and/or Publisher.

Don't worry...it'll never happen © 2011 The Estate of G.E. Johnson. T-Rex photograph taken by G.E. Johnson in the Chicago Field Museum of Natural History. *Anthropocene Era* facts courtesy of Worldwatch Institute.

...thump...thump... © 2011 The Estate of G.E. Johnson. *Oil Well Pump* courtesy of U.S. Geological Survey USGS Fact Sheet FS-019-97. Public Domain. Text added by G.E. Johnson.

...thump...thump...thump... © 2011 The Estate of G.E. Johnson. *Coal Train* photograph taken by G.E. Johnson in Lincoln, Nebraska.

Chainsaws whine and forests cry © 2011 The Estate of G.E. Johnson. *Deforestation in Santa Cruz Boliva* from the *NASA Earth Observatory.* Image courtesy NASA/GSFC/MITI/ERSDAC/JAROS, and U.S./Japan ASTER Science Team: These images from 1986 and 2001 are for an area of tropical dry forest lying east of Santa Cruz de la Sierra, Bolivia. Since the mid-1980s, the resettlement of people from the Altiplano (the Andean high plains) and a large agricultural development effort (the Tierras Baja project) has lead to this area's deforestation. Soybean production began in earnest in the early 1970s following a substantial increase in the crop's world price. The pie or radial patterned fields are part of the San Javier resettlement scheme. At the center of each unit is a small community that includes a church, bar/café, school, and soccer field. The rectangular, light colored areas are fields of soybeans cultivated for export, mostly funded by foreign loans. The dark strips running through the fields are windbreaks, which are advantageous because the soils in this area are fine and prone to wind erosion. The 1986 Landsat image (left) was acquired on August 4, 1986, and the ASTER image (right) on August 11, 2001. Public Domain. Text added by G.E. Johnson.

THUMP © 2011 The Estate of G.E. Johnson *Clover Leaf Interchange* photograph taken by G.E. Johnson from the viewing platform of the Chicago Sears Tower.*Automobile* photograph by G.E. Johnson.

Spinning Earth twists space by Mark Peplow from Nature News Service 20 October 2004 © Author and/or Publisher.

Milankovitch cycle climatic forcings graphic derived from Berger and Loutre, 1991 graph courtesy of NASA. Public Domain. These graphs show calculated values for 300,000 years of orbital variation for changes in the Earth's precession, obliquity, and eccentricity. The line labeled "0" represents today, while "-200" indicates 200,000 years in the past and "100" indicates 100,000 years from now. Font altered by G.E. Johnson.

Orbital Eccentricity; Obliquity; Precession images by Robert Simmon, NASA/GSFC. Courtesy of NASA/GSFC Earth Observatory. Public Domain. Font altered by G.E. Johnson.

Hello Seven Sisters © 2011 The Estate of G.E. Johnson *The Pleiades Star Cluster* courtesy of NASA. Public Domain. *Mayan Sun Stone* from the Mexico Anthropological Museum, Mexico City, Mexico photograph by G.E. Johnson.

No cosmic ray effect by Alex Kirby from BBC News Online 23 January 2004 © Author and/or Publisher.

Ancient supernova may have triggered eco-catastrophe by Michael Purdy from Johns Hopkins University Press Release 8 January 2002 © Author and/or Publisher.

Sunspot cycle 23 image courtesy of NASA. Sunspot counts, plotted against an x-ray image of the sun based on February 2003 data. Public Domain. Font altered by G.E. Johnson.

Sun 'minor player' in climate change by Alex Kirby from BBC News Online 3 May 2000 © Author and/or Publisher.

Climate feels the sun's effects by Alex Kirby from BBC News Online 3 October 2003 © Author and/or Publisher.

Viewpoint: the Sun and climate change by Paal Brekke from BBC News Online 16 November 2000 © Author and/or Publisher.

The truth about global warming-it's the sun that's to blame by Michael Leidig and Roya Nikkah from The Telegraph. 18 July 2004 © Author and/or Publisher.

Greenhouse gases, not solar activity, cause of global warming © Environment News Service 3 August 2004.

The Earth's Magnetosphere Sun and Earth Magnetosphere image courtesy of NASA Goddard Space Flight Center Living with a Star Program. Public Domain. Font altered by G.E. Johnson.

Scientists eye whirlpool in Earth's core by O. Baker from Science News Service 13 November 1999 © Author and/or Publisher.

Antarctic craters reveal asteroid strike by Paul Brown from Guardian 19 August 2004 © Author and/or Publisher.

Past shifts of the earth's magnetic field from *Long-term asymmetric saw-toothed pattern* by the American Geophysical Union 1999 in Eos, Vol. 76, July 18, 1995 p. 289. Reprinted with permission from the AGU. Image caption: Long-term asymmetric saw-toothed pattern: (above) Variations in relative paleointensity (expressed as VADMs, virtual axial dipole moments) across the normal-to-reverse (N -> R) Upper Jaramillo and reverse-to-normal (R -> N) Matuyama-Brunhes reversals obtained from ODP Leg 138 sediments from the equatorial Pacific. Open and solid circles are U-channel and discrete sample measurements, respectively (Courtesy of J.-P. Valet and L. Meynadier; (below) Schematic diagram of the general intensity pattern suggesting that field revitalization occurs not only during successful reversals, but also during unsuccessful, or aborted reversal attempts. Font altered by G.E. Johnson

Sun's rays to roast Earth as poles flip by Robin McKie from The Observer 10 November 2002 © Author and/or Publisher.

Hi cousin © 2011 The Estate of G.E. Johnson. *Chimpanzee* drawing courtesy of the United States Fish and Wildlife Service. Public Domain. Text added by G.E. Johnson.

Evolution on fast forward: Finches adapt to climates by Bijaal P. Trivedi from National Geographic Today 10 January 2002 © Author and/or Publisher.

Biodiversity: Buffer against climate change © Environmental News Network 5 May 2001.

Climate change will unbalance ecosystems, study says © Reuters 10 April 2002.

Biodiversity crucial to Earth's ecosystems by Catherine Gianaro from University of Chicago Press Release 24 April 2004 © Author and/or Publisher.

The domino effect © 2011 The Estate of G.E. Johnson.

Higher animals suffering major declines © Environmental News Network 5 May 2001.

World's reptile populations running thin from Environmental News Network 11 August 2000 © Environmental News Network.

Orangutans edging closer to brink of extinction by Hillary Mayell from National Geographic News 24 October 2000 © Author and/or Publisher.

British butterflies 'face extinction' © BBC News Online 2 March 2001.

World plants near extinction close to 50 percent-study by Christopher Doering from Reuters 31 October 2002 © Author and/or Publisher.

Lions 'close to extinction' © BBC News Online 18 September 2003.

Amphibians face a bleak future by Emma Marris from Nature News Service 14 October 2004 © Author and/or Publisher.

Human impact triggers massive extinctions © Environment News Service 2 August 1999.

Coral reefs will be gone in 20 years, scientists say © Associated Press 23 October 2000.

Humans moving closer to extinction, study says by Robert McClure from Seattle Post-Intelligencer Online 5 January 2001 © Author and/or Publisher.

Third of primates 'risk extinction' by Alex Kirby from BBC News Online 7 October 2002 © Author and/or Publisher.

Study: only 10 percent of big ocean fish remain by Marsha Walton from CNN News Online 14 May 2003 © Author and/or Publisher.

Decline in oceans' phytoplankton alarms scientists by David Perlman from San Francisco Chronicle Online 6 October 2003 © Author and/or Publisher.

Climate warming spells species wipeout by Jeremy Lovell from Reuters 2 February 2005 © Author and/or Publisher.

Positive Feedback: Biodiversity Loss © 2011 The Estate of G.E. Johnson. Image segment of *Bald eagle chicks* by Dave Menke courtesy of U.S. Fish and Wildlife Service.

Earth faces sixth mass extinction by Anil Ananthswamy from New Scientist 18 March 2004 © Author and/or Publisher.

Butterfly Effect © 2011 The Estate of G.E. Johnson. Center Image: *Sunspot Average for 1985-1995 from 30°S-30°N. Daily Sunspot Area Averaged Over Individual Solar Rotations* courtesy of NASA. Public Domain. Left and Right Image: *Lotis Blue Butterflies* courtesy of United States Fish and Wildlife Service. Public Domain.

3 HUM OF HUMANITY

Monkey's uncle © 2011 The Estate of G.E. Johnson. Photograph of Hans Wanamaker and Kristofer Johnson by G.E. Johnson.
U.N. says great apes in danger worldwide by Angela Doland from Associated Press 28 November 2003 © Author and/or Publisher.
Humans did come out of Africa, says DNA by Jeremy Thomson from Nature News Service 7 December 2000 © Author and/or Publisher.
Scientists find fossils of man's earliest relative by Patricia Reaney from Reuters © Author and/or Publisher.
Skull may shift evolution theories by Mark Evans from Associated Press © Author and/or Publisher.
Basically, we all have the same brain frame © 2011 The Estate of G.E. Johnson. *Skull Photo* courtesy of US National Oceanic and Atmospheric Administration.
Africa from Space as rendered for the Blue Marble project by the NASA Earth Observatory. Image courtesy of NASA. Astronaut photograph AS17-148-22727 NASA Johnson Space Center. Public Domain.
Noah walked with God © 2011 The Estate of G.E. Johnson. *Story of Noah* from Genesis in The New Testament. *Storm* image courtesy of National Oceanic and Atmospheric Administration. Retrieved from http://www.photolib.noaa.gov courtesy of: NOAA Photo Library, NOAA Central Library; OAR/ERL/National Severe Storms Laboratory (NSSL). Public Domain.
Finding Noah's flood: evidence of ancient disaster is linked to biblical legend by Hannah Fairfield from Columbia University Press Release November 1999 © Author and/or Publisher.
Meteor showers blotted out first civilizations by Rajeev Syal from The Times Online 14 December 1997 © Author and/or Publisher.
Meteor clue to end of Middle East civilizations by Robert Matthews from The Telegraph 4 November 2001 © Author and/or Publisher.
Giant crater found here © 2011 The Estate of G.E. Johnson. *Isreal and Iraq from space* courtesy of NASA. Public Domain. Text added by G.E. Johnson.
Segment from the *Epic of Gilgamesh* from Sumerien history. Public Domain.
Segment from the *Admonitions of Ipuwer* from Egypt's Old Kingdom. Public Domain.
Tikal © 2011 The Estate of G.E. Johnson. *Large image of Central America* courtesy of NASA Mesoamerican Visualization and Monitoring System. Public Domain. *Small image of Tikal* courtesy of NASA/MSFC. Public Domain.
hmmm...i wonder if they'll get it © 2011 The Estate of G.E. Johnson. Image of Maoi at Rano Raraku, source unknown.

4 MODERN DARK AGES

Modern Dark Ages Boy © 2011 The Estate of G.E. Johnson. Photograph taken in Lincoln, Nebraska of Dietrich Wanamaker.
Portrait of Albert Einstein taken by Yousuf Karsh on 11 February 1948 from The Yousuf Karsh collection of the Library and Archives Canada. Public Domain.
Albert Einstein Quote source and date unknown.
Rene Descartes Quotes from *Discourse on the Method for Conducting One's Reason Well and for Seeking Truth in the Sciences.* Fourth Edition Donald A. Cress (Trans.). © 1998 Hackett Publishing Company.
The Earth moves most for humans by Philip Ball from Nature News Service 7 March 2005 © Author and/or Publisher.
Study: Earth can't meet human demand for resources by Christopher Doering from Reuters 24 June 2002 © Author and/or Publisher.
A message from our descendants © 2011 The Estate of G.E. Johnson. Photograph taken on the northern coast of Germany of G.E. Johnson by Martin Schroeder.
Six senses of sickness © 2011 The Estate of G.E. Johnson.
Hamburger Konzerne © 1989 Klaus Staeck. Edition Staeck http://www.staeck.com. Reprinted with permission. English translation by G.E. Johnson.
Earthrise from Moon. Taken on July 20, 1969 from the orbiting Command Module of the Apollo mission. NASA photo ID AS11-44-6552. Public Domain.

5 GAIA'S WARM AND PROTECTIVE EMBRACE

Greenhouse effect said to be proved © Associated Press 14 March 2001.
No doubts global warming is real, U.S. experts say © Reuters 3 December 2003.
Abrupt climate change likely by Tom Clarke from Nature News Service 13 December 2001 © Author and/or Publisher.
Global cooperative carbon dioxide network courtesy of NOAA. Public Domain. Font altered by G.E. Johnson
Mauna Loa carbon dioxide concentration by Robert Simmon and the NOAA Goddard Institute for Space Studies. Public Domain. Font altered by G.E. Johnson.
Global temperature anomaly by Robert Simmon and the NOAA Goddard Institute for Space Studies. Font altered by G.E. Johnson.
Wildfires add carbon to the atmosphere from Environment News Service 9 December 2002 © Environment News Service.
The greenhouse effect © 2011 The Estate of G.E. Johnson. *Composite image of the sun and earth relationship* courtesy of NASA Living with a Star Program. Public Domain.
Atmospheric altitudes © 2011 The Estate of G.E. Johnson. Adapted from *The Earth's Atmosphere* from NASA/JPL with image of *Africa from Space* courtesy of NASA. Public Domain.
Global warming potential/lifetime in years © 2011 The Estate of G.E. Johnson.
Variations of atmospheric concentrations of methane, carbon dioxide and temperature © 2011 The Estate of G.E. Johnson based on data from Petit, et al. in Nature 399, 429 (1999) Sowers and Bender, 1994; Blunier et al., 1998; Fischer et al., 1999.
Scientists tie ancient warming to 'belch' © Reuters 3 June 2004.
Planting new forests can't match saving old ones in cutting greenhouse gases, study finds by Andrew C. Revkin from New York Times 22 March 2000 © Author and/or Publisher.
Positive Feedback: Rainforest Reduction © G.E. Johnson. Image segment of *Deforestation in Santa Cruz, Bolivia* taken from ASTER courtesy of NASA. Public Domain.
Global warming creates grim future for forests by Kanina Holmes from Reuters 5 March 2002 © Author and/or Publisher.
Decline in oceans' phytoplankton alarms scientists by David Perlman from San Francisco Chronicle 6 October 2003 © Author and/or Publisher.
Ozone-depleting gases are not natural © Environmental News Network 2 July 1999.
Discovery of new ozone-destroying chemical © CSIRO Australia 31 August 1998.
Scientists find ozone-destroying molecule © NASA 9 February 2004.
Positive Feedback: Increased UV Exposure © G.E. Johnson. Image segment of *Diatom* courtesy of NASA/GSFC Distributed Active Archive Center. Public Domain.

Ozone layer depletion © 2011 The Estate of G.E. Johnson. Images in collage: *A Hole in the Gossamer Veil: data from the GOME Ozone Monitoring Experiment* courtesy of the Deutsche Zentrum fuer Land und Raumfahrt. Public Domain. *CFC, Chlorine and Oxygen molecules* courtesy of NASA Goddard Space Flight Center Scientific Visualization Studio. Public Domain.

Global warming may increase ozone hole © Environmental News Network 29 March 1999.

Arctic ozone loss more sensitive to climate change than thought © NASA-JPL 23 April 2004.

Ozone hole over Chile puts southern cities on sunblock alert by Jan McGirk from The Independent 11 October 2000 © Author and/or Publisher.

Positive Feedback: Ozone Layer Depletion and Greenhouse Effect © G.E. Johnson. Image segment *Projected ice coverage for 2075 in Greenland* courtesy of NASA/Goddard Space Flight Center Scientific Visualization Studio. Public Domain.

September 2004 ozone hole over Antarctica by Greg Shirah courtesy of NASA/Goddard Space Flight Center Scientific Visualization Studio. Public Domain. Font altered by G.E. Johnson.

Huge 2004 stratospheric ozone loss tied to solar storms, Arctic winds by Cara Randall from University of Colorado 1 March 2005 © Author and/or Publisher.

Earth losing air-cleansing abilities, study says by Gary Polakovic from Los Angeles Times Online 4 May 2001 © Author and/or Publisher

Methane explosion warmed prehistoric earth, possible again © NASA-GSFC News Release 10 December 2001.

Study: Earth to warm even if greenhouse gases cut © Reuters 19 September 2002.

6 EMOCEANAL BREAKDOWN

Greenhouse gases 'do warm oceans' by Paul Rincon from BBC News Online 17 February 2005 © Author and/or Publisher.

Positive Feedback: Ocean Temperature Increase © 2011 The Estate of G.E. Johnson. Image segment *Phytoplankton bloom off Newfoundland, Canada* by Jacques Descloitres, MODIS Rapid Response Team, NASA/GSFC courtesy of NASA/GSFC. Public Domain.

Positive Feedback: Ocean and Air Temperature Increase © G.E. Johnson. Image segment Global *Land-Sea Surface Temperature Index 1900-2000* courtesy National Climatic Data Center and NOAA. Public Domain.

Plummeting plankton linked to warmer oceans by Richard Stegner from CNN News Online © Author and/or Publisher.

Most important greenhouse gas goes way up in past 50 years by Vince Stricherz from UniSci.com 25 April 2001 © Author and/or Publisher.

Tropical waters in northern hemisphere heating up, data shows © CNN News Online 28 July 2000.

In polar waters, a surge in temperatures takes scientists by surprise by Robert C. Cowen from Christian Science Monitor 21 February 2002 © Author and/or Publisher.

Ocean is warming, study finds by John Roach from Environmental News Network 24 March 2000 © Author and/or Publisher.

The rising oceans by Claude Morgan from Environmental News Network 14 August 2000 © Author and/or Publisher.

Sea level rises 'underestimated' by Jonathon Amos from BBC News Online 17 February 2002 © Author and/or Publisher.

Melting glaciers in Antarctica are raising oceans, experts say by Kenneth Chang from New York Times 11 December 2001 © Author and/or Publisher.

Oceans to rise one meter by 2100-Arctic expert by Alister Doyle form Reuters 9 November 2004 © Author and/or Publisher.

Are you gettin' this? © 2011 The Estate of G.E. Johnson. Photograph of Corey Priesman by G.E. Johnson.

NASA study finds rapid changes in polar ice sheets © NASA-JPL News Release 31 August 2002.

Glacier Retreat Facts courtesy of Worldwatch Institute.

Melting glaciers in Antarctica are raising oceans, experts say by Kenneth Chang from New York Times © Author and/or Publisher.

Mountain of Melt © 2011 The Estate of G.E. Johnson. *South Cascade Glaciers* 1928 image courtesy of the U.S. Forest Service, 2000 image courtesy of the U.S. Geological Survey. Public Domain.

Global warming melts Australia's glaciers by Michael Perry from Reuters 31 May 2001 © Author and/or Publisher.

Small glaciers of the Andes may vanish in 10-15 years by Rancou Bernard from Unisci.com 17 January 2001 © Author and/or Publisher.

Scientists: Patagonian glaciers melting by Alexa Stanard from Associated Press 17 November 2003 © Author and/or Publisher.

South American glaciers' big melt © BBC News Online 17 October 2003.

Patagonia, Argentina and Southern Chile image courtesy MODIS Land Rapid Response Project at NASA/GSFC. Public Domain. Text added by G.E. Johnson.

Record retreat in Swiss glaciers in 2003 due to climate change: scientists © Agence France-Presse 13 January 2004.

Melting Swiss glaciers threatening the Alps © Reuters 16 November 2004.

Melting snows of Kilimanjaro © NASA News February 2000.

Global warming blamed for melting Everest glacier © Reuters 6 June 2002.

Mount Kilimanjaro: Tanzania, Africa. Images from 17 February 1993 and 21 February 2000 captured by the Landsat 5 and Landsat 7 satellites, courtesy of Jim Williams and the NASA-GSFC Scientific Visualization Studio and the Landsat 7 Science Team. Public Domain. Text added by G.E. Johnson.

Many Alaskan glaciers are thinning, USGS study says by Diane Noserale from UniSci Online 11 December 2001© Author and/or Publisher.

Glaciers on thin ice: expert says melting to be faster than expected by Sabin Russell from San Francisco Chronicle 17 February 2002 © Author and/or Publisher.

85% of Alaskan glaciers melting at 'incredible rate' by Tim Radford from Guardian 19 July 2002 © Author and/or Publisher.

Muir glacier 1899-2003. Images courtesy of USGS Photo Library and R.D. Karpilo, NPS. Public Domain. Text added by G.E. Johnson.

Bering glacier 1986-2002. Images courtesy from Landsat 5 and Landsat 7 satellites NASA/USGS. Public Domain.Text added by G.E. Johnson.

Carroll glacier 1906-2003. Images courtesy of USGS Photo Library and B.F. Molina, USGS. Public Domain. Text added by G.E. Johnson.

World's glaciers continue to shrink © Environmental News Network 27 May 1998.

Mountain glaciers receding, AGU told by Rory McGee from UniSci Online 31 May 2001 © Author and/or Publisher.

Glaciers melting worldwide, study finds by Robert S. Boyd from Contra Costa Times 30 September 2002 © Author and/or Publisher.

Decline of world's glaciers expected to have global impacts over this century © NASA-GSFC News 29 May 2002.

Grass flourishes in warmer Antarctic by Jonathon Leake from The Times Online 26 December 2004 © Author and/or Publisher.

Ice flow direction change in interior West Antarctica by M.J. Siegert et al from Science Magazine 24 September 2004 © Author and/or Publisher.

Antarctic ice shelf crumbling by Molly O'Meara from Environmental News Network 10 June 1998 © Author and/or Publisher.

Antarctic ice crumbling rapidly by Alex Kirby from BBC News Online 8 April 1999 © Author and/or Publisher.

Climate change: Hundreds of Antarctic glaciers in retreat, says study © Agence France-Presse 21 April 2005.

Study: global warming linked to melting ice cap © Reuters 21 February 2001.

Antarctica's ice sheet melting naturally © BBC News Online 3 January 2003.

Retreat of Antarctic ice shelves not new by Sue Leeman from Associated Press 24 February 2005 © Author and/or Publisher.

West Antarctic ice sheet not in jeopardy © Environmental News Network 1 December 1998.

Polar ice sheet shows shrinkage by Paul Racer from Associated Press 1 February 2001 © Author and/or Publisher.

Rapid Antarctic warming puzzle by Alex Kirby from BBC News Online 6 September 2001 © Author and/or Publisher.

Antarctic experts warn of global warming meltdown by Jeremy Lovell from Reuters 28 December 2001 © Author and/or Publisher.

West Antarctic ice getting thicker by Randolph Schmid from Associated Press 17 January 2002 © Author and/or Publisher.

Antarctic ice shelf collapses in larges event of last 30 years by Ted Scambos from University of Colorado 18 March 2002 © Author and/or Publisher.

Antarctic glaciers accelerating in response to 2002 ice sheet collapse by Ted Scambos from National Snow and Ice Data Center 21 September 2004 © Author and/or Publisher.

Disintegration of the Larsen b Ice Shelf in Antarctic © 2011 The Estate of G.E. Johnson. Images of the Breakup of the Larsen Ice Shelf, Antarctica courtesy Ted Scambos, National Snow and Ice Data Center, University of Colorado, boulder, based on data from MODIS. Public Domain. Image of the Antarctica courtesy of NASA/GSFC. Public Domain. Text added by G.E. Johnson.

Antarctic ice melt poses worldwide threat by Michael Byrnes from Reuters 14 May 2002 © Author and/or Publisher.

Antarctic ice shelf is melting rapidly, scientists warn by Tim Radford from Guardian 31 October 2003 © Author and/or Publisher.

Positive Feedback: Accelerated Ice Melt © 2011 The Estate of G.E. Johnson. Image segment of the Front of a Melting Glacier courtesy of National Oceanic and Atmospheric Administration; Photographer: Giuseppe Zibordi; Credit: Michael Van Woert, NOAA NESDIS, ORA; Image ID: corp2420, NOAA Corps Collection. Public Domain.

Positive Feedback: Polar Fringe Melt © 2011 The Estate of G.E. Johnson. Image segment of Larsen Ice Shelf, Antarctica courtesy of Landsat 7 Science Team and NASA/GSFC. Public Domain.

Antarctic ice fringe 'melting faster' by Alex Kirby from BBC News Online 13 June 2002 © Author and/or Publisher.

Glaciers are flowing faster by Philip Ball from Nature News Service 23 September 2004 © Author and/or Publisher.

Study shows potential for Antarctic climate change © NASA-GSFC 6 October 2004.

Fringe melt acceleration image courtesy of NASA-GSFC. Public Domain. Text added by G.E. Johnson.

Ice albedo image courtesy of NASA-Goddard Space Flight Center Conceptual Image Lab. Public Domain. Text added by G.E. Johnson.

Study blames Arctic ice decline on humans by Margot Higgins from Environmental News Network 6 December 1999 © Author and/or Publisher.

North Pole faces a major meltdown by Walter Gibb from New York Times 11 July 2000 © Author and/or Publisher.

U.N. warns global warming is melting Arctic soil © Reuters 7 February 2001.

Substantial contribution to sea-level rise during the last interglacial from Greenland ice sheet by Kurt Cuffey et al from Nature Magazine 6 April 2000 © Author and/or Publisher.

Global View of the Arctic Ocean image courtesy of NASA-JPL, University of Alaska-Fairbanks from the RADARSAT Satellite. Public Domain Text added by G.E. Johnson.

Arctic sea ice gets thinner © BBC News Online 16 November 1999.

Arctic ice 'melting from below' by Alex Kirby from BBC News Online 27 March 2002 © Author and/or Publisher.

Arctic sea ice declines again in 2004, according to CU-Boulder researchers by Mark Sevreza et al from National Snow and Ice Data Center 4 October 2004 © Author and/or Publisher.

Arctic Sea Ice Melt © 2011 The Estate of G.E. Johnson. Northern Hemisphere images courtesy of NASA and show decreasing thickness and extent of Arctic sea ice from January 1, 1990 and January 1, 1999, respectively. Public Domain. Images were created using data from the Defense Meteorological Satellite Program's Special Scanning Microwave Image. Sea Ice Area Graph created with data from the Hadley Centre for Climate Prediction and Research.

NASA scientists detect rapid thinning of Greenland's coastal ice © NASA 20 July 2000.

Scientists alarmed at increase in melt rate of ice by Hamish MacDonell from The Scotsman 4 August 2004 © Author and/or Publisher.

Greenland is a canary by Mitzi Perdue from Scripps Howard News Service 23 October 2000 © Author and/or Publisher.

Tell me little birdy… © 2011 The Estate of G.E. Johnson. Summer Sea Ice, North Pole 31 August 1999 image by James W. Williams courtesy of NASA-Goddard Space Flight Center Scientific Visualization Studio. Public Domain. Miner from 1928 courtesy of the Mine Safety and Health Administration, U.S. Department of Labor. Public Domain.

Icelandic glacier in rapid breakup © United Press International 22 October 2000.

Europe's largest glacier melting away-expert © Reuters 17 January 2002.

Arctic ice cap melting at worrying rate: NASA © Agence France-Press 24 October 2003.

Canada's shrinking ice caps by Katie Lorentz from NASA-GSFC News 4 March 2005 © Author and/or Publisher.

Record melt in Arctic and Greenland: ice sheets shrank by 1 million square kilometers this summer by Tom Clarke from Nature News Service 9 December 2002 © Author and/or Publisher.

Positive Feedback: Permafrost Thaw © 2011 The Estate of G.E. Johnson. Image segment of Methane collection in the atmosphere courtesy of NASA/Goddard Space Flight Center Scientific Visualization Studio. Public Domain.

Thawing Subarctic permafrost increases greenhouse gas emissions © American Geophysical Union-Lund University 24 February 2004.

Researchers find frozen North may accelerate climate change © NASA-GSFC 12 October 2004.

El Niño and La Niña courtesy of NASA-JPL image from TOPEX/POSEIDON November 1997 image of El Niño and February 1999 image of La Niña. Text added by G.E. Johnson.

Global warming may worsen El Niño effects © Environmental News Network 23 February 1999.

Cycle lock causes quick change by Philip Ball from Nature News Service 8 June 2001 © Author and/or Publisher.

Night temperature soar © Australian Associated Press 22 February 2005.

NAO-AO Positive and Negative Mode © 2011 The Estate of G.E. Johnson. Earth Image by Reto Stöckli, enhancements by Robert Simmon (ocean color, compositing courtesy of NASA Goddard Space Flight Center. Public Domain.

Thousands still without power in eastern Canada © Reuters 15 November 2004.

Violent storm causes chaos across Europe © Agence-France Presse 20 November 2004.

The North Atlantic Oscillation and greenhouse-gas forcing by Svetlana Kuzmina et al from Geophysical Research Letters 11 February 2005 © Author and/or Publisher.

The Arctic Oscillation: A key to this winter's cold-and a warmer planet © National Center for Atmospheric Research 25 February 2004.

Hmmm…I wonder if we'll get it © 2011 The Estate of G.E. Johnson. Photograph of Jonathon Henning taken in Heidelberg, Germany by G.E. Johnson.

The Great Ocean Conveyor Belt of Planet Earth in 2005 © 2011 The Estate of G.E. Johnson. Earth Image by Reto Stöckli, enhancements by Robert
Simmon courtesy of NASA Goddard Space Flight Center. Public Domain. Water circulation based on Rahmstorf 2002 and Broecker 1991.

Lamont's Broecker warns gases could alter climate: oceans' circulation could collapse by Laurence Lipsett from Columbia University Record 5 December
1997 © Author and/or Publisher.

Subpolar Gyre and Thermohaline Pumps © 2011 The Estate of G.E. Johnson. Arctic image courtesy of Paul Chang NOAA/NESDIS/ORA from QuikSCAT
Ice Page. Public Domain. Water movement based on graphic by Jack Cook from Woods Hole Oceanographic Institution. Location of Thermohaline
pumps after data from Broecker 1989.

Temperature over Greenland © 2011 The Estate of G.E. Johnson. Graphic based on data as calculated by Cuffey and Clow 1997 from the data of
Grootes and Stuiver 1997.

UC study sheds new light on climate change processes by Matthew Schmidt from University of California-Davis 10 March 2004 © Author and/or Publisher.

Abrupt climate change: new research supports hypothesis that ocean currents redistributed heat during rapid warming and cooling by John Toon from
Georgia Institute of Technology 24 June 2004 © Author and/or Publisher.

Abrupt climate changes at the North Pole © 2011 The Estate of G.E. Johnson. Graph based on calculations of Cuffey and Clow 1997 and Alley 2000.

Strong thc = good, weak thc = not so good © 2011 The Estate of G.E. Johnson. Photograph of Morwenna Borden by Evan Green.

Rapid changes in the mechanism of ocean convection during the last glacial period by Trond M. Dokken, et al from Nature Magazine 30 September 1999
© Author and/or Publisher.

Warming of the tropical Atlantic ocean and slowdown of the thermohaline circulation during the last deglaciation by Carsten Ruehlemann, et al from Nature
Magazine 2 December 1999 © Author and/or Publisher.

Changes in deep-water formation during the Younger Dryas event inferred form ^{10}BE and ^{14}C records by Raimund Muscheier, et al from Nature Magazine
30 November 2000 © Author and/or Publisher.

Rapid changes of glacial climate simulated in a coupled climate model by Andrey Ganopolski, et al from Nature Magazine 11 January 2001 © Author
and/or Publisher.

Rapid environmental change in southern Europe during the last glacial period by Judy R. M. Allen, et al from Nature Magazine 19 August 1999 © Author
and/or Publisher.

Weaker Gulf Stream in the Florida Straits during the last glacial maximum by Jean Lynch-Stieglitz, et al from Nature Magazine 9 December 1999 © Author
and/or Publisher.

An abrupt climate event in a coupled ocean-atmosphere simulation without external forcing by Alex Hall, et al from Nature Magazine 11 January 2001 ©
Author and/or Publisher.

Collapse and rapid resumption of Atlantic meridional circulation linked to deglacial climate changes by J.F. McManus, et al from Nature Magazine 22 April
2004 © Author and/or Publisher.

Researchers fear decline in sea ice is changing climates by Lee Bowman from Scripps Howard News Service 24 October 2003 © Author and/or Publisher.

Satellites record weakening North Atlantic current © NASA-GSFC 15 April 2004.

Arctic rivers discharge more freshwater into ocean, reflecting changes to hydrologic cycle by Harvey Leifert from American Geophysical Union 19 January
2005 © Author and/or Publisher.

Rate of ocean circulation directly linked to abrupt climate change in North Atlantic region by Shelley Dawicki from Woods Hole Oceanographic Institution
21 April 2004 © Author and/or Publisher.

Shutdown of circulation pattern could be disastrous by James E. Kloeppel from University of Illinois at Urbana Champagne 14 December 2004 © Author
and/or Publisher.

Ocean drift disruption may chill Europe © BBC News Online 25 November 1999.

Global warming could trigger ocean current collapse © Environmental News Network 9 December 1997.

Melting glaciers diminished gulf stream, cooled western Europe, during last ice age © National Science Foundation 19 November 2001.

Are you getting' this? © 2011 The Estate of G.E. Johnson. Photograph of Evan Green and Jonathon Henning by Evan Green.

Everybody ready or not? © 2011 The Estate of G.E. Johnson. Photographs of Jonathon Henning, Morwenna Borden, Evan Green, and G.E. Johnson.

Report: climate change causing jump in natural disasters © Reuters 29 September 2000.

Natural disasters at record level in 2000-insurer © Reuters 29 December 2000.

World disasters seen as global warming outcome by Robert Evans from Reuters 19 February 2001 © Author and/or Publisher.

Extreme weather events might increase © World Meteorological Organization 2 July 2003.

Global Economic Losses from Weather Related Natural Disasters 1980-2003 © G.E. Johnson. Based on data from Worldwatch Institute http://www.
worldwatch.org

Global warming may spur more wildfires by Don Thompson from Associated Press 12 November 2003 © Author and/or Publisher.

Positive Feedback: Forest Fire Frequency © 2011 The Estate of G.E. Johnson. Image segment of *Fire in the boreal forest of Canada* taken by B.J. Stocks,
Canadian Forest Service courtesy of NASA. Public Domain.

Stormy weather buffeted U.S. in 1999 by Robinson Shaw from Environmental News Network 4 January 2000 © Author and/or Publisher.

Record number of tornadoes reported in U.S. during August © NOAA 2 September 2004.

Can you feel the winds of change? © 2011 The Estate of G.E. Johnson. Image *Tornado near end of life* courtesy of NOAA Photo Library, NOAA Central
Library; OAR/ERL/National Severe Storms Laboratory. Public Domain. Text added by G.E. Johnson.

Future heat waves will be hotter, longer, more frequent © Environment News Service 13 August 2004.

Drought's growing reach: NCAR points to global warming as key factor © NCAR-UCAR 10 January 2005.

Climate studies point to more floods in this century by Hillary Magell from National Geographic News 30 January 2002 © Author and/or Publisher.

Record cold grips much of the nation in November and December: Two-month period is the coldest on record in the United States © NOAA 5 January
2001.

Hurricanes and climate change: is there a connection? by Nicole Gordon from The National Center for Atmospheric Research and the UCAR Office of
Programs Staff Notes monthly October 2004 © Author and/or Publisher.

Eye can't wait to see you again. © 2011 The Estate of G.E. Johnson. Double Image of *Hurricane Ivan* courtesy of NASA. Hurricane Ivan lay south of the
western end of Cuba in this September 12th SeaWiFS view. The Yucatan Peninsula is in the foreground and the Florida Keys are just visible at the left edge
of the image. Public Domain. Text added by G.E. Johnson.

Hydro-Quebec cuts power to 500,000 after forest fire nears transmission line © Canadian Press 5 July 2002.

California again sets power alert amid record heat © Reuters 10 July 2002.

Ice, snow leave 335,000 in South powerless for 4th day © Reuters 28 December 2001.

Warming trend will decimate arctic peoples, report warns by Stephen Leahy from InterPress Service 9 September 2004 © Author and/or Publisher.

A quarter of China's population at risk as glaciers start melting © Agence France-Presse 13 May 2004.

Millions menaced by China floods © BBC News Online 20 August 2002.

Europe's flood part of global deluge by Arie Farnam from Christian Science Monitor © Author and/or Publisher.

Floods submerge most of Bangladesh, hit millions © Sydney Morning Herald 26 July 2004.

Global warming blamed as Australia's biggest city gets water curbs © Agence France-Presse 11 September 2003.

Sea engulfing Alaskan village by David Willis from BBC News Online 30 July 2004 © Author and/or Publisher.

Global warming to affect water supply: more rainfall means smaller sierra snowpack by David Perlman from San Francisco Chronicle 15 June 2001 © Author and/or Publisher.

Global warming to cost $300 billions a year © Reuters 3 February 2001.

Killer snowstorm traps 1.35 million people © Agence France-Presse 17 January 2001.

Effects of climate warming are here and now © Scripps Howard News Service 5 May 2004.

Disasters will outstrip aid effort as world heats up by Peter Capella from Guardian 29 June 2001 © Author and/or Publisher.

Climate 'will lead to hungry century' by Alex Kirby from BBC News Online 19 February 2001 © Author and/or Publisher.

World forum scientists: 'Climate change will become a security issue' © Reuters 2 February 2002.

Global warming triggers public health warning © Environmental News Network 8 May 2001.

Europe told there is no choice but to adapt by Paul Brown from Guardian 2 November 2000 © Author and/or Publisher.

Civilisations 'destroyed by climate change' by Roger Highfield from The Telegraph 26 January 2001 © Author and/or Publisher.

Climate related perils could bankrupt insurers © Environment News Service 7 October 2002.

Refugees, disease, water and food shortages to result from global warming © Agence France-Presse 2 February 2005.

Global warming hits species all over world-study by Ed Cropley from Reuters 27 March 2002 © Author and/or Publisher.

Study warns of global warming extinctions by Rick Callahan from Associated Press 7 January 2004 © Author and/or Publisher.

Minute shift in temperature has had a major effect on Earth, studies show by Usha Lee McFarling from Los Angeles Times 2 January 2003 © Author and/or Publisher.

Hundreds of species pressured by global warming © Environment News Service 2 January 2003.

Oblivion threat to 12,000 species by Alex Kirby from BBC News Online 18 November 2003 © Author and/or Publisher.

Warming may threaten 37% of species by 2050 by Guy Gugliotta from Washington Post 8 January 2004 © Author and/or Publisher.

Antarctica becomes too hot for the penguins: decline of 'Dinner Jacket' species is a warning to the world by Geoffrey Lean from The Independent 3 February 2002 © Author and/or Publisher.

Your pollution in my blood © 2011 The Estate of G.E. Johnson. *Emperor Penguin with Nathaniel B. Palmer* taken by Michael Van Woert, NOAA/NESDIS/ ORA image courtesy of NOAA. Public Domain. Text added by G.E. Johnson.

Thanks a lot you jerky apes © 2011 The Estate of G.E. Johnson. *Polar Bears* image courtesy of U.S. Fish and Wildlife Service.

Polar bear 'extinct within 100 years' by Helen Briggs from BBC News Online 9 January 2003 © Author and/or Publisher.

INTERMISSION

I'm dreaming of being. © 2011 The Estate of G.E. Johnson. Photograph of Evan Green and Jonathon Henning in a train from Heidelberg to Strasbourg. *Eastern Hemisphere as rendered for the Blue Marble project* by the NASA Earth Observatory at www.earthobservatory.nasa.gov courtesy of NASA/GSFC. NASA Goddard Space Flight Center Image by Reto Stöckli , enhancements by Robert Simmon. Public Domain. Smiley Face and Text added by G. E. Johnson.

7 READING THE RINGS OF TIME

Triggers to climate catastrophe still poorly understood © Agence France-Presse 2 February 2005.

No doubts global warming is real, U.S. experts say © Reuters 3 December 2003.

Michael Crichton: State of Denial © 2011 The Estate of G.E. Johnson. Based on the cover for *State of Fear* by Michael Crichton © 2004 Harper Collins.

U.S. Study links human activity to global warming by Tim Dodgett from Reuters 26 August 2004 © Author and/or Publisher.

Accept it: Kyoto won't stop global warming © The Globe and Mail 9 September 2002.

Scientists claim final proof of global warming by Mark Henderson from The Times Online 6 May 2004 © Author and/or Publisher.

Climate modeling must consider all 'greenhouse' gases © TerraDaily.com 6 December 2000.

Supercruncher image of NASA New 512 courtesy of NASA. Public Domain. Text added by G.E. Johnson

NASA stakes out world leadership with new climate supercomputer © Environmental News Network 23 July 2001.

Computer models generate extreme climate events © TerraDaily.com 16 January 2001.

NCAR releases new version of premier global climate model © NCAR-UCAR 22 June 2004.

CCSM3, The Community Climate System Model, version 3 image courtesy of the National Center for Atmospheric Research. Caption: The Community Climate System Model, version 3, incorporates data about phenomena ranging from clouds to sea ice in order to simulate Earth's complex climate system. The model is so complex that it requires about three trillion computer calculations to simulate a single day of global climate, and it produces far more information about regional climate variations than the previous version. Public Domain. Text added by G.E. Johnson.

Crystal Ball Earth © 2011 The Estate of G.E. Johnson. *Earth* image courtesy of Blue Marble 2000 project Visual Analysis Lab NASA Goddard Space Flight Center. Public Domain. *Hands* image courtesy of U.S. Department of Transportation. Public Domain.

Borehole temperatures confirm global warming by John Roach from Environmental News Network 17 February 2000 © Author and/or Publisher.

Borehole temperatures confirm global warming pattern by Rob Harris and Lee Siegel from Unisci.com 27 February 2001 © Author and/or Publisher.

Ice core reveals a worrying truth about Earth's climate by Steve Connor from The Independent 10 June 2004 © Author and/or Publisher.

Hard Core Bore image courtesy of National Ice Core Laboratory USGS. Public Domain. Caption: A person collecting a core of ice for analysis of physical properties such as salinity and crystal structure that indicate the age of the ice. The coring device is turned by hand or small engine to cut downward into the ice, surrounding a core of ice that can be pulled and stored for later analysis in a small laboratory. Text added by G.E. Johnson.

Fossils provide evidence of a warm Arctic from Environmental News Network 6 January 1999 © Environmental News Network.

Scientists say ice meteors a sign of climate change by Emma Ross-Thomas from Reuters 27 September 2002 © Author and/or Publisher.

Earth's rocks show signs of global warming © CBC News Online 18 February 2002.

Coral provides clues to climate change by Robinson Shaw from Environmental News Network 6 February 2000 © Author and/or Publisher.

Antarctic mud reveals ancient evidence of global climate change by Mark Schwartz from Stanford University 14 December 2001 © Author and/or Publisher.

Major temperature rise recorded in Arctic this year: German scientists © Agence France-Presse 28 August 2004.

Earth's ancient atmosphere trapped in rocks by John Roach from Environmental News Network 13 July 2000 © Author and/or Publisher.

Clues to bronze age comet strike © BBC News Online 25 May 1998.

Research offers explanation for Earth's bulging waistline © NASA-JPL 9 December 2002.

I'm feeling waterlogged image courtesy of NASA. Public Domain. Text added by G.E. Johnson.

Scientists mine clues of early global warmings by Timothy Bralower from Environmental News Network 10 November 1997 © Author and/or Publisher.

Ice covered volcano yields clues to ancient climate by Earle Holland from Environmental News Network 25 July 1997 © Author and/or Publisher.

Oldest ice core from the tropics recovered, new ice age evidence by Earle Holland from Ohio State University 3 December 1998 © Author and/or Publisher.

Ice cores show record of climate dating back 20,000 years © Ohio State University 26 June 1995.

Researchers in Himalayas retrieve highest ice core ever drilled by Earle Holland from Ohio State University 21 November 1997 © Author and/or Publisher.

Himalayan ice revels climate warming, catastrophic drought by Earle Holland from Ohio State University 14 September 2000 © Author and/or Publisher.

Researchers date Chinese ice core to 500,000 Years by Earle Holland from Ohio State University 29 June 1997 © Author and/or Publisher.

The Himalayas as Viewed from the International Space Station image from *On top of the world: Everest and Makalu* courtesy of NASA. Astronaut photograph ISS008-E-13304 was taken from the International Space Station using a Kodak DCS760 digital camera and a 800-mm lens on January 28, 2004 from the Earth Observations Laboratory, Johnson Space Center. Public Domain. Text added by G.E. Johnson

Early 'ice-off' may be sign of global warming © Environmental News Network 22 April 1999.

Ice records hold cold truth about climate change by Brian Mattmiller from Environmental News Network 9 September 2000 © Author and/or Publisher.

Permafrost studied for global warming clues © Environmental News Network 15 December 1998.

Tracking climate change on the tundra by Marianne Kunnen-Jones from Earth Observatory Media Alert 15 December 2001 © Author and/or Publisher.

As ice thaws, Arctic peoples at loss for words by Alister Doyle from Reuters 18 November 2004 © Author and/or Publisher.

Explorer says Arctic ice thinning noticeably by David Ljunggren from Reuters 27 May 2001 © Author and/or Publisher.

Signs of thaw in a desert of snow: scientists begin to heed Inuit warnings of climate change in Arctic by Deneen L. Brown from Washington Post 28 May 2002 © Author and/or Publisher.

Warming climate disrupts Alaska natives' lives by Yareth Rosen from Reuters 16 April 2004 © Author and/or Publisher.

Greenland ice yields climate change clues © Environmental News Network 13 October 1998.

Antarctica's climate clues by Christine McGourty from BBC News Online 26 December 2001 © Author and/or Publisher.

Satellites act as thermometers in space, show Earth has a fever © NASA-GSFC News Release 21 April 200.

NASA AQUA image courtesy of NASA-GSFC AQUA Satellite Project. Text added by G.E. Johnson

NASA launches weather satellite © Associated Press 4 May 2002.

New way upcoming to test climate change predictions © UniSci.com 12 June 2002.

Meet Europe's gigantic environment satellite © European Space Agency 13 November 2001.

Russia-NASA share Earth data © Space.com 20 December 2001.

Oom Pah Oom Pah © G.E. Johnson. Image courtesy of NASA-GSFC. Public Domain. Text added by G.E. Johnson

Smell the Dryas © 2011 The Estate of G.E. Johnson. *Image of the northern hemisphere flower dryas octopetalia* courtesy of US National Park Service. Public Domain. Text added by G.E. Johnson.

8 PREDICTING APOCALYPSE NOW...OR MAYBE LATER

Go Big Red © 2011 The Estate of G.E. Johnson. *View of Jupiter* taken by Voyager 1 courtesy of NASA-JPL. Public Domain. *Eastern Hemisphere* as rendered for the Blue Marble project by the NASA Earth Observatory. Public Domain. Text added by G.E. Johnson.

Planetary attraction © 2011 The Estate of G.E. Johnson. Image of *Magnetotail* adapted from Europe's SOHO Satellite data received 1997. Images of *Venus* and *Earth* courtesy of NASA Lunar and Planetary Institute. Public Domain. Text added by G.E. Johnson.

Planet's tail of the unexpected by Jeff Hecht from New Scientist 31 May 1997 © Author and/or Publisher.

Global warmth up, ozone hole at record, Arctic ice down for 2003-UN © United Nations 17 December 2003.

It's a warmer world, but does that mean Armageddon? by Alister Doyle from Reuters 11 February 2005 © Author and/or Publisher.

Greenhouse gases building quickly © Environmental News Network 25 June 2001.

Global growth of carbon dioxide still rising by Simon Torok from CSIRO Australia 29 March 2004 © Author and/or Publisher.

Increase in greenhouse gases seen from space by Patricia Reaney from Reuters 14 March 2001 © Author and/or Publisher.

Greenhouse gas jump spurs global warming fears by Alister Doyle from Reuters 11 October 2004 © Author and/or Publisher.

Global warming gas seen increasing dramatically by Jeff Franks from Retuers 19 November 2003 © Author and/or Publisher.

Further evidence found of global warming © Environmental News Network 23 April 1998.

20th century global warming is unprecedented © Environmental News Network 22 December 1998.

Twentieth century 'warmest in 500 years' by Alex Kirby from BBC News Online 16 February 2000 © Author and/or Publisher.

Earth hits '2,000 –year warming peak' by Alex Kirby from BBC News Online 1 September 2003 © Author and/or Publisher.

Earth gets a warm feeling all over by Rob Gutro from NASA-GISS 8 February 2005 © Author and/or Publisher.

Satellite data shows 1998 warmest this century © Environmental News Network 18 January 1999.

1999 closes warmest decade and warmest century of the millennium © Environment News Service 22 December 1999.

World temperature second highest on record © Reuters 18 December 2001.

Scientists: '02 second hottest as warming speeds by Stephanie Nebehay from Reuters 17 December 2002 © Author and/or Publisher.

2004 was fourth warmest year ever recorded by Andrew C. Revkin from New York Times 10 February 2005 © Author and/or Publisher.

2005 could be warmest year recorded-NASA by Timothy Gardner from Reuters 10 February 2005 © Author and/or Publisher.

Pace of global warming 'could double' by Charles Clover from The Telegraph 23 January 2001 © Author and/or Publisher.

Scientists alarmed by continued warming trend by Usha Lee McFarly from Los Angeles Times 12 December 2002 © Author and/or Publisher.

US experts say global warming faster than thought © Agence France-Press 24 June 2004.

Studies: Global warming to worsen © CNN News Online 18 April 2002.

Humans could trigger dramatic climate change © Environmental News Network 28 May 1997.

Climate change 'inevitable': CSIRO Chief © CSIRO Australia 13 October 2000.

Global warming will persist at least a century even if emissions curbed now © NASA Earth Observatory 17 February 2002.

Climate of change is here for good by Steve Connor from The Independent 20 March 2002 © Author and/or Publisher.

Study hints at extreme climate change by Robinson Shaw from Environmental News Network 28 October 1999 © Author and/or Publisher.

Catastrophic climate change '90% certain' by Steve Connor from The Independent 20 July 2001 © Author and/or Publisher.

The past says abrupt climate change in our future by A'ndrea Elyse Messer from Pennsylvania State University © Author and/or Publisher.

Climate could come fast and furious © Environment News Service 9 December 2002.

Earth warned on 'tipping points' by Alex Kirby from BBC News Online 26 August 2004 © Author and/or Publisher.

Global warming could trigger cascade of climatic changes © Oregon State University 13 March 2003.

Study highlights global decline by Jonathon Amos from BBC News Online 30 March 2005 © Author and/or Publisher.

Global warming already disrupting climate: scientists tell conference © Agence France-Presse 1 February 2005.

Alarm at new climate warning by Richard Black from BBC News Online 26 January 2005 © Author and/or Publisher.

The Shepherd's Boy by Aesop. Story adapted by G.E. Johnson.

Climate change already affecting global environment © The Chief Engineer Online 3 March 2005.

Climate change: countdown to global catastrophe by Michael McCarthy 24 January 2005 © Author and/or Publisher.

Reports point to proof of global warming by Matt Crenson from Associated Press 13 November 2004 © Author and/or Publisher.

'Global warming real' say new studies by Clive Cooksen from The Financial Times 18 February 2005 © Author and/or Publisher.

Did you get it? © 2011 The Estate of G.E. Johnson. Photograph of Morwenna Borden, Evan Green, and Jonathon Henning in front of Rijksmuseum, Amsterdam, Netherlands.

EPILOGUE-INTERVIEW WITH THE AUTHOR PART II

Interview Pictures of Evan Green and G.E. Johnson taken in Amsterdam, Netherlands © 2011 The Estate of G.E. Johnson.

Climate change 'entering the unknown' by Frank D. Roylance from The Baltimore Sun 3 December 2003. © Author and/or Publisher.

Visit America © 2001 Klaus Staeck. Edition Staeck http://www.staeck.com. Reprinted with permission. English translation by G.E. Johnson.

NASA: Bush stifles global warming evidence by Chuck Schoffner from Associated Press 27 October 2004 © Author and/or Publisher.

Climate 'bigger threat than terror' © The Australian 27 October 2004.

'US climate policy bigger threat to world than terrorism' by Steve Connor from The Independent 9 January 2004 © Author and/or Publisher.

Scientist 'gagged' by No 10 after warning of global warming threat by Steve Connor and Andrew Grice from The Independent 8 March 2004 © Author and/or Publisher.

Bush attacks environment 'scar stories': Secret email gives advice on denying climate change by Antony Barnett from The Observer 4 April 2004 © Author and/or Publisher.

Oil firms fund climate change 'denial' by David Adam from The Guardian 27 January 2005 © Author and/or Publisher.

Forget climate change, that's the least of our worries, say Nobel winners by John Vidal from The Guardian 21 October 2004 © Author and/or Publisher.

Retiring climate change by Ana Unruh Cohen from Center for American Progress 16 February 2005 © Author and/or Publisher.

No stopping global warming, studies predict by Maggie Fox from Reuters 17 March 2005 © Author and/or Publisher.

Geroge W. Bush Jr. quote from President Participates in Conversation on Social Security Reform 11 January 2005. Public Domain.

Chuck Hagel quote from The Omaha World Herald 6 March 2005. © Author and/or Publisher.

Now the Pentagon tells Bush: Climate change will destroy us by Mark Townsend and Paul Harris from The Observer 22 February 2004 © Author and/or Publisher.

An abrupt climate change scenario and its implications for United States national security-Executive Summary by Peter Schwartz and Doug Randall October 2003 © Author and/or Publisher.

Key findings of the Pentagon Report on Abrupt Climate change excerpts taken from *An abrupt climate change scenario and its implications for United States national security-Executive Summary* by Peter Schwartz and Doug Randall October 2003 © Author and/or Publisher.

Global warming inevitable for decades to come, science conference told © Agence France-Presse 1 February 2005.

Blair warns of climate change's threat by Ed Johnson from Associated Press 14 September 2004 © Author and/or Publisher.

Global warming threat central to policy-Britain by Jeremy Lovell from Reuters 15 March 2005 © Author and/or Publisher.

Hans Blix quote from MTV News Interview 12 March 2003 © 2003 MTV News.

Support Oil Dependency © 2011 The Estate of G.E. Johnson.

The end of oil is closer than you think by John Vidal from The Guardian 21 April 2005 © Author and/or Publisher.

5% won't git ,er dun © 2011 The Estate of G.E. Johnson.

Climate talks shift focus to how to deal with changes by Andrew C. Revkin from New York Times 2 November 2002 © Author and/or Publisher.

It's much too late to sweat global warming by Mark Hertsgaard from San Francisco Chronicle 13 February 2005 © Author and/or Publisher.

Buckle up Children of the Earth © 2011 The Estate of G.E. Johnson. Photograph of Morwenna Borden, Evan Green, and Jonathon Henning in space pod.

Remember...our children will want the world back © 2011 The Estate of G.E. Johnson. Photograph of G.E. Johnson age 8 superimposed on image of Galaxy NGC 4414 taken by the Hubble Space Telescope courtesy of NASA. Public Domain.

FONTS USED

Futura AND Times New Roman
USED UNDER LISCENSE.
ALL OTHERS ARE FREEWARE FONTS.

SPACE AGE © JUSTIN CALLAGHAN
WELTRON URBAN © FONT-A-LICIOUS FONTS
Futura Lt Bt and *Futra Lt Italic* © Bitstream
Adler copyright Author Unknown
Andromeda © Randy Everett
GOOD TIMES © RAY LARABIE
Times New Roman © Monotype
SAVED BY ZERO © RAY LARABIE

REFERENCES

Adam, David (27 January 2005). Oil firms fund climate change 'denial'. Guardian [online]. Retrieved from http://www.guardian.co.uk

Agence France-Presse (26 December 2000). 2000 among France's hottest years. Retrieved from http://www.climateark.org

Agence France-Presse (1 February 2005). Global warming already disrupting climate: Scientists tell conference. Retrieved from http://news.yahoo.com

Agence France-Presse (1 February 2005). Global warming inevitable for decades to come, science conference told. Retrieved from http://news.yahoo.com

Agence France-Presse (11 September 2003). Global warming blamed as Australia's biggest city gets water curbs. Retrieved from http://news.yahoo.com

Agence France-Presse (13 January 2004). Record retreat in Swiss glaciers in 2003 due to climate change: scientists. Retrieved from http://news.yahoo.com

Agence France-Presse (13 May 2004). A quarter of China's population at risk as glaciers start melting. Retrieved from http://news.yahoo.com

Agence France-Presse (17 January 2001). Killer snowstorm traps 1.35 million people. Retrieved from http://www.hk-imail.com

Agence France-Presse (2 February 2005). Refugees, disease, water and food shortages to result from global warming. Retrieved from http://news.yahoo.com

Agence France-Presse (2 February 2005). Triggers to climate catastrophe still poorly understood. Retrieved from http://news.yahoo.com

Agence France-Presse (20 June 2001). Cooler Europe seen as global warming hits. Retrieved from http://news.yahoo.com

Agence France-Presse (20 November 2004). Violent storm causes chaos across Europe. Retrieved from http://news.yahoo.com

Agence France-Presse (21 April 2005). Climate change: Hundreds of Antarctic glaciers in retreat, says study. Retrieved from http://news.yahoo.com

Agence France-Presse (24 June 2004). US experts say global warming faster than thought. Retrieved from http://news.yahoo.com

Agence France-Presse (24 October 2003). Arctic ice cap melting at worrying rate: NASA. Retrieved from http://news.yahoo.com

Agence France-Presse (28 August 2004). Major temperature rise recorded in Arctic this year: German scientists. Retrieved from http://www.news.yahoo.com

Agence France-Presse (29 March 2004). Australians measure alarming rise in global greenhouse gases. Retrieved from http://news.yahoo.com

Aguilera, Mario (11 June 2001). Coral records Pacific climate changes over 100 years. UniSci Online. Retrieved from http://unisci.com

Ahlstrom, Dick (2 November 2001). Warm October may signify global changes. Irish Times [online]. Retrieved from http://www.ireland.com

All About 2012 (no date). Retrieved from http://www.greatdreams.com

Allen, Jim (25 September 1999). Plato's Atlantis and the Rectangular Plain near Lake Poopo, Bolivia. Retrieved from http://www.net-zone.co.uk

Allen, Judy R. et al (1999). Rapid environmental changes in southern Europe during the last glacial period. Nature: Vol. 400; 740-743.

Alley, Richard B. et al (1993). Abrupt increase in Greenland snow accumulation at the end of the Younger Dryas event. Nature: Vol. 362; 527-529.

Alley, Richard B. (14 December 2001). Sharper warming on chances of abrupt climate change. UniSci Online. Retrieved from http://unisci.com

Alley, Richard B. (1998). Paleoclimatology: Icing the North Atlantic. Nature: Vol. 392; 335-337.

Alley, Richard B. (2000). The Two-Mile Time Machine: Ice Cores, Abrupt Climate Change, and Our Future. Princeton University Press. Princeton.

American Geophysical Union (31 July 2002). Cosmic rays said linked to global warming. Retrieved from http://www.spaceflightnow.com

American Geophysical Union–Lund University (24 February 2004). Thawing subarctic permafrost increases greenhouse gas emissions. Retrieved from http://www.agu.org

Amos, Jonathan (17 February 2002). Sea level rises 'underestimated'. BBC News Online. Retrieved from http://news.bbc.co.uk

Amos, Jonathan (19 January 2001). Seawater 'overstates' climate warming. BBC News Online. Retrieved from http://news.bbc.co.uk

Amos, Jonathan (4 May 2001). Arctic's big melt challenged. BBC News Online. Retrieved from http://news.bbc.co.uk

Amos, Jonathan (20 March 2005). Study highlights global decline. BBC News Online. Retrieved from http://news.bbc.co.uk

Amos, Jonathan (29 June 2001). Stalagmite has climate warming. BBC News Online. Retrieved from http://news.bbc.co.uk

Ananthaswamy, Anil (18 March 2004). Earth faces sixth mass extinction. New Scientist [online]. Retrieved from http://www.newscientist.com

Anisimov, Oleg (12 July 2001). The Changing Cryosphere: Impacts of Global Warming in the High Latitudes. Challenges of a Changing Earth Conference. Retrieved from http://sciconftalks.igbp.sr.unh.edu

Arctic Climate Impact Assessment Secretariat (2004). Impacts of a Warming Arctic. Cambridge University Press. Cambridge, UK.

Arctic Monitoring and Assessment Programme (10 August 2001). A State of the Arctic Environment Report. Retrieved from http://www.amap.no

Arctic Monitoring and Assessment Programme (no date). Climate change, ozone depletion and ultraviolet radiation. Retrieved from http://www.amap.no

Arguelles, Jose (no date). Mayan Timing Universal Laws of Nature. Retrieved from http://www.2012.com

Arnold, David (1 October 2002). Could global warming produce a big chill? Boston Globe [online]. Retrieved from http://www.boston.com

Asher, David (1998). Why the Earth impactor influx has varied over human history. Retrieved from http://www.spaceguard.or.jp

Asimov, Isaac and Frederik Pohl (1991). Our Angry Earth. Tor Books. New York.

Associated Press (11 December 2001). Scientists warn of climate change. Retrieved from http://news.yahoo.com

Associated Press (14 May 2001). Ocean cycle changes U.S. rainfall. Retrieved from http://www.enn.com

Associated Press (15 July 2004). Monsoon rains bring more destruction to South Asia, 326 dead. Retrieved from http://www.heatisonline.com

Associated Press (16 November 2001). Solar cycle drives ocean temperatures, study says. Retrieved from http://www.usatoday.com

Associated Press (19 October 1999). Warming trends. Retrieved from http://abcnews.go.com

Associated Press (2 November 2000). Scientists find seafloor storms. Retrieved from http://www.enn.com

Associated Press (22 March 2004). Carbon dioxide levels sky high. Retrieved from http://www.cbsnews.com

Associated Press (22 January 2001). Report warns of disaster from global warming. New York Times [online]. Retrieved from http://www.nytimes.com

Associated Press (22 September 2002). Panel says health of seas in peril. Retrieved from http://news.yahoo.com

Associated Press (23 August 2000). Coral reefs will be gone in 20 years, scientists say. Associated Press. Retrieved from http://news.yahoo.com

Associated Press (29 January 2005). Glaciers melting in warming world, scientists say. Retrieved from http://www.ctv.ca

Associated Press (30 July 2002). Experts warn of disasters from climate changes. Retrieved from http://news.yahoo.com

Associated Press (4 May 2002). NASA launches weather satellite. Retrieved from http://news.yahoo.com

Atkinson, T.C. et al (1987). *Seasonal temperatures in Britain during the past 22,000 years, reconstructed using beetle remains.* Nature: 325, 587-592.

Austin, Duncan et al (October 1998). *Contributions to climate change: Are conventional metrics misleading the debate?* World Resources Institute: Climate Notes.

Australian Associated Press (22 February 2005). *Night temperatures soar.* Retrieved from http://news.ninemsn.com.au

Bacon, Sheldon (1998). *Decadal variability in the outflow from the Nordic seas to the deep Atlantic Ocean.* Nature: Vol. 394; 871-874.

Baker, O. (13 November 1999). *Scientists eye whirlpool in Earth's core.* Science News Online. Retrieved from http://www.sciencenews.org

Ball, Philip (23 September 2004). *Glaciers are flowing faster.* Nature News Service [online]. Retrieved from http://www.nature.com

Ball, Philip (31 March 2005). *Black holes 'do not exist'.* Nature News Service [online]. Retrieved from http://www.nature.com

Ball, Philip (7 March 2005). *The Earth moves most for humans.* Nature News Service [online]. Retrieved from http://www.nature.com

Ball, Philip (8 June 2001). *Cycle lock causes quick change.* Nature News Service [online]. Retrieved from http://www.nature.com

Ballingrud, David (8 July 2002). *On alert for asteroids.* St. Petersburg Times [online]. Retrieved from http://www.sptimes.com

Banke, Jim (4 May 2002). *NASA's Aqua research spacecraft reaches polar orbit.* Cape Canaveral Bureau. Retrieved from http://www.space.com

Barker, Terry (December 2001). *Representing the Integrated Assessment of Climate Change, Adaptation, and Mitigation.* Tyndall Centre Working Paper No. 11. University of Cambridge.

Barlow, Jim (21 March 2002). *Sunlight, PCB exposure enhance skin cancer chances.* University of Illinois at Urbana-Champaign. Retrieved from http://www.eurekalert.org

Barnett, Antony (4 April 2004). *Bush attacks environment 'scare stories': secret email give advice on denying climate change.* Observer [online]. Retrieved from http://www.guardian.co.uk

Barnola, J. M., D. Raynaud, Y. S. Korotkevich and C. Lorius (1987). *Vostok ice core provides 160,000-year record of atmospheric CO2.* Nature: Vol. 329; 408-414.

Bates, Jason (10 July 2002). *Hunt for potentially deadly asteroids underfunded, panel says.* Space News Online. Retrieved from http://www.space.com

BBC News Online (11 December 2001). *Earth's equatorial 'obesity': The first image from the Grace spacecraft.* Retrieved from http://news.bbc.co.uk

BBC News Online (13 July 2004). *Flood havoc mounts in South Asia.* Retrieved from http://news.bbc.co.uk

BBC News Online (14 September 2000). *Himalayan ice tells warming story.* Retrieved from http://news.bbc.co.uk

BBC News Online (15 July 2004). *Ocean CO2 may 'harm marine life'.* Retrieved from http://news.bbc.co.uk

BBC News Online (16 November 1999). *Arctic sea ice gets thinner.* Retrieved from http://news.bbc.co.uk

BBC News Online (17 June 2004). *Shell boss 'fears for the planet'.* Retrieved from http://news.bbc.co.uk

BBC News Online (17 October 2003). *South American glaciers' big melt.* Retrieved from http://news.bbc.co.uk

BBC News Online (18 December 2000). *Researchers to drill into dinosaur crater.* Retrieved from http://news.bbc.co.uk

BBC News Online (18 September 2003). *Lions 'close to extinction'.* Retrieved from http://news.bbc.co.uk

BBC News Online (19 March 1999). *Greenland ice warning.* Retrieved from http://news.bbc.co.uk

BBC News Online (2 March 2001). *British butterflies 'face extinction'.* Retrieved from http://news.bbc.co.uk

BBC News Online (20 August 2002). *Millions menaced by China floods.* Retrieved from http://news.bbc.co.uk

BBC News Online (25 July 2002). *Asteroids 'could trigger nuclear war'.* Retrieved from http://news.bbc.co.uk

BBC News Online (25 May 1998). *Clues to Bronze Age comet strike.* Retrieved from http://news.bbc.co.uk

BBC News Online (25 November 1999). *Ocean drift disruption may chill Europe.* Retrieved from http://news.bbc.co.uk

BBC News Online (28 October 2003). *Grim picture painted for 2020.* Retrieved from http://news.bbc.co.uk

BBC News Online (3 January 2003). *Antarctica's ice sheet melting naturally.* Retrieved from http://news.bbc.co.uk

BBC News Online (3 October 2001). *Dinosaurs felt the heat.* Retrieved from http://news.bbc.co.uk

BBC News Online (4 September 2002). *Seasons 'becoming muddled'.* Retrieved from http://news.bbc.co.uk

BBC News Online (6 December 1999). *Humanity blamed for ice loss.* Retrieved from http://news.bbc.co.uk

BBC News Online (8 August 2002). *Photos show glacier's decline.* Retrieved from http://news.bbc.co.uk

Beers, Stephen (14 December 1998). *Blame it on El Niño: Are we ignoring the early warning signs of global warming?* Los Angeles Times [online]. Retrieved from http://www.lats.com

Bell, Art and Strieber, Whitley (2001). *The Coming Global Superstorm.* Pocket Star Books: Simon and Schuster. New York.

Bellingka, Pier (no date). *Flip-flop to catastrophe.* Retrieved from http://www.ourplanet.com

Belluck, Pam and Andrew Revkin (23 December 2001). *A Mitheless Autumn, for Better and Worse.* New York Times[online]. Retrieved from http://www.nytimes.com

Bender, M., T. Sowers, M.L Dickson, J. Orchado, P. Grootes, PA. Mayewski and D.A. Meese (1994). *Climate connection between Greenland and Antarctica during the last 100,000 years.* Nature: 372, 663-666.

Bentley, Molly (9 December 2002). *Record ice loss in Arctic.* BBC News Online. Retrieved from http://news.bbc.co.uk

Berger, A. and Loutre, M.F. (1991). *Insolation values for the climate of the last 10 million years.* Quaternary Science Reviews, 10: 297-317.

Berger, John J. (2000). *Beating the Heat: Why and How We Must Combat Global Warming.* Berkeley Hills Books. Albany, CA.

Berman, Morris (1984). *The Reenchantment of the World.* 2nd Edition. Bantam. New York.

Bianchi, Giancarlo G. and I. Nicholas McCave (1999). *Holocene periodicity in North Atlantic climate and deep-ocean flow south of Iceland.* Nature: Vol. 397, 515-517.

Black, Richard (11 January 2002). *El Niño 'Could return in months'.* BBC News Online. Retrieved from http://news.bbc.co.uk

Black, Richard (26 January 2005). *Alarm at new climate warning.* BBC News Online. Retrieved from http://news.bbc.co.uk

Blunier T., J. Chappellaz, J. Schwander, A. Dallenbach, B. Stauffer, T. Stocker, D. Raynaud, J. Jouzel, H.B. Clausen, C.U. Hammer and S.J. Johnsen (1998). *Asynchrony of Antarctic and Greenland climate change during the last glacial period.* Nature: Vol. 394; 739-743.

Borenstein, Seth (18 January 2003). *2002 the second-hottest year.* Free Press Washington. Retrieved from http://www.freep.com

van Boven, Theo (1982). *People Matter: View on International Human Rights Policy.* Muehlenhoff. Amsterdam.

Bowen, Jerry (29 August 2002). *Clues in the Arctic.* CBS News [online]. Retrieved from http://www.cbsnews.com

Bowman, Lee (24 October 2003). *Researchers fear decline in sea ice is changing climates.* Scripps Howard News Service. Retrieved from http://seattlepi.nwsource.com

Boyd, Robert S. (1 March 2002). Oceans help unlock puzzles of climate. Free Press Washington [online]. Retrieved from http://www.freep.com

Boyd, Robert S. (21 August 2002). Glaciers melting worldwide, study finds. Contra Costa Times [online]. Retrieved from http://news.nationalgeographic.com

Boyd, Robert S. (30 September 2002). Glaciers slip sliding away. The Philadelphia Inquirer [online]. Retrieved from http://www.philly.com

Bralower, Timothy (10 November 1997). Scientists mine clues of early global warmings. Environmental News Network. Retrieved from http://www.enn.com

Branigan, Tania (30 October 2001). October warmest on record. Guardian [online]. Retrieved from http://www.guardian.co.uk

Brekke, Paal (16 November 2000). Viewpoint: The Sun and climate change. BBC News Online. Retrieved from http://www.news.bbc.co.uk

Bridges, Andrew (10 December 2001). Three Antarctican glaciers shrinking. The Associated Press. Retrieved from http://news.yahoo.com

Bridges, Andrew (24 July 2002). Astronomers: Asteroid hit unlikely. The Associated Press. Retrieved from http://news.yahoo.com

Bridges, Andrew (28 December 2001). Enormous icebergs imperil penguins. The Associated Press. Retrieved from http://news.yahoo.com

Briggs, Helen (25 September 2001). How reptiles survived the big one. BBC News Online. Retrieved from http://news.bbc.co.uk

Briggs, Helen (9 January 2003). Polar bears 'extinct' within 100 years. BBC News Online. Retrieved from http://news.bbc.co.uk

Britt, Robert Roy (07 February 2002). Mini ozone holes over Europe pose growing health risk. Space.com. Retrieved from http://www.space.com

Britt, Robert Roy (13 November 2001). Comets, meteors and myth: New evidence for toppled civilisations and biblical tales. Space.com. Retrieved from http://www.space.com

Britt, Robert Roy (17 January 2001). Meltdown: Satellites show accelerated polar ice threat. Space.com. Retrieved from http://www.space.com

Britt, Robert Roy (24 April 2001). Reinventing Darwin again: How asteroids impacted human evolution. Space.com. Retrieved from http://www.space.com

Britt, Robert Roy (5 December 2002). Earth gets fatter thanks to faster glacial melting. Space.com. Retrieved from http://www.space.com

Broecker, W.S. et al (1989). Routing of meltwater from the Laurentide Ice Sheet during the Younger Dryas cold episode. Nature: 341, 318-321.

Broecker, W.S. (16 November 1998). What if the conveyor were to shut down? University of Washington. Retrieved from http://faculty.washington.edu

Broecker, W.S. (1995). The Glacial World According to Wally. 2nd edition. Eldigio Press. Lamont-Doherty Earth Observatory of Columbia University. New York.

Broecker, W.S. (May 1997). Will our ride into the greenhouse future be a smooth one? GSA Today: 7(5): 1-7.

Brown, DeNeen L. (28 May 2002). Signs of thaw in a desert of snow. Washington Post [online]. Retrieved from http://www.washingtonpost.com

Brown, Paul (19 August 2004). Antarctic craters reveal asteroid strike. Guardian [online]. Retrieved from http://www.guardian.co.uk

Brown, Paul (2 November 2000). Europe told there is no choice but to adapt. Guardian [online]. Retrieved from http://www.guardian.co.uk

Brown, Paul (26 April 2002). World's weather hotter than ever. Guardian [online]. Retrieved from http://www.guardian.co.uk

Brown, Paul (7 August 2003). World to warm by 8C, says thinktank. Guardian [online]. Retrieved from http://www.guardian.co.uk

Brown, Ronald (1 May 1998). Antarctic ice reveals past, may foretell future. Environmental News Network. Retrieved from http://www.enn.com

Browne, Anthony (20 August 2000). First ice-free North Pole in 50m years. Observer [online]. Retrieved from http://www.observer.co.uk

Browne, Irene (1 August 2002). Giant crater found undersea. Discovery News Online. Retrieved from http://dsc.discovery.com

Byrnes, Michael (14 May 2002). Antarctic ice melt poses worldwide threat. Retrieved from http://news.yahoo.com

Bryant, Greg (September 1999). The Dark Ages: were they darker than we imagined? Universe Magazine. Retrieved from http://gchbryant.tripod.com

Bunyard, Peter (March/April 1999). How climate change could spiral out of control. The Ecologist: Vol. 29; No. 2; 68-74.

Bunyard, Peter (March/April 1999). How global warming could cause northern Europe to freeze. The Ecologist: Vol. 29; No. 2; 79-80.

Bunyard, Peter (March/April 1999). How ozone-depletion increases global warming. The Ecologist: Vol. 29; No. 2; 85.

Burroughs, William James (2001). Climate Change: A Multidisciplinary Approach. Cambridge University Press. New York.

Callahan, Rick (7 January 2004). Study warns of global warming extinctions. Associated Press. Retrieved from http://news.yahoo.com

Calvin, William H. (2002). A Brain for All Seasons: Human Evolution and Abrupt Climate Change. The University of Chicago Press. Chicago.

Canadian Press (5 July 2002). Hydro-Quebec cuts power to 500,000 after forest fire nears transmission line. Retrieved from http://ca.news.yahoo.com

Capella, Peter (29 June 2001). Disasters will outstrip aid effort as world heats up. Guardian [online]. Retrieved from http://www.guardian.co.uk

Carrington, Damian (7 October 1999). Global warming can make sea level plunge. BBC News Online. Retrieved from http://news.bbc.co.uk

CBC News Online (12 April 2002). Earth's rocks show signs of global warming. Retrieved from http://www.cbc.ca

CBC News Online (18 February 2001). Scientists spell out risks from global warming. Retrieved from http://www.cbc.ca

CBS News Online (25 November 2002). Arctic meltdown. Retrieved from http://www.cbsnews.com

Center for International Climate and Environmental Research-Oslo (no date). Climate Change: What's at stake? Retrieved from http://www.cicero.uio.no

Chang, Kenneth (11 December 2001). Melting glaciers in Antarctica are raising oceans, experts say. New York Times [online]. Retrieved from http://www.nytimes.com

Chang, Kenneth (12 December 2001). Drastic shifts in climate are likely, experts warn. New York Times [online]. Retrieved from http://www.nytimes.com.

Chang, Kenneth (2 April 2002). The melting (freezing) of Antarctica; deciphering contradictory climate patterns is largely a matter of ice. New York Times [online]. Retrieved from http://www.nytimes.com

Chang, Kenneth (3 May 2002). Ozone hole is now seen as a cause for Antarctic cooling. New York Times [online]. Retrieved from http://www.nytimes.com

Chappellaz, J., J.-M. Barnola, D. Raynaud, Y. S. Korotkevich and C. Lorius (1990). Atmospheric CH4 record over the last climatic cycle revealed by the Vostok ice core. Nature: Vol. 345; 127-131.

Chatterjee, Neil (8 February 2002). Tropical air thins European ozone layer by 30%. Reuters. Retrieved from http://news.yahoo.com

Chrichton, Michael (2004). Sate of Fear. Harper Collins. New York.

Christensen, T. R., T. Johansson, H. J. Åkerman, M. Mastepanov, N. Malmer, T. Friborg, P. Crill, and B. H. Svensson (2004). Thawing sub-arctic permafrost: Effects on vegetation and methane emissions. Geophysical Research Letters: 31, L04501, doi:10.1029/2003GL018680.

Christianson, Gale E. (2000). Greenhouse: The 200-Year Story of Global Warming. Penguin Books USA. New York.

Citizens for a Sound Economy Foundation (2001). Climate of Hype. CSE Foundation. Washington, DC.

Clark, P. et al (2001). Freshwater forcing of abrupt climate change during the last glaciation. Science: Vol. 293; 283-287.

Clark, Maureen (25 October 2001). *Alaska ice contest helps scientists.* Associated Press. Retrieved from http://news.yahoo.com

Clarke, Tom (13 December 2001). *Abrupt climate change likely.* Nature News Service [online]. Retrieved from http://www.nature.com

Clarke, Tom (23 May 2003), *Feedback could warm climate fast.* Nature News Service [online]. Retrieved from http://www.nature.com

Clarke, Tom (9 December 2002). *Record melt in Arctic and Greenland.* Nature News Service [online]. Retrieved from http://www.nature.com

Climate Change Science Program and the Subcommittee on Global Change Research (2004). *Our Changing Planet: The U.S. Climate Change Science Program for Fiscal Years 2004 and 2005.* Retrieved from http://www.usgcrp.gov

Clover, Charles (23 January 2001). *Pace of global warming 'could double.'* The Telegraph [online]. Retrieved from http://www.telegraph.co.uk

Clube, Victor (1990). *Meteor stream key to earth's climate.* Retrieved from http://www.globalideasbank.org

CNN News Online (12 July 2001). *Climate change speeds up, says UN.* Retrieved from http://www.cnn.com

CNN News Online (15 May 2002). *New Antarctic iceberg bigger than Delaware.* Retrieved from http://www.cnn.com

CNN News Online (25 January 2002). *Scientific winds blow hot and cold in Antarctica.* Retrieved from http://www.cnn.com

CNN News Online (27 November 1997). *Global warming: no day at the beach.* Retrieved from http://www.cnn.com

CNN News Online (28 July 2000). *Tropical waters in Northern Hemisphere heating up, data shows.* Retrieved from http://www.cnn.com

CNN News Online (4 October 2000). *NASA finds largest-ever ozone hole.* Retrieved from http://www.cnn.com

CNN News Online (18 April 2002). *Studies: Global warming to worsen.* Retrieved from http://www.cnn.com

Cohen, Ana Unruh (16 February 2005). *Retiring climate change.* Center for American Progress. Retrieved from http://www.americanprogress.org

Collett-White, Mike (20 March 2002). *Ice shelf collapse reignites global warming fears.* Retrieved from http://news.yahoo.com

Colman S. (2002). *Paleoclimate: A fresh look at glacial floods.* Science: Vol. 296; 1251-1253.

Columbia University (24 May 1996). *Lamont scientist cites water vapor in climate shifts.* Columbia University Record: Vol. 21; No. 28.

Commoner, Barry (1967). *Science and Survival.* Viking Press. New York.

Connor, Steve (10 June 2004). *Ice core reveals a worrying truth about Earth's climate.* The Independent [online]. Retrieved from http://news.independent.co.uk

Connor, Steve (20 July 2001). *Catastrophic climate change '90% certain'* The Independent [online]. Retrieved from http://news.independent.co.uk

Connor, Steve (20 March 2002). *Climate of change is here for good.* The Independent [online]. Retrieved from http://news.independent.co.uk

Connor, Steve (21 November 2003). *Mother of all extinctions 'took place before the dinosaurs died.'* The Independent [online]. Retrieved from http://news.independent.co.uk

Connor, Steve (9 January 2004). *'US climate policy bigger threat to world than terrorism'.* The Independent [online]. Retrieved from http://news.independent.co.uk

Connor, Steve (19 February 2005). *The final proof: Global warming a man-made disaster.* The Independent [online]. Retrieved from http://news.independent.co.uk

Connor, Steve, and Andrew Grice (8 March 2004). *Scientist 'gagged' by No 10 after warming of global warming threat.* The Independent [online]. Retrieved from http://news.independent.co.uk

Cooksen, Clive (18 February 2005). *'Global warming real' say news studies.* The Financial Times [online]. Retrieved from http://news.ft.com

Coupland, Douglas (1991). *Generation X: Tales for an Accelerated Culture.* St. Martin's Press. New York.

Cowen, Robert C. (14 December 2001). *Be prepared: sudden shifts in climate coming.* Christian Science Monitor [online]. Retrieved from http://www.csmonitor.com

Cowen, Robert C. (21 February 2002). *In polar waters, a surge in temperatures takes scientists by surprise.* Christian Science Monitor [online]. Retrieved from http://www.csmonitor.com

Cowen, Robert C. (26 September 2002). *Into the cold?* Christian Science Monitor [online]. Retrieved from http://www.csmonitor.com

Cowen, Robert C. (4 October 2001). *A new understanding of ocean-heat transfer.* Christian Science Monitor [online]. Retrieved from http://www.csmonitor.com

Cowen, Robert C. (8 August 2002). *Wind flows fear at ozone layer.* Christian Science Monitor [online]. Retrieved from http://www.csmonitor.com

Craig, Andrew (30 May 2001). *Arctic 'getting greener'.* BBC News Online. Retrieved from http://news.bbc.co.uk

Crenson, Matt (13 November 2004). *Reports point to proof of global warming.* Associated Press. Retrieved from http://news.yahoo.com

Crenson, Matt (16 February 2002). *El Niño pattern may be forming.* Associated Press. Retrieved from http://news.yahoo.com

Cropley, Ed (27 March 2002). *Global warming hits species all over world-study.* Reuters. Retrieved from http://news.yahoo.com

Crowley, Thomas J. (14 July 2000). *Causes of climate change over the past 1000 years.* Science: Vol. 289; 270-277.

CSIRO Australia (13 October 2000,). *Climate change 'inevitable': CSIRO chief.* Retrieved from http://www.csiro.au

CSIRO Australia (31 August 1998) *Discovery of new ozone-destroying chemical.* Retrieved from http://www.csiro.au

Cuffey Kurt M., Marshall Shawn J. (6 April 2000) *Substantial contribution to sea-level rise during the last interglacial from Greenland ice sheet.* Nature: Vol. 404; Number 6778 591-594.

Cuffey, K.M.., and G.D. Clow (1997). *Temperature, accumulation, and ice sheet elevation in central Greenland through the last deglacial transition.* Journal of Geophysical Research 102 (26):383-396.

Daily, Matt (13 July 2001). *Scientists say future climate change could be sudden.* Reuters. Retrieved from http://www.enn.com

Darwin, Charles (1982 edition). *Origin of the Species.* Penguin Books. New York.

David, Leonard (20 December 2001). *Russia - NASA Share Earth Data.* Space.com. Retrieved from http://www.space.com

Davidson, Keay (4 February 2002). *Media goofed on Antarctic data: Global warming interpretation irks scientists.* The Chronicle. Retrieved from http://www.sfgate.com

Davidson, Keay (7 January 2003). *The sky is slowly rising, scientists say: Upward movement of atmospheric layer points to global warming.* San Francisco Chronicle [online]. Retrieved from http://sfgate.com

Davis, Robert E (September 2000). *Ocean depth and salinity can counter warming.* Environment and Climate News. Retrieved from http://www.heartland.org

Dawicki, Shelley (21 April 2004). *Rate of ocean circulation directly linked to abrupt climate change in North Atlantic region.* Woods Hole Oceanographic Institution. Retrieved from http://www.whoi.edu

De Siqueira, Marinez et.al (2004). *Extinction risk from climate change.* Nature: Vol. 427, pp 145-148.

Derbyshire, David (22 February 2002). *The cold snap the civilised the world.* The Telegraph [online]. Retrieved from http://www.telegraph.co.uk

Descartes, René (1637). *Discourse on the Method for Rightly Conducting One's Reason and for Seeking Truth in the Sciences.* Donald A. Cress-Translator. Fourth Edition 1998. Hackett Publishing Company. Indianapolis.

Deutsche Presse (14 July 2001). *UN-Wissenschaftsbericht sagt dramatische Folgen der Klimaerwaermung voraus.* Retrieved from http://www.habito.de

Deutsche Presse Agentur (12 January 2001). *Erstmals Erklaerung fuer rapide Klimawechsel*. Hannoversche Allgemeine Zeitung. Pg. 6.

Deutsche Presse Dienst (1 November 2000). *UN-Klimabericht sieht Erhoehung der Temperaturen*. Retrieved from http://de.news.yahoo.com

Deutsche Presse Dienst (13 August 2001). *Klimaexperten: CO2-Ausstoss auf Null zurueckfuehren*. Retrieved from http://de.news.yahoo.com

Dickinson, Robert (20 February 2002). *Global warming will continue for the next century*. UniSci Online. Retrieved from http://unisci.com

Discovery Channel Online (13 September 2002). *Ice reveals ancient climate change*. Retrieved from http://dsc.discovery.com

Doering, Christopher (24 June 2002). *Study: Earth can't meet human demand for resources*. Reuters. Retrieved from http://news.yahoo.com

Doering, Christopher (31 October 2002). *World plants near extinction close to 50 percent-study*. Reuters. Retrieved from http://news.yahoo.com

Doggett, Tom (26 August 2004). *U.S. study links human activity to global warming*. Reuters. Retrieved from http://news.yahoo.com

Dokken, Trond M. and Eystein Jansen (1999). *Rapid changes in the mechanism of ocean convection during the last glacial period*. Nature: Vol. 401; 456-461.

Doland, Angela (28 November 2003). *U.N. says great apes in danger worldwide*. Associated Press. Retrieved from http://news.yahoo.com

Dominguez, Alex (14 March 2000). *Greenhouse effect said to be proved*. Associated Press. Retrieved from http://news.yahoo.com

Doyle, Alister (11 February 2005). *It's a warmer world, but does that mean Armageddon?* Reuters. Retrieved from http://news.yahoo.com

Doyle, Alister (11 October 2004). *Greenhouse gas jump spurs global warming fears*. Reuters. Retrieved from http://news.yahoo.com

Doyle, Alister (18 November 2004). *As ice thaws, Arctic peoples at loss for words*. Reuters. Retrieved from http://news.yahoo.com

Doyle, Alister (28 January 2003). *Shrinking ice open new trade route*. Reuters. Retrieved from http://news.yahoo.com

Doyle, Alister (24 January 2005). *Panel urges U.S. to do mor on climate 'Time Bomb'*. Reuters. Retrieved from http://news.yahoo.com

Doyle, Alister (29 January 2005). *'Dangerous' global warming possible by 2026-WWF*. Reuters. Retrieved from http://news.yahoo.com

Doyle, Alister (9 November 2004). *Oceans to rise one meter by 2100-Arctic expert*. Reuters. Retrieved from http://news.yahoo.com

Dunham, William (25 January 2002). *Antarctic island called a unique climate-change lab*. Reuters. Retrieved from http://www.enn.com

Dunham, William (28 November 2001). *Scientists find hot activity under Arctic ice cap*. Reuters. Retrieved from http://news.yahoo.com

Earth Observatory News (February 2000). *Melting snows of Kilimanjaro*. NASA Earth Observatory. Retrieved from http://earthobservatory.nasa.gov

Easterbrook, Gregg (1995). *A Moment on the Earth*. Penguin Books-USA. New York.

Easterling, David (18 July 1997). *Clouds may be a culprit in global warming*. Environmental News Network. Retrieved from http://www.enn.com

ECES (5 March 2001). *Researchers find worrisome drop in global levels of atmospheric hydroxl compound*. Retrieved from http://www.eces.org

Egan, Timothy (13 June 2002). *Alaska, no longer so frigid, starts to crack, burn and sag*. New York Times [online]. Retrieved from http://www.nytimes.com

Ehrlich, Paul R. (1968). *The Population Bomb*. Ballantine Books. New York.

Eicki, Olaf (05 April 2000). *Aktische Oekosysteme durch Klimaveraenderung bedroht*. Retrieved from http://www.wcmc.org

Eliott, Susannah (19 February 2001). *Andean ice cap, Kilimanjaro ice fields disappearing*. Unisci Online. Retrieved from http://unisci.com

Emmerich, Roland (2004). *The Day After Tomorrow (movie)*. 20th Century Fox.

Enever, Andrew (10 December 2002). *Bolivian glaciers shrinking fast*. BBC News Online. Retrieved from http://news.bbc.co.uk

Environmental News Network (1 December 1998). *West Antarctic ice sheet not in jeopardy*. Retrieved from http://www.enn.com

Environmental News Network (10 April 1998). *Ozone hole may worsen through 2020*. Retrieved from http://www.enn.com

Environmental News Network (10 July 2001). *Arctic Oscillation causes climate change*. Retrieved from http://www.enn.com

Environmental News Network (11 August 2000). *World's reptile population running thin*. Retrieved from http://www.enn.com

Environmental News Network (11 May 1998). *Antarctic ice reveals past, may foretell future*. Retrieved from http://www.enn.com

Environmental News Network (12 December 2001). *Earth's cold regions five evidence of global warming*. Retrieved from http://www.enn.com

Environmental News Network (12 July 1999). *Global warming is for real, NASA says*. Retrieved from http://www.enn.com

Environmental News Network (13 June 2001). *Mountain glaciers shrinking worldwide*. Retrieved from http://enn.com

Environmental News Network (13 October 1998). *Greenland ice yields climate change clues*. Retrieved from http://www.enn.com

Environmental News Network (14 November 1997). *Arctic warming continues, study shows*. Retrieved from http://www.enn.com

Environmental News Network (14 November 2001). *Satellite data confirms warming of Earth's climate*. Retrieved from http://www.enn.com

Environmental News Network (15 December 1998). *Permafrost studied for global warming clues*. Retrieved from http://www.enn.com

Environmental News Network (15 June 1998). *Guatemala may repeat Mayan history*. Retrieved from http://www.enn.com

Environmental News Network (16 December 1997). *Satellite images show changes in Greenland's snow cover*. Retrieved from http://www.enn.com

Environmental News Network (18 February 1999). *Climate change linked to cultural evolution*. Retrieved from http://www.enn.com

Environmental News Network (18 January 1999). *Satellite data shows 1998 warmest this century*. Retrieved from http://www.enn.com

Environmental News Network (18 September 1996). *Researchers find warming waters melting glaciers*. Retrieved from http://www.enn.com

Environmental News Network (19 October 1998). *El Nino offers peak at global warming*. Retrieved from http://www.enn.com

Environmental News Network (2 July 1999). *Ozone-depleting gases are not natural*. Retrieved from http://www.enn.com

Environmental News Network (2 November 1998). *It's been a roller coaster of global warming*. Retrieved from http://www.enn.com

Environmental News Network (20 June 1997). *Ice core provides ancient climate record*. Retrieved from http://www.enn.com

Environmental News Network (20 November 2001). *Global warming occurred millions of years ago*. Retrieved from http://www.enn.com

Environmental News Network (21 January 1998). *December coldest ever in stratosphere*. Retrieved from http://www.enn.com

Environmental News Network (21 November 2001). *Geologists use lichens to track climate changes*. Retrieved from http://www.enn.com

Environmental News Network (22 April 1999). *Early 'ice-off' may be sign of global warming*. Retrieved from http://www.enn.com

Environmental News Network (22 December 1998). *20th Century global warming is unprecedented*. Retrieved from http://www.enn.com

Environmental News Network (23 April 1998). *Further evidence found of global warming*. Retrieved from http://www.enn.com

Environmental News Network (17 April 1998). *Large Antarctic ice shelf disintegrating*. Retrieved from http://www.enn.com

Environmental News Network (23 February 1999). *Global warming may worsen El Niño effects*. Retrieved from http://www.enn.com

Environmental News Network (23 July 2001). *NASA stakes out world leadership with new climate supercomputer*. Retrieved from http://www.enn.com

Environmental News Network (24 January 1998). *Last ice age likely extended around the globe*. Retrieved from http://www.enn.com

Environmental News Network (24 July 1998). *Evidence found of ancient climate swings*. Retrieved from http://www.enn.com

Environmental News Network (24 July 2001). *Researchers forecast rapid, irreversible climate warming*. Retrieved from http://www.enn.com

Environmental News Network (25 July 1997). *Ice covered volcano yields clues to ancient climate*. Retrieved from http://www.enn.com

Environmental News Network (25 June 2001). *Greenhouse gases building quickly*. Retrieved from http://www.enn.com

Environmental News Network (26 March 1999). *Warm Arctic may enhance global warming*. Retrieved from http://www.enn.com

Environmental News Network (27 May 1998). *Higher animals suffering major declines*. Retrieved from http://www.enn.com

Environmental News Network (27 May 1998). *World's glaciers continue to shrink*. Retrieved from http://www.enn.com

Environmental News Network (28 May 1997). *Humans could trigger dramatic climate change*. Retrieved from http://www.enn.com

Environmental News Network (28 October 1998). *El Niño influenced melting glaciers*. Retrieved from http://www.enn.com

Environmental News Network (29 March 1999). *Global warming may increase ozone hole*. Retrieved from http://www.enn.com

Environmental News Network (30 March 1998). *Current El Niño called a strange one*. Retrieved from http://www.enn.com

Environmental News Network (5 June 1998). *El Niño's warm waters linked to trade winds*. Retrieved from http://www.enn.com

Environmental News Network (5 March 1999). *Greenland's glaciers thinning, NASA finds*. Retrieved from http://www.enn.com

Environmental News Network (5 May 2001). *Biodiversity: Buffer against climate change*. Retrieved from http://www.enn.com

Environmental News Network (6 April 1998). *Global warming could flood New York City*. Retrieved from http://www.enn.com

Environmental News Network (6 January 1999). *Fossils provide evidence of a warm Arctic*. Retrieved from http://www.enn.com

Environmental News Network (7 May 2001). *Dust particles have global impact*. Retrieved from http://www.enn.com

Environmental News Network (7 October 1998). *Antarctic ozone hole sets record*. Retrieved from http://www.enn.com

Environmental News Network (8 May 2001). *Global warming triggers public health warning*. Retrieved from http://www.enn.com

Environmental News Network (9 December 1997). *Global warming could trigger ocean current collapse*. Retrieved from http://www.enn.com

Environmental News Network (9 December 1998). *El Niño spurs rise in global average sea level*. Retrieved from http://www.enn.com

Environment News Service (13 August 2004). *Future heat waves will be hotter, longer, more frequent*. Retrieved from http://ens-newswire.com

Environment News Service (2 August 1999). *Human impact triggers massive extinctions*. Retrieved from http://ens-newswire.com

Environment News Service (2 January 2003). *Hundreds of species pressured by global warming*. Retrieved from http://ens-newswire.com

Environment News Service (7 October 2002). *Climate related perils could bankrupt insurers*. Retrieved from http://ens-newswire.com

Environment News Service (9 December 2002). *Climate change could come fast and furious*. Retrieved from http://ens-newswire.com

Environment News Service (9 December 2002). *Wildfires add carbon to the atmosphere*. Retrieved from http://ens-newswire.com

Environment News Service (4 October 2000). *Another large iceberg splits off Antarctic ice shelf*. Retrieved from http://ens-newswire.com

Environment News Service (1 December 1998). *1998 record year for weather related disasters*. Retrieved from http://ens-newswire.com

Environment News Service (1 May 2003). *Fresh analysis of satellite data reveals global warming*. Retrieved from http://ens-newswire.com

Environment News Service (21 May 2004). *Scientists, religious leaders urge Congress to take action on global warming*. Retrieved from http://ens-newswire.com

Environment News Service (22 December 1999). *1999 closes warmest decade and warmest century of the millennium*. Retrieved from http://ens-newswire.com

Environment News Service (26 February 2003). *Methane eruptions could fuel global warming*. Retrieved from http://ens-newswire.com

Environment News Service (28 September 2000). *Red List of Threatened Species reveals global extinction crisis*. Retrieved from http://ens-newswire.com

Environment News Service (3 August 2004). *Greenhouse gases, not solar activity, cause of global warming*. Retrieved from http://ens-newswire.com

Erickson, Jim (5 March 2000). *Yucatan crater holds clues to complex role of impacts*. The Arizona Daily Star [online]. Retrieved from http://www.azstarnet.com

Erickson, Jim (7 December 2002). *Ice makes case for global warming*. Rocky Mountain News [online]. Retrieved from http://www.rockymountainnews.com

European Space Agency (13 November 2001). *Meet Europe's gigantic environment satellite*. Retrieved from http://spaceflightnow.com

European Space Agency (27 February 2002). *Sea level anomaly in the tropical Pacific Ocean*. Retrieved from http://www.esa.int

Evans, Mark (10 July 2002). *Skull may shift evolution theories*. Associated Press. Retrieved from http://news.yahoo.com

Evans, Robert (19 February 2001). *World disasters seen as global warming outcome*. Reuters. Retrieved from http://news.yahoo.com

Fabi, Randy (07 March 2002). *NOAA sees El Niño trend through early 2003*. Reuters. Retrieved from http://news.yahoo.com

Fahey, David W. (2002). *Twenty Questions About the Ozone Layer*. WMO-NOAA-UC-UNEP-NASA. http://www.wmo.ch/web/arep/ozone.html

Fairfield, Hannah (November 1999). *Finding Noah's flood: Evidence of ancient disaster is linked to Biblical legend*. Columbia University. Retrieved from http://www.columbia.edu

Farnam, Arie (15 August 2002). *Europe's flood part of global deluge*. Christian Science Monitor [online]. Retrieved from http://www.csmonitor.com

Fischer, Daniel (1999). *Showdown in the Desert, The Leonid Storm of 1999: How theory met reality at the Al Azraq meteor camp*. Retrieved from http://www.astro.uni-bonn.de

Fischer, H., M. Wahlen, J. Smith, D. Mastoiani and B. Deck (1999). *Ice core records of atmospheric CO$_2$ around the last three glacial terminations*. Science: Vol. 283; 1712-1714.

Fitton, Laura (1998). *Cassandra's Prophesies for Troy*. Paper submitted to Images of Women in the Ancient World: Issues of Interpretation and Identity. Retrieved from http://www.arthistory.sbc.edu

Foley, Grover (March/April 1999). The Threat of Rising Seas. The Ecologist: Vol. 29; No. 2; 76-78.

Fordahl, Matthew (16 December 2000). Ozone layer over Arctic will recover slower than expected, scientists say. Associated Press. Retrieved from http://www.climateark.org

Fowler, Jonathan (19 December 2001). Earth's 2001 temperatures to be high. Associated Press. Retrieved from http://news.yahoo.com

Fox, Maggie (17 March 2005). No stopping global warming, studies predict. Reuters. Retrieved from http://news.yahoo.com

Fox, Maggie (19 July 2002). Melting Alaskan glaciers raise sea level. Reuters. Retrieved from http://news.yahoo.com

Fox, Maggie (2 February 2001). Study: Melting Antarctic glacier raising sea level. Reuters. Retrieved from http://news.yahoo.com

Fox, Maggie (26 October 2000). Draft report shows world getting even warmer. Reuters. Retrieved from http://news.yahoo.com

Francou, Bernard (17 January 2001). Small glaciers of the Andes may vanish in 10-15 years. UniSci Online. Retrieved from http://unisci.com

Franke, Klaus (2000). Ueberleben im Treibhaus. Der Spiegel 28/2000.

Franks, Jeff (19 November 2003). Global warming gas seen increasing dramatically. Reuters. Retrieved from http://news.yahoo.com

Fry, Tom (28 August 2001). Ozone depletion increases skin cancer risk. Environmental News Network. Retrieved from http://www.enn.com

Fryer, Joanne (17 August 2000). Carbon dioxide level higher than for millions of years. University of Bristol. Retrieved from http://www.bris.ac.uk

Ganopolski, Andrew (1998). Simulation of modern and glacial climates with a coupled global model of intermediate complexity. Nature: Vol. 391; 351-356.

Ganopolski, Andrey and Stefan Rahmstorf (11 January 2001). Rapid changes of glacial climate simulated in coupled climate model. Nature: Vol. 409; 153-158.

Gardner, Timothy (10 February 2005). 2005 could be warmest year recorded–NASA. Reuters. Retrieved from http://news.yahoo.com

Gelbspan, Ross (1997). The Heat is On. Addison Wesley. New York.

Gelbspan, Ross (September, October 2000). Hot and bothered: The global warming debate is over. It's real, inexorable, and headed our way. EMagazine [online] Retrieved from http://www.emagazine.com

Gentry, Phillip (16 January 2002). Raw empirical data not everything in climate study. Unisci Online. Retrieved from http://unisci.com

Geological Society of America (28 June 2001). Wheels within wheels: Orbital anomaly may have caused global cooling 23 million years ago. Retrieved from http://www.geosociety.org

German Aerospace Center (2000). An Antarctic ice shelf under stress. Retrieved from http://www.dfd.dlr.de

Gianaro, Catherine (24 April 2004). Biodiversity crucial to Earth's ecosystems. University of Chicago. Retrieved from http://www.eurekalert.org

Gibb, Walter (11 July 2000). North Pole faces a major meltdown. New York Times [online]. Retrieved from http://www.nytimes.com

Gilber, Adrian and Maurice Cotterell (1995). The Mayan Prophecies. Retrieved from http://knowledge.co.uk

Glanz, James (11 January 2001). Droughts might speed climate changes. New York Times [online]. Retrieved from http://www.nytimes.com

Global Change Online (1999). Slowdown ahead for Atlantic conveyor belt? Pacific Institute for Studies in Development, Environment, and Security. Retrieved from http://www.globalchange.org

Globe and Mail [online] (4 August 2003). Arctic lakes deemed warmest in history. Retrieved from http://www.globeandmail.com

Goldman, Jane (10 January 2001). NOAA Scientist and colleague find extreme climate event...without even trying. NOAA. Retrieved from http://www.publicaffairs.noaa.gov

Goldsmith, Edward, Robert Allen, Michael Allaby, John Davoll, Sam Lawrence (1972). Blueprint for Survival. Signet Books. New York.

Gonzalez, Julieta (4 March 2002). Asteroids rained down during final formation of Earth. Retrieved from http://www.spacedaily.com

Gordon, Nicole (October 2004). Hurricanes and climate change: Is there a connection? NCAR-UCAR Office of Programs Staff Notes. Retrieved from http://clinton2.nara.gov

Gore, Al (no date). The dangers of carbon pollution on the environment. Retrieved from http://www.ucar.edu

Gore, Al (1993). Earth in the Balance: Ecology and the Human Spirit. Plume Books. New York.

Gorman, Jessica (June 2000). How dry we were. Discover: Vol. 21; No. 6.

Gould, Stephen Jay (1977). Velikovsky in collision. Retrieved from http://www.freethought-web.org

Greenpeace Germany (2000). Dangerous Interference with the Climate System: Implications of the IPCC Third Assessment Report for Article 2 of the Climate Convention. Bonn.

Gribbin, John (1990). Hothouse Earth. Grove Weidenfeld. New York.

Griggs, Kim (16 October 2001). Antarctic cores reveal ice history. BBC News Online. Retrieved from http://news.bbc.co.uk

Groots, P.M., and M. Stuiver (1997). Oxygen 18/16 variability in Greenland snow and ice with 10^{-3} to 10^{5}-year time resolution. Journal of Geophysical Research: 102:26, 455-26,470.

Grub, Michael et al (November 2000). Keeping Kyoto: A study of approaches to maintaining the Kyoto Protocol on Climate Change. Climate Strategies: International Network for Climate Policy Analysis.

Guardian [online] (5 January 2002). Ice field loss puts alpine rivers at risk. Retrieved from http://www.guardian.co.uk

Guardian [online] (29 November 1998). Abrupt climate change. Retrieved from http://www.guardian.co.uk

Guardian [online] (14 July 2004). Melting ice: the threat to London's future. Retrieved from http://www.guardian.co.uk

Guardian [online] (6 May 2004). Scientists claim new evidence of warming. Retrieved from http://www.guardian.co.uk

Gugliotta, Guy (14 January 2002). In Antarctic, no warming trend. Washington Post [online]. Retrieved from http://www.washingtonpost.com

Gugliotta, Guy (18 November 1999). For Noah's flood, a new wave of evidence. Washington Post [online]. Retrieved from http://www.washingtonpost.com

Gugliotta, Guy (8 January 2004). Warming may threaten 37% of species by 2050. Washington Post [online]. Retrieved from http://www.washingtonpost.com

Gutro, Rob (8 February 2005). Earth gets a warm feeling all over. NASA-GISS. Retrieved from http://www.giss.nasa.gov

Hall, Alex and Ronald J. Stouffer (11 January 2001). An abrupt climate event in a coupled ocean-atmosphere simulation without external forcing. Nature: Vol. 409; 171-174.

Hall, Carl T. (25 June 2002). Humanity is taking more than Earth can give. San Francisco Chronicle [online]. Retrieved from http://www.sfgate.com

Hannover Algemeine Zeitung (22 August 2000). Schmelze am Nordpol alarmiert Berlin. Retrieved from http://www.haz.de

Hansen, J. et al (2005). Earth's energy imbalance: Confirmation and implications. Science: doi:10.1126/science.1110252.

Hansen, J., M. Sato, J. Glascoe, and R. Ruedy (1998). A common-sense climate index: Is climate changing noticeably? Proc. Natl. Acad. Sci. 95, 4113-4120.

Hansen, J., R. Ruedy, J. Glascoe, and M. Sato (1999). GISS analysis of surface temperature change. Journal of Geophysical Research 104, 30997-31022.

Hansen, J., R. Ruedy, M. Sato, and R. Reynolds (1996). Global surface air temperature in 1995: Return to pre-Pinatubo level. Geophysical Research Letters 23, 1665-1668.

Hansen, J. et al (2001). A closer look at United States and global surface temperature change. Journal of Geophysical Research: 106, 23947-23963.

Harris, Rob and Lee Siegel (27 February 2001). Borehole temperatures confirm global warming pattern. UniSci Online. Retrieved from http://unisci.com

Harvey, Brian and John D. Hallett (1977). Environment and Society: An Introductory Analysis. The MIT Press. Cambridge, MA.

Hauser, Rachel (no date). Polar Paradox. Retrieved from http://earthobservatory.nasa.gov

Hawkes, Nigel (8 March 1997). Bronze Age cities may have been destroyed by comet. The Times Online. Retrieved from http://www.timeonline.co.uk

Heap, Tom (2 December 2000). Caves reveal clues to UK weather. BBC News Online. Retrieved from http://news.bbc.co.uk

Hebert, H. Josef (24 March 2000). Oceans getting warmer. Associated Press. Retrieved from http://abcnews.go.com

Hebert, H. Josef (25 October 2000). Pollution adds to global warming. Associated Press. Retrieved from http://www.ncar.ucar.edu

Hecht, Jeff (31 May 1997). Planet's tail of the unexpected. New Scientist [online]. Retrieved from http://www.newscientist.com

Heilprin, John (22 September 2002). Panel Says Health of Seas in Peril. Associated Press. Retrieved from http://news.yahoo.com

Heilprin, John (7 June 2001). Scientists see global warming rise. Associated Press. Retrieved from http://news.yahoo.com

Heilprin, John (9 November 2001). CO2 Emissions Climb in 2000. Associated Press. Retrieved from http://news.yahoo.com

Henderson, Mark (15 February 2003). 'Don't tell public of doomsday asteroid'. The Times Online. Retrieved from http://www.timesonline.co.uk

Henderson, Mark (2 February 2001). Antarctic glacier is losing 4 ft of ice each year. The Times Online. Retrieved from http://www.timesonline.co.uk

Henderson, Mark (4 December 2000). Ozone hole will heal, say scientists. The Times Online. Retrieved from http://www.timesonline.co.uk

Henderson, Mark (6 May 2004). Scientists claim final proof of global warming. The Times Online. Retrieved from http://www.timesonline.co.uk

Herbert, H. Josef (26 December 2000). Global warming theory affirmed. Associated Press. Retrieved from http://www.washingtonpost.com

Hertsgaard, mark (13 February 2005). It's much too late to sweat global warming. San Francisco Chronicle [online]. Retrieved from http://www.sfgate.com

Hickman, Hayes (20 January 2003). Arctic ice shelf shrinking. Knoxville News-Sentinel [online]. Retrieved from http://www.knoxnews.com

Higgins, Margot (2 November 2000). Vicious cycle: Global warming feeds fire potential. Environmental News Network. Retrieved from http://www.enn.com

Higgins, Margot (6 December 1999). Study blames Arctic ice decline on humans. Environmental News Network. Retrieved from http://www.enn.com

Higgins, Margot (13 September 2000). Monkey's extinction may be a sign. Environmental News Network. Retrieved from http://www.enn.com

Highfield, Roger (26 January 2001). Civilisations 'destroyed by climate change'. The Telegraph [online]. Retrieved from http://www.telegraph.co.uk

Hoegh-Guldberg, O. et al (October 2000). Pacific in Peril: Biological, economical, and social impacts of climate change on Pacific coral reefs. Greenpeace. Retrieved from http://www.greenpeace.org

Holland, Earle (14 September 2000). Himalayan ice reveals climate warming, catastrophic drought. Ohio State University. Retrieved from http://researchnews.osu.edu

Holland, Earle (17 October 2001). Antarctic seafloor core suggests Earth's orbital oscillations may be the key to what controlled ice ages. Ohio State University. Retrieved from http://researchnews.osu.edu

Holland, Earle (18 February 2001). Ice caps in Africa, tropical South America likely to disappear within 15 years. Ohio State University. Retrieved from http://researchnews.osu.edu

Holland, Earle (21 November 1997). Researchers in Himalayas retrieved highest ice core ever drilled. Ohio State University. Retrieved from http://researchnews.osu.edu

Holland, Earle (29 June 1997). Researchers date Chinese ice core to 500,000 years. Ohio State University. Retrieved from http://researchnews.osu.edu

Holland, Earle (3 December 1998). Oldest ice core from the tropics recovered, new ice age evidence. Ohio State University. Retrieved from http://researchnews.osu.edu

Holland, Earle (30 November 1992). Chinese ice cores provide climate records of four ice ages. Ohio State University. Retrieved from http://researchnews.osu.edu

Holmes, Kanina (5 March 2002). Global warming creates grim future for forests. Reuters. Retrieved from http://news.yahoo.com

Hotz, Robert Lee (26 October 2000). Scientists increase estimate of global warming's severity. Los Angeles Times [online]. Retrieved from http://www.climateark.org

Hughes, Katherine (22 September 1997). What is El Niño? Environmental News Network. Retrieved from http://www.enn.com

Hunt, Katherine (14 June 2001). Polluted clouds may give patchy cooling to world. Reuters. Retrieved from http://www.dailynews.yahoo.com

Imperial College (20 November 2000). New Insight offered into formation of Yucatan crater. Retrieved from http://spaceflightnow.com

Imperial College (no date). Hydrocode modeling of crater collapse. Retrieved from http://www.ese.ic.ac.uk

Intergovernmental Panel on Climate Change (J.T. Houghton et al, Eds). (2001). Climate Change 2001: The Scientific Basis. Cambridge University Press. New York.

Intergovernmental Panel on Climate Change (James J. McCarthy et al, Eds) (2001). Climate Change 2001: Impacts, Adaptations, and Vulnerability. Cambridge University Press. New York.

Intergovernmental Panel on Climate Change (Bert Metz et al, Eds.) (2001). Climate Change 2001: Mitigation. Cambridge University Press. New York.

International Climate Change Taskforce (2005). Meeting the Climate Challenge: Recommendations of the International Climate Change Taskforce. Retrieved from http://www.igbp.kva.se

International Geosphere-Biosphere Programme (not date). How have past climate changes affected human societies? Retrieved from http://www.igbp.kva.se

International Geosphere-Biosphere Programme (2004). Global Change and the Earth System: A planet under pressure-Executive Summary. Retrieved from http://www.igbp.kva.se

Japanese Space Exploration Agency (2003). Traces of meteor collision with Earth. Retrieved from http://spaceinfo.jaxa.jp

Jaura, Ramesh (11 February 2005). Climate Change: 'A period of uncertainty has closed'. Inter Press Service. Retrieved from http://www.ipsnews.net

Jenkins, John Major (23 May 1994). The how and why of the Mayan end date in 2012 AD. Mountain Astrologer [online]. Retrieved from http://www.alignment2012.com

Johansen, Dag Ove (21 December 2001). Death of a jaguar. Retrieved from http://www.geocities.com

Johnson, Ed (14 September 2004). Blair warns of climate change's threat. Associated Press. Retrieved from http://news.yahoo.com

NCAR-AMS (25 September 2001). Atmosphere, not oceans, carries most heat to the poles from the equator. Retrieved from http://www.ucar.edu

Jouzel J. et al (2001). A new 27 ky high resolution East Antarctic climate record. Geophysical Research Letters: Vol. 28; No. 16; 3199-3202.

Joyce, Terrence (18 April 2002). Policy matters: The heat before the cold. New York Times [online]. Retrieved from http://www.aei.brookings.org

Kaplan, George (no date). The seasons and the Earth's orbit - Milankovitch Cycles. Retrieved from http://aa.usno.navy.mil

Kennedy, Barbara (10 August 2001). Plausible biological cause for major climate events. UniSci Online. Retrieved from http://unisci.com

Kerr, Richard A (16 January 1998). Sea floor dust shows drought felled Akkadian Empire. Science: Vol. 279; No. 5349; 325-326.

Kettle, Martin (11 September 2000). Ice retreats to open North-West passage. Guardian [online].Retrieved from http://www.guardian.co.uk

Kettlewell, Julianna (12 July 2004). *Climate warning from the deep.* BBC News Online. Retrieved from http://news.bbc.co.uk

Kettlewell, Julianna (9 June 2004). *Ice cores unlock climate secrets.* BBC News Online. Retrieved from http://news.bbc.co.uk

Kightlinger, Diane (3 December 2000). *Equatorial waters hold undercurrent to global warming.* Environmental News Network. Retrieved from http://www.enn.com

Kilian, Michael (6 April 2000). *Europeans possibly first Americans.* Chicago Tribune [online]. Retrieved from http://www.chicagotribune.com

King James Version of the Bible. *Book of Genesis, Chapter 1.* Retrieved from http://sun.science.wayne.edu

Kirby, Alex (1 February 2001). *Antarctic ice sheet shrinks.* BBC News Online. Retrieved from http://news.bbc.co.uk

Kirby, Alex (1 September 2003). *Earth hits '2,000-year warming peak.* BBC News Online. Retrieved from http://www.bbc.co.uk.

Kirby, Alex (10 September 1999). *Climate disaster possible by 2100.* BBC News Online. Retrieved from http://news.bbc.co.uk.

Kirby, Alex (11 February 2003). *Greenhouse gases 'at record levels'.* BBC News Online. Retrieved from http://news.bbc.co.uk

Kirby, Alex (12 May 2004). *Climate film "flawed but useful".* BBC News Online. Retrieved from http://news.bbc.co.uk

Kirby, Alex (13 June 2002). *Antarctic ice fringe 'melting faster'.* BBC News Online. Retrieved from http://news.bbc.co.uk

Kirby, Alex (13 October 2004). *Carbon 'reaching danger levels'.* BBC News Online. Retrieved from http://news.bbc.co.uk

Kirby, Alex (14 December 2000). *Climate model shows dual cause.* BBC News Online. Retrieved from http://news.bbc.co.uk

Kirby, Alex (16 April 2002). *Himalayan warming 'may trigger floods'.* BBC News Online. Retrieved from http://news.bbc.co.uk

Kirby, Alex (16 February 2000). *Twentieth century 'warmest in 500 years'.* BBC News Online. Retrieved from http://news.bbc.co.uk

Kirby, Alex (17 August 2000). *Carbon at 20 million year high.* BBC News Online. Retrieved from http://news.bbc.co.uk

Kirby, Alex (18 December 2001). *2001 'warm, but no record'.* BBC News Online. Retrieved from http://news.bbc.co.uk

Kirby, Alex (18 November 2003). *Oblivion threat to 12,000 species.* BBC News Online. Retrieved from http://news.bbc.co.uk

Kirby, Alex (19 February 2001). *Climate 'will lead to hungry century'.* BBC News Online. Retrieved from http://news.bbc.co.uk

Kirby, Alex (21 July 2000). *Greenland's coastal ice thins fast.* BBC News Online. Retrieved from http://news.bbc.co.uk

Kirby, Alex (22 January 2001). *Climate change outstrips forecasts.* BBC News Online. Retrieved from http://news.bbc.co.uk

Kirby, Alex (23 December 2003). *Soot 'makes global warming worse'.* BBC News Online. Retrieved from http://news.bbc.co.uk

Kirby, Alex (23 February 2004). *Arctic ice 'melting from below'.* BBC News Online. Retrieved from http://news.bbc.co.uk

Kirby, Alex (23 January 2004). *'No cosmic ray climate effects.'* BBC News Online. Retrieved from http://news.bbc.co.uk

Kirby, Alex (24 January 2000). *West warned on climate refugees.* BBC News Online. Retrieved from http://news.bbc.co.uk

Kirby, Alex (26 April 2001). *Records 'show strong recent warming'.* BBC News Online. Retrieved from http://news.bbc.co.uk

Kirby, Alex (26 April 2002). *Record warm start to 2002.* BBC News Online. Retrieved from http://news.bbc.co.uk

Kirby, Alex (26 August 2004). *Earth warned on 'tipping points'.* BBC News Online. Retrieved from http://news.bbc.co.uk

Kirby, Alex (27 March 2002). *Arctic ice 'thins by almost half'.* BBC News Online. Retrieved from http://news.bbc.co.uk

Kirby, Alex (3 May 2000). *Sun 'minor player' in climate change.* BBC News Online. Retrieved from http://news.bbc.co.uk

Kirby, Alex (3 November 2004). *Climate gas cuts 'are affordable'.* BBC News Online. Retrieved from http://news.bbc.co.uk

Kirby, Alex (3 October 2000). *Fossils nag at carbon's climate role.* BBC News Online. Retrieved from http://news.bbc.co.uk

Kirby, Alex (7 December 2001). *Forests 'only temporary carbon absorbers'.* BBC News Online. Retrieved from http://news.bbc.co.uk

Kirby, Alex (7 February 2001). *Arctic 'now adding to global warming'.* BBC News Online. Retrieved from http://news.bbc.co.uk

Kirby, Alex (7 June 2002). *Greenland's warming ice flows faster.* BBC News Online. Retrieved from http://news.bbc.co.uk

Kirby, Alex (6 August 2002). *Cosmic rays 'explain climate conundrum'.* BBC News Online. Retrieved from http://news.bbc.co.uk

Kirby, Alex (6 September 2000). *Global warming 'a reality'.* BBC News Online. Retrieved from http://news.bbc.co.uk

Kirby, Alex (6 September 2001). *Rapid Antarctic warming puzzle.* BBC News Online. Retrieved from http://news.bbc.co.uk

Kirby, Alex (7 March 2000). *Earth enters big thaw.* BBC News Online. Retrieved from http://news.bbc.co.uk

Kirby, Alex (7 March 2002). *Alaska's oil 'melts its ice'.* BBC News Online. Retrieved from http://news.bbc.co.uk

Kirby, Alex (7 October 1999). *Nature blamed for melting ice.* BBC News Online. Retrieved from http://news.bbc.co.uk

Kirby, Alex (7 October 2002). *Third of primates 'risk extinction'.* BBC News Online. Retrieved from http://news.bbc.co.uk

Kirby, Alex (8 April 1999). *Antarctic ice crumbling rapidly.* BBC News Online. Retrieved from http://news.bbc.co.uk

Kloeppel, James E. (14 December 2004). *Shutdown of circulation pattern could be disastrous, researchers say.* University of Illinois at Urbana-Champaign. Retrieved from http://www.news.uiuc.edu

Kloeppel, James E. (3 December 2001). *Stratospheric polar vortex influences winter cold, researchers say.* University of Illinois at Urbana-Champaign. Retrieved from http://www.eurekalert.org

Kloeppel, James E. (10 October 2001). *High uncertainty of climate change from global warming.* University of Illinois at Urbana-Champaign. Retrieved from http://unicsci.com

Knight, Daniel (7 May 2000). *Researchers warn Earth's ice is melting away.* InterPress Third World News Agency [online]. Retrieved from http://www.climattreark.org

Koebert, Christian and Virgil L. Sharpton (1992). *Terrestrial impact craters.* Lunar and Planetary Institute. Retrieved from http://www.hawastsoc.org

Korotkevich, A. (1985). *150,000-year climatic record from Antarctic ice.* Nature: Vol.316; 591-595.

Kudrass, H.R. et al (1991). *Global nature of the Younger Dryas cooling event inferred from oxygen isotope data from Sulu Sea cores.* Nature: Vol. 349; 406-409.

Kuffey, Kurt M. and Shawn J. Marshal (6 April 2000). *Substantial contribution to sea-level rise during the last interglacial from Greenland ice sheet.* Nature: Vol. 404 (6778); 591-4.

Kunnen-Jones, Marianne (15 December 2001). *Tracking climate change on the tundra: Team devising new tools.* Earth Observatory Media Alert. Retrieved from http://earthobservatory.nasa.gov

Kuzmina, S. I., L. Bengtsson, O. M. Johannessen, H. Drange, L. P. Bobylev, and M. W. Miles (2005). *The North Atlantic Oscillation and greenhouse-gas forcing.* Geophysical Research Letters: 32, L04703, doi:10.1029/2004GL021064.

Lange, Larry (19 December 2002). *Cascade glaciers are shrinking, posing threat to everything below.* Seattle Post [online]. Retrieved from http://seattlepi.nwsource.com/

Lavender, Kara L. et al (2000). *Mid-depth recirculation observed in the interior Labrador and Irminger seas by direct velocity measurements.* Nature: Vol. 407; 66-69.

Lazaroff, Cat (3 December 1999). *Melting Arctic ice linked to human combustion of oil and gas.* Environment News Service. Retrieved from http://www.ens.lycos.com

Lazaroff, Cat (21 July 2000). *Greenland ice sheet melting away.* Environment News Service. Retrieved from http://news.yahoo.com

Lazaroff, Cat (30 September 2002). *Soot linked to flooding, drought, global warming.* Environment News Service. Retrieved from http://news.yahoo.com

Leahy, Stephen (9 September 2004) *Warming trend will decimate Arctic peoples, report warns.* Inter Press Service. Retrieved from http://news.yahoo.com

Leake, Jonathon (26 December 2004). *Grass flourishes in warmer Antarctic.* The Times Online. Retrieved from http://www.timesonlin.co.uk

Leakey, Richard and Roger Lewin (1996). *The Sixth Extinction: Patterns of life and the future of humankind.* Doubleday, New York.

Lean, Geoffrey (25 January 2004). *Global warming will plunge Britain into new ice age 'within decades'.* The Independent [online]. Retrieved from http://news.independent.co.uk

Lean, Geoffrey (28 March 2004). *Global warming spirals upwards.* The Independent [online]. Retrieved from http://news.independent.co.uk

Lean, Geoffrey (3 February 2002). *Antarctica becomes too hot for the penguins.* The Independent [online]. Retrieved from http://news.independent.co.uk

Lean, Geoffrey (6 February 2005). *Apocalypse Now: How mankind is sleepwalking to the End of the Earth.* The Independent [online]. Retrieved from http://www.independent.co.uk

Leeman, Sue (24 February 2005). *Retreat of Antarctic ice shelves not new.* Associated Press. Retrieved from http://news.yahoo.com

Lehman S. & L. Keigwin (1992). *Sudden changes in North Atlantic circulation during the last deglaciation.* Nature: Vol. 356; 757-762.

Leidig, Michael and Roya Nikkah (18 July 2004). *The truth about global warming-it's the Sun that's to blame.* The Telegraph [online]. Retrieved from http://www.telegraph.co.uk

Leifert, Harvey (19 January 2005). *Arctic rivers discharge more freshwater into ocean, reflecting changes to hydrologic cycle.* American Geophysical Union. Retrieved from http://www.agu.org

Leopold, Aldo (1949). *A Sand County Almanac.* Oxford University Press. Oxford.

Lichtl, Martin (1999). *Ecotainment: Der neue Weg im Umweltmarketing.* Ueberreuter. Frankfurt a.M.

Liepins, Larissa (15 November 2004). *Thousands still without power in eastern Canada.* Reuters. Retrieved from http://news.yahoo.com

Lilley, Ray (25 March 2002). *Scientist: Ice shelves face breakup.* Associated Press. Retrieved from http://news.yahoo.com

Lippsett, Laurence (5 December 1997). *Lamont's Broecker warns gases could alter climate: Oceans' circulation could collapse.* Columbia University Record: Vol. 23; No. 11.

Ljunggren, David (27 May 2001). *Explorer says Arctic ice thinning noticeably.* Reuters. Retrieved from http://news.yahoo.com

Lockwood, Deirdre (30 March 2005). *World Bank pinpoints disaster hot spots.* Nature News Service [online]. Retrieved from http://www.nature.com

Lomborg Bjorn (2001). *The Skeptical Environmentalist: Measuring the Real State of the World.* Cambridge University Press. Cambridge, UK.

Lorentz, Katie (4 March 2005). *Canada's shrinking ice caps.* NASA-GSFC News. Retrieved from http://www.nasa.gov

Los Angeles Times [online] (5 October 2001). *Colorado glacier finds put a chill on global warming trends.* Retrieved from http://www.latimes.com

Lovell, Jeremy (15 March 2005). *Global warming threat central to policy-Britain.* Reuters. Retrieved from http://news.yahoo.com

Lovell, Jeremy (28 December 2001). *Antarctic Experts Warn of Global Warming Meltdown.* Reuters. Retrieved from http://news.yahoo.com

Lovelock, James (2000). *Gaia: A New Look at Life on Earth.* Oxford University Press. Oxford.

Lovelock, James (1991). *Healing Gaia: Practical Medicine for the Planet.* Harmony Books. New York.

Lovelock, James (1988). *The Ages of Gaia: A Biography of Our Living Earth.* Oxford University Press. Oxford.

Lowell, Jeremy (2 February 2005). *Climate warming spells species wipeout.* Reuters. Retrieved from http://news.yahoo.com

Lozan, Jose L.; Hartmut Grassl, and Peter Hupfer (Eds.) (2001). *Climate of the 21st Century: Changes and Risks.* Wissenschaftliche Auswertungen. Hamburg.

Lynas, Mark (November 2000). *Too hot for Heidi.* The Ecologist: Vol. 30; No. 8.

Lynch-Stieglitz, Jean et al (1999). *Weaker Gulf Stream in the Florida Straits during the last Glacial Maximum.* Nature: Vol. 402, pp. 644-648.

Macdonell, Hamish (4 August 2004). *Scientists alarmed at increase in melt rate of ice.* The Scotsman [online]. Retrieved from http://thescotsman.scotsman.com

Magell, Hillary (30 January 2002). *Climate studies point to more floods in this century.* National Geographic News [online]. Retrieved from http://news.nationalgeographic.com

Malthus, Thomas (1798). *An Essay on the Principle of Population: An Essay on the Principle of Population, as it Affects the Future Improvement of Society with Remarks on the Speculations of Mr. Godwin, M. Condorcet, and Other Writers.* J. Johnson in St. Paul's Church-Yard. London.

Mann, Michael E. et al (1999). *A paleoperspective on global warming.* Retrieved from http://www.ngdc.noaa.gov

Mann, Michael E. et al (1999). *Global temperature patterns in past centuries: An interactive presentation.* Retrieved from http://www.ngdc.noaa.gov

Mann, Michael E. et al (1999). *Northern hemisphere temperatures of the last six centuries.* Retrieved from http://www.ngdc.noaa.gov

Mapes, Jennifer (6 February 2001). *U.N. scientists warn of catastrophic climate changes.* National Geographic News [online]. Retrieved from http://news.nationalgeographic.com

Mardyks, Ray (no date). *Keys under the Sphinx decoding the Hall of Records.* Retrieved from http://www.floweroflife.org/keys.htm

Marris, Emma (14 October 2004). *Amphibians face a bleak future.* Nature News Service [online]. Retrieved from http://www.nature.com

Maslin M. and S. Burns (2001). *Reconstruction of the Amazon basin effective moisture availability over the past 14,000 years.* Science: Vol. 290; 2285-2287.

Mason, Betsy (10 December 2003). *Man has been changing climate for 8,000 years.* Nature News Service [online]. Retrieved from http://www.nature.com

Mason, Margie (30 May 2001). *Increased shrubbery found in Arctic.* The Associated Press. Retrieved from http://news.yahoo.com

Master, Sharad (26 February 2002). *A possible Holocene impact structure in the Al'Amarah marshes, near the Tigris-Euphrates confluence, southern Iraq.* University of Witwatersrand. Johannesburg, South Africa. Retrieved from http://abob.libs.uga.edu

Mastny, Lisa (6 March 2002). *Melting of Earth's ice cover reaches new high.* Worldwatch News Brief. Retrieved from http://www.worldwatch.org

Matthews, Robert (11 November 2001). *First evidence meteor impact ended The Bronze Age.* The Telegraph [online]. Retrieved from http://www.telegraph.co.uk

Matthews, Robert (4 November 2001). *Meteor clue to end of Middle East civilisations*. The Telegraph [online]. Retrieved from http://www.telegraph.co.uk

Matthews, Robert (6 November 2001). *Slam, bang, thanks Saddam: new meteor theory*. The Telegraph [online]. Retrieved from http://www.telegraph.co.uk

Mattmiller, Brian (9 September 2000). *Ice records hold cold truth about climate change*. Environmental News Network. Retrieved from http://www.enn.com

Mayell, Hillary (18 December 2001). *Is warming causing Alaskan meltdown?* National Geographic News [online]. Retrieved from http://news.nationalgeographic.com

Mayell, Hillary (24 October 2000). *Orangutans edging closer to brink of extinction*. National Geographic News [online]. Retrieved from http://news.nationalgeographic.com

Mayell, Hillary (27 December 2001). *Did planetary 'belch' cause prehistoric warming?* National Geographic News [online]. Retrieved from http://news.nationalgeographic.com

McCarthy, Michael (30 January 2002). *Climate studies point to more floods in this century*. National Geographic News [online]. Retrieved from http://news.independent.co.uk

McCarthy, Michael (12 July 2001). *How the world is threatened by massive change*. The Independent [online]. Retrieved from http://news.independent.co.uk

McCarthy, Michael (21 October 2004). *Climate change threatens world aid effort*. The Independent [online]. Retrieved from http://news.independent.co.uk

McCarthy, Michael (24 January 2005). *Climate change: Countdown to global catastrophe*. The Independent [online]. Retrieved from http://news.independent.co.uk

McCarthy, Michael (30 March 2004). *'Dead zones' in world's oceans are growing, say alarmed UN scientists*. The Independent [online]. Retrieved from http://news.independent.co.uk

McCarthy, Michael (12 July 2001). *Global warming much worse than predicted, say scientists*. The Independent [online]. Retrieved from http://news.independent.co.uk

McClure, Robert (5 January 2001). *Humans moving closer to extinction, study says*. Seattle Post Intelligencer [online]. Retrieved from http://seattlepi.nwsource.com

McDonald, Joe (22 January 2001). *Report warns of disaster from global warming*. Associated Press. Retrieved from http://news.yahoo.com

McFarling, Usha Lee (12 December 2002). *Scientists alarmed by continued warming trend*. Los Angeles Times [online]. Retrieved from http://www.latimes.com

McFarling, Usha Lee (13 July 2001). *Fear growing over a sharp climate shift*. Los Angeles Times [online]. Retrieved from http://www.latimes.com

McFarling, Usha Lee (2 January 2003). *Minute shift in temperature has had a major effect on Earth, study says*. Los Angeles Times [online]. Retrieved from http://www.latimes.com

McFarling, Usha Lee (25 September 2002). *Glacial melting takes human toll*. Los Angeles Times [online]. Retrieved from http://www.latimes.com

McFarling, Usha Lee (3 May 2002). *Wind theory may clear up warming mysteries*. Los Angeles Times [online]. Retrieved from http://www.latimes.com

McGee, Rory (31 May 2001). *Mountain glaciers around the world receding, AGU told*. UniSci Online. Retrieved from http://unisci.com

McGirk, Jan (11 October 2000). *Ozone hole over Chile puts southern cities on sunblock alert*. The Independent [online]. Retrieved from http://news.independent.co.uk

McGivering, Jill (30 September 2000). *Floods hit 20 million*. BBC News Online. Retrieved from http://news.bbc.co.uk

McGourty, Christine (17 January 2002). *Ice 'thickens' in West Antarctica*. BBC News Online. Retrieved from http://news.bbc.co.uk

McGourty, Christine (18 December 2001). *Warmth puts penguins under pressure*. BBC News Online. Retrieved from http://news.bbc.co.uk

McGourty, Christine (26 December 2001). *Antarctica's climate clues*. BBC News Online. Retrieved from http://news.bbc.co.uk

McGourty, Christine (27 December 2001). *Low probability of ice collapse*. BBC News Online. Retrieved from http://news.bbc.co.uk

McKibben, Bill (1989). *The End of Nature*. Anchor Books, Random House. New York.

McKie, Robin (10 November 2002). *Sun's rays to roast Earth as poles flip*. Observer [online]. Retrieved from http://observer.guardian.co.uk

McKie, Robin (22 October 2000). *Now Europe's biggest glacier falls to global warming*. Guardian [online]. Retrieved from http://www.guardian.co.uk

McKie, Robin (3 March 2002). *Melting ice will open sea trail to the East*. Observer [online]. Retrieved from http://www.observer.guardian.co.uk

McMahon, Barbara and Paul Brown (23 September 2005). *900,000-year-old ice may destroy US case on Kyoto*. Guardian [online]. Retrieved from http://www.guardian.co.uk

McManus, J.F. et al (22 April 2004). *Collapse and rapid resumption of Atlantic meridional circulation linked to deglacial climate changes*. Nature: Vol. 428, pp. 834-837.

Meadows, Donella H., Dennis L. Meadows, Joergen Rander, William W. Behrens III (1972). *The Limits to Growth: A report for the Club of Rome's project on the predicament of mankind*. Universe Books. New York.

Meek, James (23 August 2000). *The heat is on*. Guardian [online]. Retrieved from http://www.guardian.co.uk

Melville, Nancy A. (23 September 2000). *Ocean oscillations may mask global warming*. Retrieved from http://news.yahoo.com

Mesarovic, Mihajlo and Eduard Pestel (1974). *Mankind at the Turning Point: The second report to the Club of Rome*. Signet Books. New York.

Messer, Andrea Elyse (13 December 2001) *The past says abrupt climate change in our future*. Pennsylvania State University. Retrieved from http://www.eurakaalert.org

Met Office (18 April 2002). *Understanding climate change*. Retrieved from http://www.met-office.gov.uk

Met Office (31 July 2002). *Record-breaking January to June temperatures*. Retrieved from http://www.met-office.gov.uk

Mileti, Dennis S. (Ed) (1999). *Disasters by Design: A Reassessment of Natural Hazards in the United States*. Joseph Henry Press. Washington, D.C.

Millenium Ecosystem Assessment: Synthesis Report (2005). Retrieved from http://www.millenniumassessment.org

Millenium Ecosystem Assessment: Living Beyond Our Mean-Natural Assets and Human Well Being (2005). Retrieved from http://www.millenniumassessment.org

Miller, G. Tyler Jr. (1998). *Living in the Environment: Principles, Connections, and Solutions*. Tenth Edition. Wadsworth Publishing Company. Belmont, CA.

Milmo, Cahal (9 January 2002). *Huge asteroid narrowly misses Earth*. The Independent [online]. Retrieved from http://www.independent.co.uk

Moore, Thomas (1998). *Climate of Fear: Why We Shouldn't Worry about Global Warming*. The Cato Institute. Washington, DC.

Morgan, Claude (14 August 2000). *The rising oceans*. Environmental News Network. Retrieved from http://www.enn.com

Morison, James (July 2001). *Recent changes in the Arctic, the study of environmental Arctic change* (abstract). Challenges of a Changing Earth Conference 10-13 July 2001. Retrieved from http://sciconftalks.igbp.sr.unh.edu

Mosley-Thompson, Ellen (20 June 1997). *Ice core provides ancient climate record*. Environmental News Network. Retrieved from http://www.enn.com

MTV News Interview with U.N. Weapons Inspector Hans Blix (12 March 2003). *Caught between Iraq and a hard place*. Retrieved from http://www.mtv.com/bands/i/iraq/news_feature_031203/index.5.jhtml

Munro, Margaret (27 March 2002). *Hotter times on planet Earth, researchers find*. National Post [online]. Retrieved from http://www.nationalpost.com

Muscheler, Raimund et al (30 November 2000). *Changes in deep-water formation during the Younger Dryas event inferred from 10Be and 14C records*. Nature: Vol. 408; 567-570.

Myth of the 5 suns (no date). Retrieved from http://www.stormpages.com

NASA (5 September 2001). *Earth is becoming a greener greenhouse.* Retrieved from http://www.spaceflightnow.com

NASA (6 December 2000). *A disintegrating glacier.* Retrieved from http://science.nasa.gov

NASA (20 July 2000). *NASA scientists detect rapid thinning of Greenland's coastal ice.* Retrieved from http://www.gsfc.nasa.gov

NASA (3 February 2005). *Satellites see ocean plants increase, coasts greening.* Retrieved from http://www.nasa.gov

NASA (9 February 2004). *Scientists find ozone-destroying molecule.* Retrieved from http://science.nasa.gov

NASA (February 2000). *Melting snows of Kilimanjaro.* Retrieved from http://earthobservatory.nasa.gov.

NASA Earth Observatory (no date). *Milutin Milankovitch.* Retrieved from http://earthobservatory.nasa.gov.

NASA Earth Observatory (no date). *Some effects of ultraviolet-B (UVB) radiation on the biosphere.* Retrieved from http://earthobservatory.nasa.gov

NASA-GISS (8 February 2005). *Earth gets warm feeling all over.* Retrieved from http://www.nasa.gov

NASA-GISS (no date). *General characteristics of the world's oceans.* Retrieved from http://icp.giss.nasa.gov

NASA-GSFC (10 December 2001). *Methane explosion warmed the prehistoric Earth, possible again.* Retrieved from http://www.gsfc.nasa.gov

NASA-GSFC (11 November 2001). *Satellites shed light on a warmer world.* Retrieved from http://spaceflightnow.com

NASA-GSFC (12 October 2004). *Researchers find frozen North may accelerate climate change.* Retrieved from http://www.gsfc.nasa.gov

NASA-GSFC (15 April 2004). *Satellites record weakening North Atlantic Current.* Retrieved from http://www.gsfc.nasa.gov

NASA-GSFC (16 July 2001). *Climate change in Atlantic larger than previously thought.* Retrieved from http://www.gsfc.nasa.gov

NASA-GSFC (21 April 2004). *Satellites act as thermometers in space, show Earth has a fever.* Retrieved from http://www.nasa.gov

NASA-GSFC (29 May 2002). *Decline of world's glaciers expected to have global impacts over this century.* Retrieved from http://earthobservatory.nasa.gov

NASA-GSFC (29 November 2001). *Ocean circulation shut down by melting glaciers after last ice age.* Retrieved from http://www.gsfc.nasa.gov

NASA-GSFC. (6 October 2004) *Study shows potential for Antarctic climate change.* Retrieved from http://www.gsfc.nasa.gov

NASA-JPL (09 December 2002). *Research offers explanation for Earth's bulging waistline.* Retrieved from http://www.spaceflightnow.com

NASA-JPL (22 June 2001). *Satellite shows no El Niño in Pacific yet, but one due.* Retrieved from http://www.spaceflightnow.com

NASA-JPL (31 August 2002). *NASA study finds rapid changes in polar ice sheets.* Retrieved from http://www.spaceflightnow.com

NASA-JPL (23 April 2004). *Arctic ozone loss more sensitive to climate change than thought.* Retrieved from http://www.earthobservatory.nasa.gov

NASA-Langley (07 February 2002). *NASA scientists find new evidence of climate change.* Retrieved from http://www.spaceflightnow.com

National Academy of Sciences (no date). *Ozone Depletion.* Retrieved from http://www.nas.edu

National Academy of Sciences (11 December 2001). *Possibility of abrupt climate change needs research and attention.* Retrieved from http://www.nas.edu

National Academy of Sciences (2001). *The Ozone Depletion Phenomenon.* Retrieved from http://www.nas.edu

National Center for Atmospheric Research-UCAR (25 February 2004). *The Arctic Oscillation: A key to this winter's cold-and a warmer planet.* Retrieved from http://www.ncar.ucar.edu

National Center for Atmospheric Research-UCAR (10 January 2005). *Drought's growing reach: NCAR study points to global warming as key factor.* Retrieved from http://www.ncar.ucar.edu

National Center for Atmospheric Research-UCAR (22 June 2004). *NCAR releases new version of premier global climate model.* Retrieved from http://www.ncar.ucar.edu

National Geographic News (no date). *Black sea legends.* Retrieved from http://news.nationalgeographic.com

National Geographic News (6 September 2001). *'Greenhouse' growing greener on patches of earth, study finds.* Retrieved from http://news.nationalgeographic.com

National Research Council (1989). *Global Change and Our Common Future: Ppaers from a Forum Committee on Global Change.* National Academies Press. Washington, DC.

National Research Council (2000). *Reconciling Observations of Global Temperature Change.* National Academies Press. Washington, DC.

National Research Council (2002) *Abrupt Climate Change: Inevitable Surprises.* National Academies Press. Washington, DC.

National Science Foundation (19 November 2001). *Melting glaciers diminished Gulf Stream, cooled western Europe, during last ice age.* Retrieved from http://www.nsf.gov

National Science Foundation (2000). *Massive iceberg peels off from Antarctic Ice Shelf.* Retrieved from http://www.nsf.gov

National Science Foundation (22 June 2004). *New Version of premier global climate model released.* Retrieved from http://www.nsf.gov

National Science Foundation (25 November 2001). *Melting glaciers shut down Gulf Stream in past.* Retrieved from http://www.nsf.gov

Natural Science Online (1 November 1997). *A new European ice age?* Retrieved from http://naturalscience.com/ns/cover/cover5.html

Nebehay, Stephanie (6 October 2000). *UN experts say ozone depletion at record level.* Reuters. Retrieved from http://news.yahoo.com

Neergaard, Lauran (24 January 2002). *Antarctic study finds warming change.* Associated Press. Retrieved from http://news.yahoo.com

Nelson, Daniel (16 April 2002). *'Catastrophic threat' from world's mountain lakes, warns UN.* One World UK [online]. Retrieved from http://news.yahoo.com

Netherlands Organization for Scientific Research (25 January 2002). *Climate change following collapse of the Maya empire.* Retrieved from http://sciencedaily.com

New Scientist [online] (31 May 1997). *Venus' tail reaches all the way to Earth.* Retrieved from http://www.newscientist.com

New Scientist News Service (26 April 2002). *First quarter of 2002 is 'warmest for a millennium'.* Retrieved from http://www.newscientist.com

New York Times [online] (13 January 2002). *Antarctic desert getting colder.* Retrieved from http://www.nytimes.com

New York Times [online] (19 June 2001). *Both sides now: New way that clouds may cool.* Retrieved from http://www.nytimes.com

New Zealand Herald [online] (11 October 2001). *Ice research projects study climate change.* Retrieved from http://www.nzherald.co.nz

Niroma, Timo (4 September 2002). *The climax of a turbulent millennium: Evidence for major impact events in the late Third Millennium BC.* Retrieved from http://personal.eunet.fi/pp/tilman/tilman2.htm

Niroma, Timo (4 September 2002). *The effect of solar variability on climate.* Retrieved from http://personal.eunet.fi/pp/tilman/tilman2.htm

Niroma, Timo (4 September 2002). *The myth history of the catastrophe events and their cultural effects.* Retrieved from http://personal.eunet.fi/pp/tilman/tilman2.htm

NOAA (10 September 2004). *NOAA announces return of El Niño.* Retrieved from http://www.noaanews.noaa.gov/

NOAA (2 September 2004). *Record number of tornadoes reported in U.S. during August.* National Oceanic and Atmospheric Administration. Retrieved from http://www.noaanews.noaa.gov

NOAA (31 March 2005). *After two large annual gains, rate of atmospheric co2 increase returns to average, NOAA reports.* Retrieved from http://www.noaanews.noaa.gov

NOAA (5 January 2001). Record cold grips much of the nation in November and December: Two-month period is the coldest on record in the United States. Retrieved from http://www.noaanews.noaa.gov

NOAA (9 September 2003). Astronomical theory of climate change. Retrieved from http://www.ngdc.noaa.gov

NOAA (no date). The Medieval Warm Period. Retrieved from http://www.ngdc.noaa.gov

NOAA (no date). The Mid-Cretaceous Period. Retrieved from http://www.ngdc.noaa.gov

NOAA (no date). The Penultimate Interglacial Period. Retrieved from http://www.ngdc.noaa.gov

NOAA (no date). The So-Called Mid-Holocene Warm Period. Retrieved from http://www.ngdc.noaa.gov

NOAA-NGDC (no date). The Greenhouse Effect. Retrieved from http://www.ngdc.noaa.gov

NOAA-NGDC (no date). What is 'Global Warming'? Retrieved from http://www.ngdc.noaa.gov

Noah's Ark: The Story (2000). Retrieved from http://www.arksearch.com

Noble, Ian (July 2001). Non-linear responses and surprises to global change (abstract). Challenges of a Changing Earth Conference July 2001. Retrieved from http://www.sciconf.igbp.kva.se

Noserale, Diane (11 December 2001). Many Alaskan glaciers are thinning. USGS study says. Unisci Online. Retrieved from http://unisci.com

Nullis, Clare (18 February 2001). Global warming risks outlined. Associated Press. Retrieved from http://news.yahoo.com

Nuttall, Nick (2002). Regional and global impacts of vast pollution cloud detailed in new scientific study. UNEP. Retrieved from http://www.rrcap.unep.org

O'Gorman, Sarah and Nduka Uche (Eds.) (2000). Cold Comfort Fire: Essays, poems and stories against climate catastrophe. A SEED Europe. Amsterdam.

O'Hanlon, Larry (12 August 2002). Cosmic rays, global warming linked. Discovery News [online]. Retrieved from http://dsc.discovery.com

O'Harra, Doug (19 July 2002). On thinning ice Alaska glaciers making biggest contribution to sea level change. Anchorage Daily News [online]. Retrieved from http://www.and.com

O'Harra, Doug (4 September 2000). North Pole is melting -- Well, it's more complicated. Anchorage Daily News [online]. Retrieved from http://www.and.com

O'Meara, Molly (10 June 1998). Antarctic ice shelf crumbling. Environmental News Network. Retrieved from http://www.enn.com

O'Neill, Brian C., Landis F. MacKeller, and Wolfgang Lutz (2001). Population and Climate Change. Cambridge University Press. Cambridge, UK.

Office of the White House Press Secretary (11 January 2005). President participates in conversation on Social Security Reform. Retrieved from http://www.whitehouse.gov

Ohio State University (1997) Latest evidence of global warming found in tropics and subtropics. Retrieved from http://researchnews.osu.edu

Ohio State University (26 June 1995) Ice cores show record of climate dating back 20,000 years. Retrieved from http://researchnews.osu.edu

Ohio State University (3 October 2002). African ice core analysis reveals catastrophic droughts, shrinking ice fields and civilization shifts. Retrieved from http://researchnews.osu.edu

Onion, Amanda (26 March 2002). More icebergs at risk. ABCNews Online. Retrieved from http://abcnews.com

Ontario Consultants on Religious Tolerance (4 October 2001). The catastrophic deluge which entered the Black Sea, circa 7,500 years ago. Retrieved from http://www.religioustolerance.org/ev_noah.htm

Oregon State University (13 March 2003). Global warming could trigger cascade of climate changes. Retrieved from http://earthobservatory.nasa.gov

Osborn, Tim (2000). The thermohaline circulation. Climatic Research Unit. Retrieved from http://www.cru.uea.ac.uk

Overpeck, J. et al (no date). Arctic temperatures of the last four centuries. Retrieved from http://www.ngdc.noaa.gov

Paillard, Didier (2001). Glacial Hiccups. Nature: Vol. 409; 147-8.

Paine, Michael (5 November 1999). Asteroids and Tsunamis. Space.com. Retrieved from http://www.space.com

Palutikof, Jean (no date). Global temperature record. Retrieved from http://www.cru.uea.ac.uk

Parry, Martin (Ed.) (2000). Assessment of Potential Effects and Adaptation for Climate Change in Europe: The Europe Acacia Project. Jackson Environment Institute, University of East Anglia. Norwich, UK.

Patterson, Thom (6 July 2001). Earth-shaking 'wobble' may have killed dinosaurs. CNN News Online. Retrieved from http://asia.cnn.com

PBS Online (no date). 160,000 BP - Global Warming. Retrieved from http://www.pbs.org

Pearce, Fred (18 July 2001). Most Predictions on global warming are probably wrong. New Scientist [online]. Retrieved from http://www.newscientist.com

Pearce, Fred (25 November 1998). Nature plants doomsday devices. Guardian [online]. Retrieved from http://www.guardian.co.uk

Pearce, Fred (2002). Global Warming: A Beginner's Guide to Our Changing Climate. Doling Kindersley. London.

Pearson, Paul N. and Martin R. Palmer (17 August 2000). Atmospheric carbon dioxide concentrations over the past 60 million years. Nature: Vol. 406, 695-699.

Peck, Grant (30 July 2002). Experts warn of disasters from climate changes. Associated Press. Retrieved from http://news.yahoo.com

Peiser, Benny (1998). Comparative analysis of late Holocene environmental and social upheaval: Evidence for a global disaster around 4000 BP. British Archaeological Report. Oxford.

Peiser, Benny (6 November 2001). Collapse of early Bronze Age civilisations: Has the smoking gun been found? Retrieved from http://abob.libs.uga.edu

Peplow, Mark (18 October 2004). Unseen comets show impact risk for Earth. Nature News Service [online]. Retrieved from http://www.nature.com

Peplow, Mark (20 October 2004). Spinning Earth twists space. Nature News Service [online]. Retrieved from http://www.nature.com

Perdue, Mitzi (23 October 2000). Greenland is a canary. Scripps Howard News Service. Retrieved from http://www.climateark.org/

Perlman, David (15 June 2001). Global warming to affect water supply: More rainfall means smaller sierra snowpack. San Francisco Chronicle [online]. Retrieved from http://www.sfgate.com

Perlman, David (17 February 2003). Earth's temperatures heating up: Averages to rise 8 degrees by end of century, climate scientist says. San Francisco Chronicle [online]. Retrieved from http://www.sfgate.com

Perlman, David (18 December 2000). Space object that killed dinosaurs broke through Earth's crust. San Francisco Chronicle [online]. Retrieved from http://www.sfgate.com

Perlman, David (20 March 2002). Antarctic ice shelf disintegrating. San Francisco Chronicle [online]. Retrieved from http://www.sfgate.com

Perlman, David (6 October 2003). Decline in oceans' phytoplankton alarms scientists. San Francisco Chronicle [online]. Retrieved from http://www.sfgate.com

Perry, Michael (31 May 2001). Global warming melts Australia's glaciers. Reuters. Retrieved from http://news.yahoo.com

Peterson, T.C., and R.S. Vose (1997). An overview of the global historical climatology network temperature database. Bulletin of the American Meteorological Society: 78; 2837-2849.

Petit, J.R. et al (1987). Vostok ice core: a continuous isotope temperature record over the last climatic cycle 160,000 years. Nature: 329; 403-407.

Petit, J.R. et al (1999). Climate and atmospheric history of the past 400,000 years from the Vostok ice core in Antarctica. Nature: 399; 429-436.

Petre, Jonathon (22 February 2002). Newton set 2060 for end of world. The Telegraph [online]. Retrieved from http://www.telegraph.co.uk

Philander, S. George (2000). Is the Temperature Rising? The Uncertain Science of Global Warming. Princeton University Press. Princeton.

Philippens, Michel (18 March 2002). *Too much sun can harm ocean life.* Netherlands Institute for Sea Research. Retrieved from http://www.spaceref.com

Pinin, Eric (12 December 2001). *A Warning on climate change.* Washington Post [online]. Retrieved from http://www.washingtonpost.com

Pitman, Nigel C.A. Pitman and Peter M. Joergensen (1 November 2002). *Estimating the size of the world's threatened flora.* Science: Vol. 298; 989.

PlanetArk.org (12 July 2004). *Millions marooned by floods in Bangladesh.* Retrieved from http://www.heatisonline.com

Polakovic, Gary (4 May 2001). *Earth losing air-cleansing ability, study says.* Los Angeles Times [online]. Retrieved from http://www.latimes.com

Pollack, Henry N. et al (no date). *Global temperatures of the last five centuries.* NOAA. Retrieved from http://www.ngdc.noaa.gov

Prokosch, Peter (21 June 2001). *Sibirien taut auf.* World Wildlife Fund Deutschland. Retrieved from http://www.wwf.de

Purdy, Michael (8 January 2002). *Ancient supernova may have triggered eco-catastrophe.* Johns Hopkins University. Retrieved from http://www.eurekalert.org

Quinn, Andrew (12 December 2001). *Scientists warn of abrupt world climate change.* Reuters. Retrieved from http://news.yahoo.com

Racer, Paul (1 February 2001). *Polar ice sheet shows shrinkage.* Associated Press. Retrieved from http://news.yahoo.com

Radford, Ceri (16 November 2004). *Melting Swiss glaciers threatening the Alps.* Reuters. Retrieved from http://news.yahoo.com

Radford, Tim (19 July 2002). *85% of Alaskan glaciers melting at 'incredible rate'.* Guardian [online]. Retrieved from http://www.guardian.co.uk

Radford, Tim (31 October 2003). *Antarctic ice shelf is melting rapidly, scientists warn.* Guardian [online]. Retrieved from http://www.guardian.co.uk

Radford, Tim and Paul Brown. *Global warming could be worst in 10,000 Years.* Guardian Weekly. Volume 164. No. 5. 2001.

Ramanujan, Krishna (22 August 2002). *Satellites show overall increases in Antarctic sea ice cover.* NASA-GSFC. Retrieved from http://www.gsfc.nasa.gov

Ramanujan, Krishna (28 August 2002). *Atmospheric wave linked to sea ice flow near Greenland, study finds.* NASA-GSFC. Retrieved from http://www.gsfc.nasa.gov

Randall, Cora (1 March 2005). *Huge 2004 stratospheric ozone loss tied to solar storms, arctic winds.* University of Colorado. Retrieved from http://www.eurekalert.org

Ray, Justin (4 May 2002). *New environmental eye on Earth launched into space.* Spaceflight Now [online]. Retrieved from http://www.spaceflightnow.com

Reancy, Patricia (12 March 2001). *Increase in greenhouse gases seen from space.* Reuters. Retrieved from http://news.yahoo.com

Reaney, Patricia (11 July 2001). *Scientists find fossils of man's earliest relative.* Reuters. Retrieved from http://news.yahoo.com

Recer, Paul (1 February 2001). *Polar ice sheet shows shrinkage.* The Associated Press. Retrieved from http://news.yahoo.com

Recer, Paul (16 November 2001). *Oceans temps, solar cycles linked.* Associated Press. Retrieved from http://www.space.com

Recer, Paul (18 July 2002). *Study: Glaciers melting faster.* Associated Press. Retrieved from http://news.yahoo.com

Recer, Paul (2 February 2001). *Scientists see a warming in melting of Antarctic ice.* Philadelphia Inquirer [online]. Retrieved from http://inq.philly.com

Recer, Paul (2 February 2001). *View from space shows melting ice sheet.* Associated Press. Retrieved from http://news.yahoo.com

Recer, Paul (3 May 2001). *Decline in natural chemical found.* Associated Press. Retrieved from http://news.yahoo.com

Recer, Paul (4 April 2002). *Asteroid could hit in 878 years.* Associated Press. Retrieved from http://news.yahoo.com

Recer, Paul (4 January 2003). *Antarctic ice may vanish in 7,000 Years.* Associated Press. Retrieved from http://news.yahoo.com

Reckess, Gila (23 December 1999). *North braces for climate changes.* Environmental News Network. Retrieved from http://www.enn.com

Rempel, Alan (31 May 2001). *Chemicals in ancient ice move, affecting ice cores.* UniSci Online. Retrieved from http://unisci.com

Renshaw, Susie (2000). *Is Mexico's impact crater linked to mass extinctions?* UniSci Online. Retrieved from http://unisci.com

Repetto, Robert et al (October 1997). *U.S. Competitiveness is not at Risk in the Climate Negotiations.* World Resources Institute: Climate Notes. Washington, DC.

Reuters (6 January 2002). *Global warming blamed for melting Everest glacier.* Retrieved from http://www.planetark.com

Reuters (1 August 2002). *World heads for warmest year yet.* Retrieved from http://news.yahoo.com

Reuters (10 April 2002). *Climate change will unbalance ecosystems, study says.* Retrieved from http://news.yahoo.com

Reuters (10 July 2002). *California again sets power alert amid record heat.* Retrieved from http://news.yahoo.com

Reuters (18 December 2001). *World temperature second highest on record.* Retrieved from http://news.yahoo.com

Reuters (19 March 2001). *Global warming could put palm trees in Swiss Alps.* Retrieved from http://news.yahoo.com

Reuters (19 September 2002). *Study - Earth to warm even if greenhouse gas cut.* Retrieved from http://news.yahoo.com

Reuters (2 February 2002). *Dinosaurs survived cataclysm 200 million years ago.* Retrieved from http://news.yahoo.com

Reuters (2 October 1999). *Global temperatures stay high in 2002, report says.* Retrieved from http://www.heatisonline.com

Reuters (21 February 2001). *Lost Atlantis found in Bolivia, explorer says.* Retrieved from http://news.yahoo.com

Reuters (12 June 2002). *World forum scientists, 'Climate change will become a security Issue'.* Retrieved from http://www.cnn.com

Reuters (12 July 2004). *Millions flee South Asia floods.* Retrieved from http://news.yahoo.com

Reuters (17 January 2002). *Peru sees El Niño return by year end, no earlier.* Retrieved from http://news.yahoo.com

Reuters (18 April 2002). *Europe's largest glacier melting away – expert.* Retrieved from http://news.yahoo.com

Reuters (22 January 2003). *Scientists firm up global climate forecasts.* Retrieved from http://news.yahoo.com

Reuters (25 March 2002). *Monsoon records show link with global climate.* Retrieved from http://news.yahoo.com

Reuters (26 October 2000). *Scientists to meet to discuss melting Antarctic ice.* Retrieved from http://news.yahoo.com

Reuters (27 March 2002). *Arctic faces ozone damage by 2020, says scientist.* Retrieved from http://www.enn.com

Reuters (28 December 1999). *Global warming hits species all over the world.* Retrieved from http://news.yahoo.com

Reuters (28 December 2001). *1999 second-warmest year century, following last year.* Retrieved from http://abcnews.go.com

Reuters (29 December 2000). *Ice, snow leave 335,000 in South powerless for 4th day.* Retrieved from http://news.yahoo.com

Reuters. *Natural disasters at record level in 2000-insurer.* Retrieved from http://www.climateark.org

Reuters (29 September 2000). Report: Climate change causing jump in natural disasters. Retrieved from http://news.yahoo.com

Reuters (3 December 2003). No doubts global warming is real, U.S. experts say. Retrieved from http://news.yahoo.com

Reuters (3 February 2001). Global warming to cost $300 billion a year. Retrieved from http://news.yahoo.com

Reuters (3 June 2004). Scientists tie ancient warming to 'belch': Emissions today are even faster, they note. Retrieved from http://www.reuters.com

Reuters (3 May 2001). Atmosphere's pollution-fighting chemical waning. Retrieved from http://news.yahoo.com

Reuters (30 October 2002). U.N. says environmental disasters cost $70 Billion. Retrieved from http://news.yahoo.com

Reuters (4 September 2001). Satellites see a greener Northern Hemisphere. Retrieved from http://www.space.com

Reuters (5 April 2000). NASA, EU report massive Arctic ozone loss. Retrieved from http://www.enn.com

Reuters (7 February 2001). U.N. warns global warming is melting Arctic soil. Retrieved from http://news.yahoo.com

Reuters (8 October 2002). More than 11,000 plants, animals face extinction. Retrieved from http://news.yahoo.com

Revkin, Andrew C. (10 February 2005). 2004 was fourth warmest year ever recorded. New York Times [online]. Retrieved from http://www.nytimes.com

Revkin, Andrew C. (13 April 2001). Studies tie rise in ocean heat to greenhouse gases. New York Times [online]. Retrieved from http://www.nytimes.com

Revkin, Andrew C. (18 January 2002). A chilling effect on the great global melt. New York Times [online]. Retrieved from http://www.nytimes.com

Revkin, Andrew C. (2 November 2002). Climate talks shift focus to how to deal with changes. New York Times [online]. Retrieved from http://www.nytimes.com

Revkin, Andrew C. (22 April 2005). Climate research faulted over missing components. New York Times [online]. Retrieved from http://www.nytimes.com

Revkin, Andrew C. (22 September 2000). Planting new forests can't match saving old ones in cutting greenhouse gases, study finds. New York Times [online]. Retrieved from http://www.nytimes.com

Revkin, Andrew C. (26 January 2001). Ancient coral may hold hint of worsening weather cycle. New York Times [online]. Retrieved from http://www.nytimes.com

Revkin, Andrew C. (4 May 2001). Study finds a decline in natural air cleanser. New York Times [online]. Retrieved from http://www.nytimes.com

Reynolds, R.W., and T.M. Smith (1994). Improved global sea surface temperature analyses. Journal of Climate: 7; 929-948.

Reynolds, R.W., N.A. Rayner, T.M. Smith, D.C. Stokes, and W. Wang (2002). An improved in situ and satellite SST analysis for climate. J. Climate: 15; 1609-1625,

Rhein, Monica (2000). Oceanography: Drifters reveal deep circulation. Nature: Vol. 497; 30-31.

Rife, Philip (no date). Those enigmatic erratics: Out-of-place artifacts or out-of-whack chronology. Retrieved from http://www.strangemag.com

Rifkin, Jeremy (1989). Entropy: Into the Greenhouse World. Bantam Books. New York

Rincon, Paul (17 February 2005). Greenhouse gases 'do warm oceans'. BBC News Online. Retrieved from http://news.bbc.co.uk

Rind, D. et al (1986). The impact of cold North Atlantic sea surface temperatures on climate: implications for the Younger Dryas cooling (11-10k). Climate Dynamics: 1; 3-33.

Roach, John (1 March 2000). Little ice age holds big climate clues. Environmental News Network.

Roach, John (13 December 2002). Freshwater Runoff Into Arctic on the Rise, Scientists Say. National Geographic News [online]. Retrieved from http://news.nationalgeographic.com

Roach, John (13 January 2000). La Niña goes with the stream flow. Environmental News Network. Retrieved from http://www.enn.com

Roach, John (13 July 2000). Earth's ancient atmosphere trapped in rocks. Environmental News Network. Retrieved from http://www.enn.com

Roach, John (17 February 2000). Borehole temperatures confirm global warming. Environmental News Network. Retrieved from http://www.enn.com

Roach, John (19 November 1999). Historic global warming linked to methane release. Environmental News Network. Retrieved from http://www.enn.com

Roach, John (22 September 1997). The impacts of El Niño. Environmental News Network. Retrieved from http://www.enn.com

Roach, John (24 March 2000). Ocean is warming, study finds. Environmental News Network. Retrieved from http://www.enn.com

Roach, John (25 August 2000). Satellite sees through Arctic ice for signs of climate change. Environmental News Network. Retrieved from http://www.enn.com

Roach, John (31 January 2000). Global-warming warnings are more than hot air. Environmental News Network. Retrieved from http://www.enn.com

Romanovsky, Vladimir et al (2001). Permafrost dynamics in Alaska and East Siberia: changes and impacts. Challenges of a Changing Earth Conference. Retrieved from http://sciconftalks.igbp.sr.unh.edu

Rosen, Yareth (16 April 2004). Warming climate disrupts Alaska natives' lives. Reuters. Retrieved from http://news.yahoo.com

Rosen, Yereth (19 July 2002). Experts confirm observation of glaciers' fast melt. Reuters. Retrieved from http://news.yahoo.com

Rosentrater, Lynn (2002). Secrets of the ice: An Antarctic expedition. Retrieved from http://www.secretsofthice.org

Ross-Thomas, Emma (27 September 2002). Scientist says ice meteors a sign of climate change. Reuters. Retrieved from http://news.yahoo.com

Rowe, Mark (8 October 2000). Global warming to leave UK out in the cold. The Independent [online]. Retrieved from http://www.independent.co.uk

Roylance, Frank D (3 December 2003) Climate change 'entering the unknown'. The Baltimore Sun [online]. Retrieved from http://www.baltimoresun.com

Rubin, Josh (4 September 2002). Dry, hot summer confirms warming trend. The Toronto Star [online]. Retrieved from http://www.thestar.com

Ruhlemann, Carsten et al (1999). Warming of the tropical Atlantic Ocean and slowdown of thermohaline circulation during the last deglaciation. Nature: Vol. 402; 511-514.

Russell, Sabin (15 February 2002). Somber opening to science meeting - Warning of future environmental havoc resulting from Western society's lifestyle. San Francisco Chronicle [online]. Retrieved from http://www.sfgate.com

Russell, Sabin (17 February 2002). Glaciers on thin ice: Expert says melting to be faster than expected. San Francisco Chronicle [online]. Retrieved from http://www.sfgate.com

Rutgers University (12 October 2000). Rise in CO2 concentrations presents continuing global challenge. Retrieved from http://ur.rutgers.edu

Samuels, Sam Hooper (15 May 2001). Ocean whitecaps may help a warming planet keep its cool. New York Times [online]. Retrieved from http://www.nytimes.com

Sanders, Jane M. (17 February 2002). Global warming will persist at least a century even if emissions curbed now. NASA Earth Observatory Media Alert. Retrieved from http://earthobservatory.nasa.gov

Sandia National Laboratories News Release (16 September 2002). Earth rings: A clue to climate change? Retrieved from http://www.spaceflightnow.com

Sawin, Janet L. (April 2005). Climate change poses greater security threat than terrorism. Worldwatch Institute Global Security Brief #3. Retrieved from http://www.worldwatch.org

Scambos, Ted (18 March 2002). Antarctic ice shelf collapses in largest event of last 30 years. University of Colorado. Retrieved from http://www.colorado.edu

Scambos, Ted (21 September 2004). Antarctic glaciers accelerating in response to 2002 ice sheet collapse. National Snow and Data Center. Retrieved from http://www.eurekalert.org

Schell, Jonathon (1982). *The Fate of the Earth*. Avon Books. New York.

Schmid, Randolph (17 January 2002). West Antarctic ice getting thicker. Associated Press. Retrieved from http://news.yahoo.com

Schmid, Randolph (23 May 2002). Antarctic icebergs seen as normal. Retrieved from http://news.yahoo.com

Schmid, Randolph (23 May 2002). Rush of Antarctic icebergs seems to be a return to normal for continent. Associated Press. Retrieved from http://news.yahoo.com

Schmidt, Matthew (10 March 2004). UC study sheds new light on climate-change processes. University of California-Davis. Retrieved from http://www.news.ucdavis.edu

Schneider, Stephen H. (1996). *Laboratory Earth: The Planetary Gamble We Can't Afford to Lose*. Weidenfield and Nicolson. Great Britain.

Schoffner, Chuck (27 October 2004). NASA: Bush stifles global warming evidence. Associated Press. Retrieved from http://news.yahoo.com

Schotterer, Ulrich and Peter Andermatt (1990). *Climate-Our Future?* Kuemmerly and Frey and the Swiss Academy of Science. Bern.

Schrader, Anne (23 January 2001). Warmer climate affecting polar ice caps. Denver Post [online]. Retrieved from http://www.denverpost.com

Schwartz, Mark (14 December 2001). Antarctic mud reveals ancient evidence of global climate change. Stanford University. Retrieved from http://www.stanford.edu/news

Schwartz, Peter and Doug Randall (October 2003). An abrupt climate change scenario and its implications for United States national security. Prepared for the United States Pentagon. Retrieved from http://www.ems.org/climate/pentagon_climatechange.pdf

Scofield, Bruce (July 2001). What really happened in 3100 BC- and where are we headed now? Retrieved from http://www.onereed.com

Scripps Howard News Service (5 May 2004). Effects of climate warming are here and now. Retrieved from http://www.heatisonline.org

Seabrook, Charles (4 August 2001). Global warming not just theory in Alaska. Atlanta Journal-Constitution [online]. Retrieved from http://www.climateark.org

Serreze, Mark et al (4 October 2004). Arctic sea ice declines again in 2004, according to CU-Boulder researchers. National Snow and Ice Data Center. Retrieved from http://www.nsidc.org

Shark, Anne (3 July 2001). Researchers determine global warming during 20th century may be slightly larger than earlier estimates. Lawrence Livermore National Laboratory. Retrieved from http://www.eurekalert.org

Sharp, David (12 November 2001). Researcher: Sea Level Rising Faster. Associated Press. Retrieved from http://news.yahoo.com

Shaw, Robinson (15 December 1999). El Niño/La Niña stirs more than ocean storms. Environmental News Network. Retrieved from http://www.enn.com

Shaw, Robinson (16 November 1999). Dramatic thinning of Arctic ice found. Environmental News Network. Retrieved from http://www.enn.com

Shaw, Robinson (21 July 2000). Iceman cometh to Greenland; sea level rises. Environmental News Network. Retrieved from http://www.enn.com

Shaw, Robinson (23 February 2000). Earth heating up faster than forecasted, study says. Environmental News Network. Retrieved from http://www.enn.com

Shaw, Robinson (25 February 2000). Global warming will stop at nothing, researcher predicts. Environmental News Network. Retrieved from http://www.enn.com

Shaw, Robinson (26 September 1999). Climate rides on ocean conveyor belt. Environmental News Network. Retrieved from http://www.enn.com

Shaw, Robinson (28 October 1999). Study hints at extreme climate change. Environmental News Network. Retrieved from http://www.enn.com

Shaw, Robinson (4 December 1999). Ozone layer could heal by 2050. Environmental News Network. Retrieved from http://www.enn.com

Shaw, Robinson (4 January 2000). Stormy weather buffeted U.S. in 1999. Environmental News Network. Retrieved from http://www.enn.com

Shaw, Robinson (6 February 2000). Coral provides clues to climate change. Environmental News Network. Retrieved from http://www.enn.com

Shaw, Robinson (6 June 2000). Polar clouds cause more ozone loss in Arctic. Environmental News Network. Retrieved from http://www.enn.com

Shaw, Robinson (7 October 2000). Clouds' role in global warming studied. Environmental News Network. Retrieved from http://www.enn.com

Shaw, Robinson (7 September 1999). Fire's role in global warming studied. Environmental News Network. Retrieved from http://www.enn.com

Shukman, David (31 March 2005). Carbon dioxide continues its rise. BBC News Online. Retrieved from http://news.bbc.co.uk

Siegel, Lee (1 September 2000). Earth's offspring? The collision theory. Space.com. Retrieved from http://www.space.com

Siegert, M.J. et al (24 September 2004). Ice flow direction change in interior West Antarctica. Science: Vol. 305; No. 5692; 1948-1951.

Simms, Andrew (2001). *An Environmental War Economy: The lessons of ecological debt and global warming*. New Economics Foundation. London.

Simpson, Sarah (2001). Triggering a Snowball. Scientific American [online]. Retrieved from http://www.sciam.com

Simpson, Shepherd (no date). Maya Galactic Alignment 2012. Retrieved from http://www.geocities.com/astrologyages/maya2012.htm

Smil, Vaclev (2000). *Cycle as Life: Civilization and the Biosphere*. WH Freeman and Co. New York.

Smith, Adam (1776). *An inquiry into the Nature and Causes of the Wealth of Nations*. Edwin Cannan (Ed). Fifth Edition 1904. Metheuen and Co. London.

Smith, T.M., R.W. Reynolds, R.E. Livesay, and D.C. Stokes (1996). Reconstruction of historical sea surface temperature using empirical orthogonal functions. J. Climate 9, 1403-1420.

Sohlman, Eva (13 May 2002). Expert warns world warming faster than expected. Reuters. Retrieved from http://news.yahoo.com

Space.com (20 December 2001). Russia-NASA share Earth data. Retrieved from http://www.space.com

SpaceDaily.com (23 January 2001). Earth may have cooled 10 degrees over past 3 million years. Retrieved from http://www.spacedaily.com

Spaceflight Now [online] (4 May 2002). New environmental eye on Earth launched into space. Retrieved from http://www.spaceflightnow.com

Sparling, Brien (no date). The Antarctic ozone hole. NASA Avanced Supercomputing Division. Retrieved from http://www.nas.nasa.gov

Sparling, Brien (no date). The ozone layer and stratosphere. NASA Advanced Supercomputing Division. Retrieved from http://www.nas.nasa.gov

Sparling, Brien (no date). Ultraviolet Radiation. NASA Advanced Supercomputing Division. Retrieved from http://www.nas.nasa.gov

Spotts, Peter (18 March 2005). How to prepare a planet for global warming. Christian Science Monitor [online]. Retrieved from http://www.csmonitor.com

Spotts, Peter (25 March 2004). Melting glaciers: unexpected boost to rising oceans. Christian Science Monitor [online]. Retrieved from http://www.csmonitor.com

Spotts, Peter (25 May 2001). Trees no savior for global warming. Christian Science Monitor [online]. Retrieved from http://www.csmonitor.com

Stanard, Alexa (17 November 2003). Scientists: Patagonian glaciers melting. Associated Press. Retrieved from http://news.yahoo.com

Stauffer, B. et al (1998). Atmospheric CO2 concentration and millennial-scale climate change during the last glacial period. Nature: Vol. 392;59-62.

Steel, Duncan (no date). A Possible Source for the 3100 BC Event. Retrieved from http://www.personal.eunet.fi

Steffan, Will et.al (2004). Executive Summary, Global Change and the Earth System: A Planet Under Pressure. IGBP Secretariat. Sweden.

Steffen, Will et.al (2004). *Abrupt changes: The Achilles' Heels of the Earth System.* Environment Magazine: 6(3). 9-20.

Stegner, Richard (14 August 2002). *Plummeting plankton linked to warmer oceans.* CNN News Online. Retrieved from http://www.cnn.com

Stegner, Richard (31 July 2002). *Report: Cosmic rays influence climate change.* CNN News Online. Retrieved from http://www.cnn.com

Steiger, Brad (1979). *Worlds Before Our Own.* Berkley Publishing Group. New York.

Steitz, David E. (22 April 2002). *Massive icebergs may affect Antarctic sea life and food chain.* NASA. Retrieved from http://www.eurekalert.org

Stephanie, Nebehay (17 December 2002). *Scientists: '02 second hottest as warming speeds.* Reuters. Retrieved from http://news.yahoo.com

Stiles, Lori (30 November 2000). *Lunar meteorite ages strongly support lunar cataclysm.* Retrieved from http://www.spacedaily.com

Stricherz, Vince (5 July 2001). *Arctic Oscillation has moderated northern winters of 1980s and '90s.* University of Washington. Retrieved from http://www.washington.edu

Stricherz, Vince (10 May 2001). *Collapse of simple life forms linked to mass extinction 200M years ago.* University of Washington. Retrieved from http://www.washington.edu

Stricherz, Vince (25 April 2001). *Most important greenhouse gas goes way up in past 50 years.* UniSci Online. Retrieved from http://unisci.com

Sueddeutsche Zeitung (9 June 2000). *Der Mensch veraendert das Klima.* Retrieved from http://www.habito.de/planeco/placli.htm

Sunday Times (1988). *Atlantis? It could just be 12,000 ft up in Bolivia.* Retrieved from http://www.bolivia.co.uk

Suplee, Curt (3 December 1999). *Study: Arctic sea ice is rapidly dwindling.* Washington Post [online]. Retrieved from http://www.washingtonpost.com

Suzuki, David (29 June 2001). *Weakened ocean currents could cause climate flip.* Environmental News Network. Retrieved from http://www.enn.com

Syal, Rajeev (14 December 1997). *Meteor showers blotted out man's first civilisations.* The Sunday Times [online]. Retrieved from http://abob.libs.uga.edu

Taipei Times [online] (21 October 2001). *Adapting to the climate's caprices.* Retrieved from http://www.taipeitimes.com

Tait, Paul (5 March 2002). *Kyoto won't help sinking Pacific Isles, U.S. says.* Reuters. Retrieved from http://news.yahoo.com

Taylor, K. et al (1997). *The Holocene-Younger Dryas transition recorded at summit, Greenland.* Science: Vol. 278, 825.

TerraDaily Online (15 August 2001). *Giant impact theory for moon formation boosted.* Retrieved from http://www.terradaily.com

TerraDaily Online (16 February 2000). *Giant impact puts moon in a twist.* Retrieved from http://www.terradaily.com

TerraDaily Online (27 March 2002). *Meteorites tell of shocking experience in planetary formation.* Retrieved from http://www.terradaily.com

TerraDaily Online (6 December 2000). *Climate modeling must consider all 'greenhouse' gases.* Retrieved from http://www.terradaily.com

The Australian [online] (27 October). *Climate 'Bigger threat than terror'.* Retrieved from http://www.theaustralian.news.com.au

The Chief Engineer [online] (3 March 2005). *Climate change already affecting global environment.* Retrieved from http://www.chiefengineer.org

The Components of the Mayan Calendar (no date). Retrieved from http://mayacalendar.com

The formation of complex craters (no date). Retrieved from http://www.ese.ic.ac.uk

The Inca and Mayan Prophecies (no date). Retrieved from http://www.web-of-light.com

The Jewish Chronicle (6 March 1998). *Catch a falling star.* Retrieved from http://abob.libs.uga.edu

The Scotsman [online] (17 November 2001). *Comment: Catastrophic consequences of climate change.* Retrieved from http://www.thescotsman.co.uk

The Sydney Morning Herald [online] (26 July 2004). *Floods submerge most of Bangladesh, hit millions.* Retrieved from http://www.smh.com.au

Thompson, Don (12 November 2003). *Global warming may spur more wildfires.* Associated Press. Retrieved from http://news.yahoo.com

Thomson, Jeremy (7 December 2000). *Humans did come out of Africa, says DNA.* Nature News Service [online]. Retrieved from http://www.nature.com

Thomson, Jerry (2001). *Climate: Great rivers of the ocean.* Nature News Service [online]. Retrieved from http://www.nature.com

Thorsell, William (9 September 2002). *Accept it: Kyoto won't stop global warming.* The Globe and Mail [online]. Retrieved from http://www.globeandmail.com

Tickell, Sir Crispin (22 October 2001). *Catastrophes.* The St. Andrews Prize Lecture, Royal Institution, London. Retrieved from www.ri.ac.uk

Tisdall, Jonathan (2 October 2002). *Highest mean temperature ever measured.* Aftenposten English Web Desk. Retrieved from http://www.aftenposten.no

Tobar, Hector (27 February 2002). *Chilean links ozone loss to skin ailments.* Los Angeles Times [online]. Retrieved from http://www.latimes.com

Toohey, Darin (28 May 2002). *Ozone losses may be speeding up at higher latitudes, according to U. of Colorado study.* Retrieved from http://www.eurekalert.org

Toon, John (24 June 2004). *Abrupt climate change: new research supports hypothesis that ocean currents redistributed heat during rapid warming and cooling.* Georgia Institute of Technology. Retrieved from http://gtresearchnews.gatech.edu

Torok, Simon (29 March 2004). *Global growth of carbon dioxide still rising.* CSIRO Australia. Retrieved from http://www.csiro.au

Townsend, Mark and Paul Harris (22 February 2004). *Now the Pentagon tells Bush: climate change will destroy us.* Observer [online]. Retrieved from http://observer.guardian.co.uk

Trivedi, Bijal P. (10 January 2002). *Evolution on fast forward: Finches adapt to climates.* National Geographic Today [online]. Retrieved from http://news.nationalgeographic.com

Union of Concerned Scientists Global Warming Website. Retrieved from http://www.ucsusa.org/global_environment/global_warming/index.cfm

Union of Concerned Scientists. *Global warming: Early warning signs* (1999). UCS, WRI. Retrieved from http://www.climatehotmap.com

UniSci Online (4 October 2001). *Dinosaurs' world heated by greenhouse effect: study.* Retrieved from http://unisci.com

UniSci Online (10 June 2002). *New way uncovered to test climate change predictions.* Retrieved from http://unisci.com

UniSci Online (14 December 2001). *Sharper warming on chances of abrupt climate change.* Retrieved from http://unisci.com

UniSci Online (17 January 2001). *Little global-scale warming seen during past 22 years.* Retrieved from http://unisci.com

UniSci Online (19 March 2002). *Depths of the southern ocean being starved of oxygen.* Retrieved from http://unisci.com

UniSci Online (20 February 2002). *Global warming will continue for the next century.* Retrieved from http://unisci.com

UniSci Online (22 February 2001). *Fingerprints link ice sheet melting to sea level rise.* Retrieved from http://unisci.com

UniSci Online (23 February 2001). *Philippine volcano helps solve climate change mystery.* Retrieved from http://unisci.com

UniSci Online (26 January 2001). *Silica not solely responsible for ice age CO2 Levels.* Retrieved from http://unisci.com

UniSci Online (31 May 2001). *Mountain glaciers receding, AGU told.* Retrieved from http://unisci.com

United Nations (15 March 2001). Satellite pictures show greenhouse effect. Retrieved from http://www.enn.com

United Nations (17 December 2003). Global warmth up, ozone hole at record, Arctic ice down for 2003-UN. Retrieved from http://www.un.org

United Nations Environment Programme (2002). Global Environment Outlook 3: Past, Present and Future Perspectives. Earthscan. London.

United Nations Environment Programme (22 May 2002). The state of the environment: Past, present, future? Retrieved from http://www.unep.org

United Press International (19 May 2000). Warmest January-April on record in U.S. Retrieved from http://www.enn.com

United Press International (22 October 2000). Icelandic glacier in rapid breakup. Retrieved from http://www.enn.com

United States Environmental Protection Agency Website. Global Warming. Retrieved from http://yosemite.epa.gov/oar/globalwarming.nsf/content/climate.html

United States Environmental Protection Agency Website. Future climate - sea level. Retrieved from http://www.epa.gov/ozone

United States Environmental Protection Agency Website. Brief history and glossary of terms and concepts for ozone. Retrieved from http://www.epa.gov/ozone

United States Environmental Protection Agency Website. Health effects from increased exposure to ultraviolet-B (UVB) radiation due to stratospheric ozone depletion. Retrieved from http://www.epa.gov/ozone

United States Environmental Protection Agency Website. Health effects of overexposure to the Sun. Retrieved from http://www.epa.gov/ozone

United States Environmental Protection Agency Website. Ocular damage from increased ultraviolet-B exposure due to ozone depletion. Retrieved from http://www.epa.gov/ozone

United States Environmental Protection Agency Website. Risk factors that influence the occurrence of skin cancer within human populations. Retrieved from http://www.epa.gov/ozone

United States Environmental Protection Agency Website. The ozone depletion process. Retrieved from http://www.epa.gov/ozone

United States Environmental Protection Agency Website. Ozone: good up high, bad nearby. Retrieved from http://www.epa.gov/ozone

University of Bristol and British Antarctic Survey (23 September 2004). New structure found deep within West Antarctic ice sheet. Retrieved from http://www.antarctica.ac.uk

University of Santa Cruz (16 April 2001). Climate change linked to anomaly in Earth's orbit. Retrieved from http://www.spaceflightnow.com

University of Washington (22 July 1999). UW scientists say Arctic oscillation might carry evidence of human-induced global warming. Retrieved from http://www.uwnews.org/

University of Washington (30 August 2002). Scientists zero in on arctic, hemisphere-wide climate swings. Retrieved from http://www.sciencedaily.com

VanDeVeer, Donald and Pierce, Christine (1994). The Environmental Ethics and Policy Book: Philosophy, Ecology, and Economics. Wadsworth Publishing Company. Belmont, California.

Veithen, Corine (July 1997). Auswirkungen von Klimaveranderungen in der Arktis. Retrieved from http://www.greenpeace.de

Velikovsky, Immanuel (1950). Worlds in Collision. McMillan. New York.

Velikovsky, Immanuel (1955). Earth in Upheaval. Doubleday. New York.

Verrengia, Joseph B (1999). La Niña, tornadoes linked. Associated Press. Retrieved from http://www.hprcc.unl.edu

Victor, David G. (2001). The Collapse of the Kyoto Protocol and the Stuggle to Slow Global Warming. Princeton University Press. Princeton.

Vidal, John (21 April 2005). The end of oil is closer than you think. Guardian [online]. Retrieved from http://www.guardian.co.uk

Vidal, John (21 October 2004). Forget climate change, that's the least of our worries say Nobel winners. Guardian [online]. Retrieved from http://www.guardian.co.uk

Vidal, John (6 August 2003). Global warming may be speeding up, fears scientist. Guardian [online]. Retrieved from http://www.guardian.co.uk

Visbeck, Martin (no date). North Atlantic Oscillation. Lamont-Doherty Earth Observatory. Retrieved from http://www.ldeo.columbia.edu

von Grafenstein, U. et al (1999). A mid-European decadal isotope-climate record from 15,500 to 5,000 years BP. Science: 284; 1654-1657.

Wagner, F. et al (1999). Century-scale shifts in Early Holocene atmospheric CO2 concentration. Science: 284; 1971-1973.

Wagner, Angie (18 June 2004). Western drought beats dust bowl, could be worst in 500 years. Associated Press. Retrieved from http://www.heatisonline.com

Walton, Marsha (14 May 2003). Study: only 10 percent of big ocean fish remain. CNN News Online. Retrieved from http://www.cnn.com

Warrick, Joby (21 April 1998). Mass extinction underway, majority of the biologists say. Washington Post. Retrieved 5 December 2000 from http://www.washingtonpost.com

Warshofsky, Fred (1977). Doomsday: The Science of Catastrophe. Simon and Schuster. New York.

Weather Channel (no date). Hot Planet. Retrieved from http://www.weather.com

Weiner, Jonathon (1986). Planet Earth. Bantam Books. New York.

Weiner, Jonathon (1990). The Next One Hundred Years: Shaping the Fate of Our Living Earth. Bantam Books. New York.

Weissert, Will (8 September 2000). Scientists uncover lost Mayan marketplace. Associated Press. Retrieved from http://www.nandotimes.com

Whitehouse, David (5 March 2001). 'Heat vent' may diminish global warming. BBC News Online. Retrieved from http://news.bbc.co.uk

Whitehouse, David (7 January 2002). Space rock hurtles past Earth. BBC News Online. Retrieved from http://news.bbc.co.uk

Whitehouse, David (17 October 2001). Hawking predicts terrestrial doom and gloom. BBC News Online. Retrieved from http://news.bbc.co.uk

Whitehouse, David (23 February 2001). Asteroid 'destroyed life 250m years ago'. BBC News Online. Retrieved from http://news.bbc.co.uk

Whitehouse, David (28 November 2000). Sun's warming influence 'under-estimated'. BBC News Online. Retrieved from http://news.bbc.co.uk

Whitehouse, David (31 July 2002). Galaxy 'may cause ice ages'. BBC News Online. Retrieved from http://news.bbc.co.uk

Wigley, Tom (14 December 2000). Global warming trend stronger than had been thought. UniSci Online. Retrieved from http://unisci.com

Wilcock, David (6 December 2000). The Shift of the Ages. Retrieved from http://www.ascension2000.com

Willis, David (30 July 2004). Sea engulfing Alaskan village. BBC News Online. Retrieved from http://news.bbc.co.uk

Wilson, E.O. (Ed.) (1988). Biodiversity. National Academy of Sciences. Washington D.C.

Wilson, Scott (9 July 2001). Warming shrinks Peruvian glaciers. Washington Post [online]. Retrieved from http://www.washingtonpost.com

Witze, Alexandra (25 December 2000). Scientists study nature's role in destroying ozone. The Dallas Morning News. Retrieved from http://www.dallasnews.com

World Meteorological Organization (16 December 1999). 1999 closes the warmest decade and warmest century of the last millennium according to WMO annual statement on the global climate. Retrieved from http://www.wmo.ch

World Meteorological Organization (19 December 2000). WMO statement on the status of the global climate in 2000. Retrieved from http://www.wmo.ch

World Meteorological Organization (2 July 2003). Extreme weather events might increase. Retrieved from http://www.wmo.ch

World Meteorological Organization (2001). WMO Statement on the status of the global climate in 2001. Retrieved from http://www.wmo.ch

Woodard, Colin (21 August 2000). Slowly but surely, Iceland is losing its ice; Global warming is prime suspect in meltdown. San Francisco Chronicle [online]. Retrieved from http://www.sfgate.com

World Bank (14 June 2001). More hunger and poverty may be enduring impact of climate change. Retrieved from http://allafrica.com

Worldwatch Institute (2000). State of the World 2000. Earthscan. London.

Worldwatch Institute (2001). State of the World 2001. Earthscan. London.

Worldwatch Institute (2002). State of the World 2002. Earthscan. London.

Worldwatch Institute (2003). State of the World 2003. Earthscan. London.

Worldwatch Institute (2004). State of the World 2004: Special focus-The Consumer Society. Earthscan. London.

Worldwatch Institute (2005). State of the World 2005: Redifining Global Security. Earthscan. London.

Wright, Karen (5 October 2000). Birth of an Antarctic Super-Berg. Discover Magazine [online]. Retrieved from http://www.ngnews.com

Wu, P., R. Wood, and P. Stott (2005). Human influences on increasing Arctic river discharges. Geophysical Research Letters: 32, L02703, doi:10.1029/2004GL 021570.

Wynn, Jeffery C. and Eugene M. Shoemaker (1 November1998). The day the sands caught fire. Scientific American [online]. Retrieved from http://www.sciam.com

Yaxk'in, Aluna Joy (1998). The Maya Cosmovision. Retrieved from http://www.lightnews.org

Zens, Josef (1 November 2000). Es wird heisser – und zwar schnell: Umweltexperten untersuchen, was der Klimawandel in Europa anrichten wird. Berliner-Zeitung [online] Retrieved from www.berlin-online.de

Zens, Josef (25 October 2000). Die Erde wird sich auf jeden Fall erwaermen: Umweltforscher Hans-Joachem Schellnhuber ueber das Kyoto-Protokoll und die anstehende Klimakonferenz. Berliner-Zeitung [online]. Retrieved from www.berlin-online.de

Zhou, et al (25 September 2001). Vegetation increasing at 40-70 Degrees North latitude. UniSci Online. Retrieved from http://unisci.com

sound of sirens cast of characters...

evan green as
the interviewer

jonathon henning as
the gauze trooper

morwenna borden as
the chanteuse

kristofer johnson as
the brother

ge johnson as
the author

corey priesman as
the doctor

porter chollet and
barton lamb green III as
the emphasizers

hans wanamaker as
the grillwalker

...and introducing dietrich wanamaker as future man

sound of sirens was brought to you with support from the following letters:

richard and winnie johnson (aka mom and dad), kristofer and dori johnson, amy and beth sennentz, todd and logan sennentz, dell and karen sennentz, matt and cory waldron, barb and ernie waldron, john, christine, and zachary sennentz, rose and bill johnson, opa, george and barbara sennentz, stevens, dittos, and quitslunds, priesmans, greens, woods, kiechels, coyles, roehrs, satories, wanamakers, sandra noetzel and family, karin mueller and family, the french cafe` crew, carrie assheuer and the unfccc, martin lichtl and lichtl sustainability communications, danyel reiche, larry johnson, university of hannover international studies office, university of nebraska-lincoln international affairs office, dee dee, tracy and the cookie company, straws, sticks and bricks, klima-buendnis, iuniverse, eg solar, unerc, leo and mario, stefan koerner, eduardo and brahim, jean-louis romane, reike, danyel, and martin, manuela, britta, and doro, scott, katja, neil, melle, julia, carsten, jonathon and mike, perry and myriam williams, markus and sandra, rueben and christiane, martina gerle, christina anhalt, leon, roxanne and dean buck, dirk and karin, bolte, socke, iwan werner, oliver koegler, tobias pietz, sandrine rommel and uwe herring, thorsten hunsicker and ulrike hartnagel, veronic and werner valentijn von daniels, lisa sock, hank, nelle, and tom woods, lars anderson, todd petersen, buck and lisa kiechel, dave and scott pittock, the burkey boys, zac roehrs, mike satorie, chris doan, kristi and brendan wamstad-evans, lili and tony merritt, tina and tommy arsiaga, roxanne jackson, gabe kaputska, carlee sherman, eric and katie leyden, amy schulze and rich gilbert, rupali, garett, sonali, and anchil, josh and jen, erika flanders and kevin sullivan, alex and marty, lisa schulze, ben and tara, katie flanagan, andy brown, marissa, corey and kristin donahue, john and caroline, julie perette, b.harris, s.hatfield, feistner, finley, hoppe and tricker, conoley and ayuko ospovat, simien bucacek, seth connard, olivia brown, elizabeth goodbrake, ben knauss, mark schroll, sarah and evan neal, kalish, garcia, fields, wieland, borno and brown, travis green, jeff and scott semrad, scott gealy, paul, annie and oakie samulsen, clifton and kate, jason and susie ortiz, dana waldrop, aaron barksdale, aaron spilker, cadi, keele, saint, bridie, sue, regan, and heather, bobby crane, bobby mcwhite, zuba, marty scheiber, kort and konen, noman mehboob, dj jaazon, tim ihrig, paul kulik.

thank you to:
°heather green, emily green, marissa and hans wanamaker, ben hunter, and ryan palmer for your editing advice.
°sara brockmeier for your research assistance.
°sheila kappeler and melton cartes for your publishing assistance.
°richard and winnie johnson, barbara sennentz, and henry cochrane woods III for your generous and continuing support.
°adam jacobs for your legal advice.

thank you all for being who you are and affecting my life along the way. you've made me what i am and this work is a part of you.
i apologize for any ommissions. you know who you are out there.

-g

COMING SOON...

SOUND OF SIRENS
21ST CENTURY GLOBAL TEMPERATURE CHANGE
BOOK II: ABSOLUTIONS

G.E. JOHNSON

NIGHT IS FALLING GREAT APES
TIME TO MAKE YOUR PEACE
AND TAKE YOUR SHELTER